Before Voltaire

Before Voltaire

The French Origins of "Newtonian" Mechanics, 1680–1715

J. B. SHANK

The University of Chicago Press
CHICAGO & LONDON

The University of Chicago Press, Chicago 60637
The University of Chicago Press, Ltd., London

Published 2018
Printed in the United States of America

27 26 25 24 23 22 21 20 19 18   1 2 3 4 5

ISBN-13: 978-0-226-50929-7 (cloth)
ISBN-13: 978-0-226-50932-7 (e-book)
DOI: 10.7208/chicago/9780226509327.001.0001

Library of Congress Cataloging-in-Publication Data

Names: Shank, John Bennett, author.
Title: Before Voltaire : the French origins of "Newtonian" mechanics, 1680–1715 / J. B. Shank.
Description: Chicago : The University of Chicago Press, 2018. | Includes bibliographical references and index.
Identifiers: LCCN 2017043824 | ISBN 9780226509297 (cloth : alk. paper) | ISBN 9780226509327 (e-book)
Subjects: LCSH: Mechanics—France—History—17th century. | Mechanics—France—History—18th century. | Mechanics, Analytic—History—17th century. | Mechanics, Analytic—History—18th century. | Newton, Isaac, 1642–1727—Influence. | Newton, Isaac, 1642–1727. Principia. | Académie royale des sciences (France)
Classification: LCC QA802 .S43 2018 | DDC 531.0944/09032—dc23
LC record available at https://lccn.loc.gov/2017043824

♾ This paper meets the requirements of ANSI/NISO Z39.48-1992 (Permanence of Paper).

Contents

Acknowledgments

This second book is in many ways older than my first, and it is one with a circuitous genealogy. In it I publish for the first time some of my first research, begun over twenty years ago. Because the book stands as a kind of prequel to my first book, *The Newton Wars and the Beginning of the French Enlightenment*, *Before Voltaire* brings into print arguments that were present, at least implicitly, in that earlier book. It also closes in reverse chronological order my overall project of research begun in the mid-1990s: the historical reinterpretation of the eighteenth-century French reception of Newton's *Principia*. This book brings that twenty-year research project to a close, and its arrival a decade after the publication of my first book has not erased the debt it owes to all the people and institutions already acknowledged in 2008. I therefore ask that you import everything and everyone thanked in *The Newton Wars* here, and treat all of those acknowledgments as expressions of gratitude applicable in every respect to this book as well.

One cluster of redundant thanks must be made explicitly, however. These speak to the different journey that has produced this second book. In 2008, I thanked the University of Minnesota McKnight Land-Grant Professorship for the support it gave me in completing my first book, yet this award played an even more significant role in making this one. It enabled me in particular to travel to Florence, Italy, in 2005–6, and to begin my indispensable friendship with Niccolò Guicciardini. My time in Italy was transformative in too many ways to enumerate, but it was especially important for this book because it was there, under the generous and supportive guidance of Niccolò, that I began the project of working seriously as a historian of mathematics. Niccolò has since become the guiding star for all that I do in this area, and the influence of his scholarship and friendship on this book cannot be overstated. He also had the brilliant idea in 2005 of bringing me into the fold of professional history of mathematics directly by inviting me to attend one of its defining institutions: the semiregular meetings held at the Mathematisches Forschungsinstitut Oberwolfach in the heart of the Black Forest in Germany. I have attended two meetings at the MFO, and each has been invaluable for my development as a historian of mathematics. My special thanks to Karine Chemla, Moritz Epple, Toni Malet, Marco Panza, Jeanne Peiffer, Helmut Pulte, David Rabouin, Patricia Radelet de Grave, David Rowe, and the many others whom I met in Oberwolfach for their stimulating dialogue and debate. Special thanks also to Henk J. M. Bos, whom I also met at the MFO. Niccolò pointed me toward his scholarship early on, and he has become a second guiding light shaping everything I do in the history of mathematics. I am also grateful for the generous and supportive conversations we have had regarding my work. I mentioned Michel Blay in *The Newton Wars*, thanking him for the generous support he offered me in Paris in 1995–96, when I was a predoctoral fellow at the Centre Alexandre Koyré. This guidance has been invaluable ever since, and the extent of his influence on this book is enormous. Although we have not had the opportunity to cross paths for many years, I owe a great deal to his generous support of me and my work. Finally, three good friends and colleagues also mentioned previously deserve special recognition because of their connection to the history of mathematics: Amir Alexander, Matthew Jones, and Massimo Mazzotti. Each has been a friend and colleague for over two decades, and each has contributed in more ways than they know to my own development as a historian of mathematics. I am also grateful for the conversations, however brief and infrequent, with my other friends and colleagues in the field, such as H. Floris Cohen, François de Gandt, and Benjamin Wardaugh. Their shared passion for discussion of all things early modern and mathematical has been uplifting.

New friends, institutions, and intellectual opportunities pursued since 2008 have also been indispensable in bringing this book to completion. A Dibner Fellowship in the History of Science at the Huntington Library allowed me important time for research and writing, as did money from the Imagine Fund and the Grant-in-Aid for Research programs at the University of Minnesota. Most recently a visiting professorship in 2016 at Utrecht University, together with a residency at the Descartes Center for the History and Philosophy of the Humanities and the Sciences, created the space needed to finish this book. My involvement the last five years with the "Baroque Science" project organized by Ofer Gal and Raz Chen-Morris has also been very influential in the development of my work. I am especially grateful to each of them for the many discussions we have had about early modern mathematics and mathematical physical science, which have enriched my thinking enormously. Also invaluable to me have been my conversations about academic science in Old Regime France with Anita Guerrini, Oded Rabinovitch, Jacob Soll, Stéphane van Damme, and especially Nick Dew, whose generosity, knowledge, and good cheer are beyond measure.

I remain very grateful for the many friends and colleagues at the University of Minnesota who continue to make "the U" and the Twin Cities a fabulous place to work and live, even during a polar vortex. This book also owes a debt to its editor at the University of Chicago Press, Karen Merikangas Darling. An unwavering supporter of the project from the outset, Karen steered the manuscript through the review process with a steady hand, especially at a crucial moment when it encountered complicated disciplinary crosswinds. The book is better because of her judicious, intelligent, and highly professional supervision. I am lastly, and most deeply, thankful for the unending support of my family in all its extensions during the long and tortuous journey involved in bringing this book to completion. My wife, Alison, who endured far too many days and nights ruined because of my frustration with "the ordeal of the f-ing book" especially has my deep and heartfelt thanks. The very long odyssey of this project would not have ended happily without all the love and distraction that she and the rest of my family have provided. I am blessed to have their unending affection and support in my life every day.

Abbreviations

Ac. Sci. Archives de l'Académie des sciences. Paris.

A.N. Archives Nationales. Paris.

BJB *Der Briefwechsel von Johann I Bernoulli.* 3 vols. Basel: Birkhauser, 1955–. Volume 1, Otto Spiess ed., *Der Briefwechsel mit Jakob Bernoulli, dem Marquis de l'Hôpital* (1955). Volume 2, Pierre Costabel and Jeanne Peiffer eds., *Der Briefwechsel mit Pierre Varignon, Teil I: 1692–1702* (1988). Volume 3, Pierre Costabel and Jeanne Peiffer eds., *Der Briefwechsel mit Pierre Varignon, Teil II: 1703–1714* (1992).

B.N. Bibliothèque Nationale. Paris.

DSB *Dictionary of Scientific Biography*, Charles Coulston Gillispie ed. 16 vols. New York, 1970–80.

HARS *Histoire de l'Académie royale des sciences. Avec les mémoires de mathématique et de physique . . . tirés des registres de cette académie (1699–1790).* 92 vols. Paris, 1702–97. Each volume comprises two separately paginated parts, referred to as *Hist.* and *Mem.*

J.S. *Journal des savants.* 90 vols. Paris, 1670–1760.

J.T. *Mémoires pour servir à l'histoire des sciences et des beaux arts.* 67 volumes. Trévoux, 1701–67. (Commonly known as the *Journal de Trévoux*).

LMS C. I. Gerhardt ed. *Leibniz: Mathematische schriften.* 7 vols. Hildesheim: Olms, 1971.

Merc. gal. *Mercure galant.* Paris, 1672–1723.

O.F. Alain Niderst ed. *Oeuvres complètes de Fontenelle.* 9 vols. Paris: Fayard, 1990–97.

O.H. *Oeuvres complètes de Christiaan Huygens.* 22 vols. The Hague: Martinus Nijhoff, 1888–1955.

O.M. André Robinet ed. *Oeuvres complètes de Malebranche.* 20 vols. Paris: J. Vrin, 1958–68.

PVARS Procès-verbaux de l'Académie royale des sciences. Archives de l'Académie des sciences. Paris.

CHAPTER 1

Introduction
Translating Newton

Every translation is a continuous interpretation.
I. BERNARD COHEN, English translator of Newton's *Principia*

Owen Gingrich echoing Arthur Koestler has called Nicolaus Copernicus's 1543 *De revolutionibus orbium coelestium* the revolutionary scientific treatise that nobody read.[1] Few, it is true, read *De revolutionibus*, but certainly Isaac Newton's *Philosophiae naturalis principia mathematica* surpasses Copernicus's work as the most influential unread book in the history of science. The claim of influence is the easier of the two to trace. From the moment that Newton's treatise appeared, and throughout the centuries that have elapsed since, the *Principia* has been treated as a work of epochal significance, one whose aura compels description in nothing less than supernatural terms. The treatise in fact entered the world accompanied by its own divinizing chorus, since Newton's scientific colleague Edmund Halley contributed a Latin ode to the text that sang of the *Principia* as a "splendid ornament of our time and our nation."[2] The poem reached its climax by situating the author alongside the immortals of Olympus.

> O you who rejoice in feeding on the nectar of the gods in heaven,
> Join me in singing the praises of Newton, who reveals all this,
> Who opens the treasure chest of hidden truth,
> Newton, dear to Muses,
> The one in whose pure heart Phoebus Apollo dwells and whose mind he has filled with all his divine power;
> No closer to the gods can any mortal rise.[3]

Halley's enthusiasm was neither isolated nor exceptional, for in the decades after 1687—the year the *Principia* appeared—a tradition was established whereby references to the book as a "prodigy," a "glory to the human spirit," an "immortal work," a treatise of "sublime genius," and, to follow Halley literally, a "near divine achievement" became commonplace. Alexander Pope demonstrated the increasing influence of this tradition in 1727, the year of Newton's death, when he proposed that Sir Isaac's tomb be graced with the following epitaph: "Nature and nature's laws lay hid in night: God said, 'Let Newton be!' and all was light." Such exuberance became ordinary during Newton's lifetime, and in the decades after his death these "honey-sauced eulogies," to use Clifford Truesdell's description,[4] also began to anchor more general stories about the new and progressive scientific age that Newton's *Principia* was said to have initiated. Few educated Europeans escaped the influence of this public canonization of Newton as the prophet and patron saint of scientific modernity, and by 1800 the *Principia* was widely held to have been the ignorance-piercing agent of a world-changing, modernity-making revolution.

To be sure, few could have explained in precise scientific detail the essence of the *Principia*'s revolutionary scientific accomplishments, or even what Newton had done exactly to effect this monumental change. Joseph-Louis Lagrange summed up the situation by saying that there was only one system of the world to discover, and Newton had found it, and less sophisticated observers largely fell in line with this expert opinion by offering comparable pocket descriptions that spoke of "the discovery of universal gravitation and the true laws of physics" or the "unraveling of the mystery of the cosmos." Yet underlying the repetition of these platitudes was the confidence, almost universally shared, that epochal accomplishments had been realized in this book, and that with it, Newton had achieved a historic dissipation of ignorance while initiating a new age of light.

In 1762, Alexandre Saverien, an obscure naval engineer and would-be French *philosophe*, demonstrated the solidification of this understanding in his grandiloquent *Histoire des philosophes modernes*.[5] To convey Newton's world-changing genius, Saverien retold a story about the reception of the *Principia* that had already become a canned tale available on the shelf whenever sugared accounts of Newton's accomplishments were needed. In the story, Newton's colleague Dr. John Arbuthnot is asked in 1696 by the French mathematician the Marquis de l'Hôpital to show him the best demonstration by an Englishman of the solid that exerts the least resistance when immersed in a fluid. Arbuthnot gave l'Hôpital the *Principia*, and the marquis is described as being awestruck by what he found therein. "Good God, what a fund of knowledge there is in that book," one account of the story has him exclaim.[6] Saverien followed other narrators in reporting that l'Hôpital was then led to inquire about the book's author, asking "Is he like other men? Does he eat and drink and sleep?" L'Hôpital reportedly expressed astonishment when he was told that Newton in fact "conversed cheerfully with his friends" and "put himself upon a level with all mankind."[7]

One can still find this story repeated in contemporary accounts of the reception of Newton's *Principia*. The persistence of it attests to the continuing influence of the eighteenth-century confectioner's discourse about Newton, along with its consecrating intentions, in our understanding of Newton's life and legacy today.[8] Yet in the story's evocation of a direct encounter between a flesh-and-bones mathematician and the actual contents of Newton's 1687 opus, the tale also points to the other part of the claim that started this book, the one that posits a disjuncture between the heroic understanding of Newton's immortal influence and the actual history of the *Principia*'s literal textual reception.

Who in fact read the "Preface to the Reader," the "Definitions," the "Axioms, or Laws of Motion," and the three abstruse books that comprised the first Latin edition of Newton's *Philosophiae naturalis principia mathematica*? L'Hôpital certainly did, and even if we will soon discover reasons to doubt the literal word of the Arbuthnot story, we can accept that he encountered the *Principia* soon after its publication and found much brilliant mathematical work in it. Yet, as an actual, informed reader of Newton's *Principia*, l'Hôpital did not exemplify the book's initial European reception. He is in fact better seen as an exception that proves other rules. For the truth of the matter is that few in Europe in Newton's lifetime, and even fewer as the *Principia* became musty with age, possessed the technical capacities and the intellectual inclination—the first generally produced the second—to digest Newton's treatise directly. A Cambridge University student may have used sarcasm to most effectively capture the predicament that the *Principia* posed when he spied the Lucasian Professor on the street and remarked, "There goes the man who hath writ the book that neither he nor anyone else understands."[9]

Some, such as l'Hôpital and his colleagues in the French Académie Royale des Sciences, were able to read and understand Newton's book, and among this cadre of highly skilled mathematicians the *Principia* received a robust reception. However, since this coterie of mathematical experts never included more than a few dozen individuals anywhere in Europe, the direct textual reception of Newton's work was also exceedingly limited and narrow in scope. If one also considers the intense stamina required for even the most highly skilled reader to work through every detail of Newton's exceedingly long and idiosyncratic treatise, the list of actual readers of the *Principia* becomes smaller still.

Abraham de Moivre illustrates the point. He was one of the rare European mathematicians suitably trained to read the *Principia* from cover to cover, and when he encountered the book in 1688, he purchased a copy and set out to do just that. But as the secretary of the Académie Royale des Sciences explained in his eulogy of de Moivre in 1754, "The young mathematician was soon forced to admit that what he had taken for simple mathematics was in fact beyond his understanding, and that the treatise was actually the beginning of a long and difficult course of study that he had yet to undertake."[10] Despite persistent effort, de Moivre ultimately struggled to master Newton's dense tome, yet he became an ardent "Newtonian" afterward nevertheless.

De Moivre's partial and labored digestion of the *Principia* was typical, as was John Locke's indirect and mediated digestion of the same. Unlike de Moivre, Locke did not possess the mathematical skills necessary to even try to comprehend the *Principia* directly, so to compensate he turned to his friend, the esteemed mathematician Christiaan Huygens, and asked him to vouchsafe Newton's work for him. When Huygens confirmed that the mathematical demonstrations were sound, Locke adopted the major conclusions of the *Principia* as his own.[11] In *Some Thoughts concerning Education*, he explained the logic of his approach, writing that since Newton has given us "so good and clear an account . . . in his admirable book, *Philosophiae naturalis Principia Mathematica* . . . of the motions, properties, and operations of the great masses of matter, in this our solar system, . . . his book will deserve to be read." Yet acknowledging that, "very few have mathematicks enough to understand its demonstrations," Locke also asked students to simply trust, as he did, that the arguments were sound. "[Newton's] conclusions . . . may be depended on as propositions well proved," Locke wrote reassuringly, since "the most accurate mathematicians who have examin'd them allow them to be such."[12] Ironically, Locke's principal authority on these matters, Huygens, went to his grave in 1695 convinced that the physics that Newton had built upon his "most accurate mathematics" was questionable at best.[13] As we will see, Huygens's reservations were not exceptional during these years.

L'Hôpital, de Moivre, Huygens, and Locke all confirm, therefore, that the *Principia* was not in fact an unread book. But Locke calling a treatise he could not understand "a good and clear account" and Huygens expressing doubts about the physical conclusions drawn from Newton's "most accurate" mathematical demonstrations also point to the historical gap separating these partial, fractured, and exceedingly narrow readings of the *Principia* from the widespread consensus about the significance of the book that would soon take hold. Recognizing and acknowledging this gap ultimately raises the following question: How did it happen that such a dauntingly recondite treatise, one that challenges the interpretive capacities of even the most expert and patient reader, and a book that was only read initially by a very small number of individuals (because only that many possible readers even existed), and a book that was only partially understood even by the few readers who did undertake the challenge (because it is an exceedingly technical and abstruse treatise that is difficult for even expert readers to comprehend)—how did it happen that *this of all books* became within a matter of decades the widely perceived agent of a revolutionary transformation in modern science and even of modernity tout court?

The only possible answer is through a process of translation, and it is this process in its initial iterations that is the focus of this book. In another book, published in 2008, I focused on the longer-term results of this process of translation, analyzing how it happened that Newton of all people become the hero of the French Enlightenment and the icon of the philosophical modernity that the eighteenth-century French *philosophes* began to polemically defend after 1750.[14] This earlier book looked at the socially embedded chains of reading, interpretation, and writing, along with those of scientific observation, inscription, and theorization, that transformed the technical claims of Newton's abstruse science into a widely held understanding of "Newtonianism," one that became a vehicle for widespread historical transformation. *Before Voltaire* is a prequel of sorts to that earlier study, but it also has a different focus. Rather than try to understand how technical mathematical science was made into the sugary metanarratives of Enlightenment, this book aspires to wipe away the sticky pink fluff that these later, celebratory accounts have spun around Newton's *Principia* in order to return directly and precisely to its initial European reception. It further strives to build from this pre-Enlightenment vantage point—a perspective before Voltaire, in other words—a new and sugar-free account of the outcomes that ensued when other mere mortals (or in this case, Frenchmen) began to study Newton's *Principia* without any awareness of the epochal significance that later interpreters would attribute to this treatise or their reading of it.

Halley's verse encomium, published alongside the recondite mathematical arguments of Newton's opus, reminds us, however, that we cannot expect to understand this initial and initiating process of translation by hoping to return to some original moment when Newton's virginal science was revealed and then disseminated to the readers of his texts. Text and context are always and forever intertwined, and while entities like "Newtonian mechanics" certainly came together after 1687 because of human encounters with the raw material of Newton's original science, they were also shaped decisively by those who read, interpreted, and put to use Newton's work in ways that he may never have intended. Newton's direct influence on these outcomes was certainly important, and in many cases his own contribution was the most decisive of all. But modern mathematical physics was not a Newtonian creation, revealed in the *Principia* and then disseminated through the reception and further dissemination of his work. Quite the contrary, Newtonian mechanics, this book argues, was a contingent historical creation like every other science, and to understand its historical beginnings we must cast aside the later, retrospective understandings that have served since the Enlightenment to canonize Newton as the heroic originator of modern mathematical physics. We do so in order to return with unfogged eyes to the actual history of the *Principia*'s initial reception.

To pursue this project, the Enlightenment narratives that Voltaire and other eighteenth-century "Newtonians" spun around their hero must be discarded. We must also cast away modernist preconceptions about the inherently progressive, rationalist development of mathematical science. The later triumphalist understanding of the Newtonian achievement poured a sticky Enlightenment goo all over the Newton archive, and today this hagiographic sludge separates us from the actual history of Newton's initial European reception. It is the project of *Before Voltaire* to wipe away this historiographical treacle so as to offer a fresh and thoroughly unsweetened reinterpretation of Newton's precise role in the historical beginnings of eighteenth-century mathematical physics. Crucial to this project is a careful reconsideration of the current historiography about the *Principia*'s European reception and legacy, and to see what is needed let us consider the current historical scholarship on the *Principia*'s eighteenth-century reception and the ongoing influence of Truesdell's "honey-sauced" reminiscences within it.

The Continuing Influence of the "Newtonian Revolution"

At the core of the Enlightenment *philosophes'* understanding of Newton's *Principia* was the claim that his treatise was a revolutionary work, one that was singularly responsible for an epochal, modernizing transformation that cast away the errors of the past and ushered in a new age of scientific light. Over a half a century ago, in the early, pioneer days of the professional history of science, the founding fathers of the discipline (and a couple of founding mothers as well) likewise made this Enlightenment understanding the foundation for the modern, postwar understanding of the development of modern science. They did so by arguing that modern mathematical physics was indeed the singular result of a rational, progressive revolution accomplished by Newton's *Principia*. H. Floris Cohen has called this body of scholarship "the Great Tradition" of the history of science,[15] and sustained by its assumptions canonical historians such as Alexandre Koyré, A. Rupert Hall, and I. Bernard Cohen wrote detailed accounts of the link tying together Newton's *Principia* with the eighteenth-century development of modern mathematical physics.[16]

For these historians, as for the Enlightenment *philosophes* they echoed, Newton's *Principia* was the epic hero in the story of modernity's triumph through science. With this one scientific treatise, they contended, Newton brought to a climax the world-historic seventeenth-century event called the Scientific Revolution, while also providing its greatest offspring: the methods and principles of modern physics. Scholarship in the history of science is no longer determined by the Great Tradition in the way that it once was, but it has not fully liberated itself from this founding narrative either. This is especially true in historical writing about mathematics, physics, and their eighteenth-century entanglement in what is now called "mathematical physics."

The history of these so-called exact sciences is still routinely written through recourse to teleological assumptions about their rational progress toward ever-greater perfection. This is especially true of the contemporary literature on the history of mathematical physics as it developed in the wake of Newton's *Principia*. This is the precise history that this book proposes to reexamine, and in one influential thread of the scholarship, the idea of an imagined "Newtonian Revolution" still remains pervasive and stubbornly persuasive to many.

Ernst Mach articulated the essential features of this understanding at the dawn of the modern historiographical era when he wrote in 1883 (before the twentieth-century transformations of physics) that work in mathematical mechanics since 1687 was little more than "the deductive, formal, and mathematical development of [the science] based on Newton's laws [as found in the *Principia*]."[17] In short, and implicitly echoing Pope's epitaph, Mach claimed that Newton had revealed all the essential features of modern mechanics in his *Principia*, and that the history of this science since 1687 was reducible to a story about the absorption, application, and further development of the divine light emanating from Newton's oracular and inaugurating treatise.

A. R. Hall echoed this understanding seventy-five years later in a chapter of his canonical textbook, *The Scientific Revolution, 1500–1800: The Formation of the Modern Scientific Attitude*. Titled "The Principate of Newton," the chapter, first published in 1954 and then reissued in several editions over the next three decades, asserted that, "With the work of Isaac Newton [in the *Principia*], the Scientific Revolution reached its climax so far as the physical sciences are concerned. . . . For the eighteenth century . . . it only remained for mathematicians to arrange the details of the Newtonian universe in somewhat more exact order."[18] In 1980, I. Bernard Cohen elaborated the same understanding in even greater detail in his book *The Newtonian Revolution*. Cohen argued that "a revolution in the sciences was wrought by and revealed in the *Principia*" such that it became "the standard by which all other science was measured."[19] George Smith, a self-professed follower of Cohen's Newtonian Revolution approach, echoed the same understanding in 2007 in his entry on Newton's *Principia* in the *Stanford Encyclopedia of Philosophy*. "By the last decades of the [eighteenth] century," Smith writes, "no one could deny that a science had emerged [from Newton's *Principia*] that, at least in certain respects, so far exceeded anything that had ever gone before that it stood alone as the ultimate exemplar of science generally." What ensued was a new project: finding "the precise nature and limits of the knowledge attained in this science" and the methods by which "this extraordinary advance had been achieved."[20]

The still widely prevalent view, articulated here by Mach, Hall, Cohen, and Smith, which holds Newton to be the father of modern mathematical physics, and which sees the development of this science after 1700 as the overdetermined consequence of the reception of the *Principia*, remains stubbornly persistent. This is especially true in the recent historical work on the *Principia* written by Anglo-American academic philosophers.[21] In the philosophical histories of the Newtonian influence, the *Principia* is taken without argument to have been a singular step forward in the making of modern science. With this legacy presumed, Newton's texts (i.e., the *Principia* and the unpublished manuscripts that illuminate it) are then mined for the "extraordinary advances," to use Smith's terms, that they offered (and continue to offer) attentive readers today.

The epistemological paradigms of contemporary academic philosophy are one reason for this approach. Professional academic philosophers, especially those addressing Anglophone audiences, tend as a group to be interested in timeless, universal legacies more than the messy contingencies of history.[22] Yet many historians of the exact sciences share the same tendencies, and in this way, the continuing influence of the Newtonian Revolution framework remains strong and widespread, at least as a governing assumption, even as the credibility of these stories as wider interpretations of Enlightenment science has all but evaporated.

Cohen's *Newtonian Revolution*, for example, is still widely cited,[23] and in its basic approach it still informs a lot of the technical literature on eighteenth-century mathematical physics. Cohen conceives of Newton's revolutionary influence in terms of what he calls the "Newtonian style" in science, a style, he argues, that was articulated in the *Principia* and then disseminated decisively as a defining feature of modern physics. As Cohen describes the accomplishment he perceives: "Newton's outstanding achievement was to show how to introduce mathematical analysis into the study of nature in a rather new and particularly fruitful way, so as to disclose *Mathematical Principles of Natural Philosophy*, as the *Principia* was titled in full." "In the *Principia*," he continues, "the science of motion is developed in a way that I have characterized as the Newtonian style. . . . This style permitted Newton to treat problems in the exact sciences as if they were exercises in pure mathematics and to link experiment and observation to mathematics in a notably fruitful manner. The Newtonian style also made it possible to put to one side, and to treat as an independent question, the problem of the cause of universal gravity and the manner of its action and transmission." Moreover, Cohen concludes, "not only did Newton exhibit a powerful means of applying mathematics to nature, he made use of a new mathematics which he himself had been forging."[24]

For Cohen, then, the Newtonian style of the *Principia* is the first instantiation of what has since become a foundational feature of modern physics as a whole: the use of mathematical analysis to account empirically and predictively for the observed phenomena of nature. In the *Principia*, Cohen contends, this new approach to physics, including the new mathematics that made it possible, the infinitesimal calculus, is articulated for the first time. The revolution that ensues, a revolution that Cohen contends was triggered by the publication and reception of Newton's work, involves the introduction and dissemination of the "Newtonian mathematical style of physics" throughout Europe, and then the initiation through its use of the new mathematicized approach to physics that is the hallmark of modern physical science today.

Like Cohen's book, *Before Voltaire* is also concerned with the innovative new entanglement of mathematical analysis and physical explanation in Europe around 1700. But unlike *The Newtonian Revolution*, this book will resist all claims for a singular "Newtonian" origin of this development, or a singular revolutionary source for it in the *Principia*. Cohen's account of the influence of the so-called Newtonian Revolution on Jacob Hermann's 1716 *Phoronomia* illustrates well the weakness of his broad-brush conception of a general Newtonian style as the source for Hermann's work.

Cohen notes, for example, that Hermann's work was "generally Leibnizian," and that the *Phoronomia* used "Leibniz's algorithm for the calculus," which was nowhere to be found in Newton's *Principia*. Newton, as we will soon see, avoided using his own calculus in his treatise, at least in the explicit and assertive way that Hermann used Leibniz's in his. Nevertheless, Cohen gives Newton credit for inventing the mathematics that made Hermann's mathematicization of physics possible. As we will soon see in detail, this elision is not trivial, for it was Leibniz, not Newton, who developed the precise mathematics that made Hermann's mechanics possible, and his decision to eschew the alternative mathematics offered in the *Principia*, a choice that many in England made differently, was no insignificant matter. Cohen also notes that Hermann offered an "effusive tribute" to Leibniz in the treatise, but rather than explore the many reasons for this praise by examining the reasons for the many Leibnizian sources for Hermann's science, Cohen simply notes the general similarities in mathematical approach between Hermann's work and Newton's, and then concludes that the *Phoronomia* shows "how overtly the Newtonian dynamics tended to infiltrate even the Continental opposition." The assumption about widespread Continental opposition to Newton's work is another pillar of the old framework that is taken for granted, and it is enough for Cohen that Hermann refers to Newton's *Principia* as a "golden work" (*aureum opus*), and that he deploys the "somewhat Newtonian style of dealing mathematically with the properties of imagined [physical] systems . . . that all writers on this subject eventually used" to make him an agent of the Newtonian Revolution. This despite Hermann's Leibnizian mathematics, his praise for Leibniz's scientific outlook, and the many other non-Newtonian tendencies in his science.

Triumphalist ahistorical interpretations such as these inevitably result when the assumption is made that modern mathematical physics flows solely from the fount of Newton's *Principia*. This is why *Before Voltaire* begins by casting aside all remnants of the Newtonian Revolution framework, and likewise parts company with the philosophical historiography that still depends on these assumptions. This book is a study of contingent historical developments, not a reconstruction of the universal rationality that is imagined to have unfolded out of Newton's treatise. *Before Voltaire* therefore breaks entirely with all the lingering rationalist and teleological tendencies of the "Great Tradition" literature so as to focus precisely and attentively on the actual historical steps that led from Newton's mathematical mechanics as published in 1687 to the later science that came to be associated with his name and legacy.

It further breaks with the scholarly program just outlined by posing for historical scrutiny the claim that eighteenth-century mathematical mechanics, and by extension modern mathematical physics, was primarily a Newtonian creation, developed and announced in the *Principia* and then completed and established through the reception and acceptance of his work.

Cutting a historiographical path explicitly across the persistent influence of this Great Tradition literature, *Before Voltaire* pays close attention to the precise science found in Newton's *Principia*, and the ways that eighteenth-century mathematical mechanics was and was not a consequence of its direct, indirect, and even oblique influence. It will further look at the emergence in France at precisely this moment of a new science of mathematical mechanics, one that is often seen as the teleological next step in the revolutionary unfolding of classical Newtonian mechanics. Yet contrary to the Newtonian Revolution understanding, *Before Voltaire* considers calculus-based mathematical mechanics as an underdetermined and contingent historical development, one produced without question through the influence of the *Principia*, an influence that will be analyzed in detail, but one that was also engendered by a host of other influences in a manner that calls into question the later labeling of it as "Newtonian mechanics," plain and simple.

To accomplish this, all remnants of the triumphalist Newtonian Revolution framework must be cast aside. This means marginalizing (or better said, legitimately ignoring) the increasingly copious philosophical historiography about Newton's *Principia* and its legacy, a literature that is still anchored in the Great Tradition paradigm. Unlike this literature, my study does not propose to understand what readers could have, or, worse yet, should have gathered from studying the *Principia*. It rather attempts to survey the different readings of the book that actually proliferated in the first two decades of its reception. What the recent scholarship on Newton's first published scientific treatise has shown us is that early readers of the book saw many other things in it besides what rigorous academic historians and philosophers have discerned to be its most valuable contributions. The project of *Before Voltaire* is to let these early interpretations stand on their own and speak for themselves, and then to allow these historical readings, not our own contemporary scholarly understanding, to show us what Newtonian science was originally understood to have been, and how it was actually used in the making of eighteenth-century science.

Especially important to this project is the rejection of the idea, central to the philosophical historiography, that the *Principia* contained some intellectual unity called "Newtonian science" out of which modern mathematical physics was born. The best recent philosophical scholarship has as its goal the establishment of precisely this "Newtonian Way," or "Newtonian Method," that sums up the essence of Newton's achievement in the *Principia* and defines for us the essence of his modern scientific legacy. I appreciate this literature as a guide for reading the *Principia* in light of modern scientific thinking today, but the goal of this book is not to read and interpret the *Principia* in twenty-first-century terms. My goal is to reconstitute the earliest readings of the book from around 1700, the ones that were actually instrumental in starting the chain of interpretations and translations that produced eighteenth-century mathematical mechanics as a contingent historical outcome. To write this history, *Before Voltaire* eschews all remnants of the Great Tradition's concept of a Newtonian Revolution and opposes all insinuations that eighteenth-century mathematical physics was a coherent scientific unity, assembled in Newton's study and at his lab bench and then delivered to the world through the publication and dissemination of his scientific masterpiece.

In the beginning was the translation, or so I argue in this book. Its objective is to look carefully, and scrupulously historically, at the initial and initiating translations that started the process whereby Newton's work in the *Principia* was read, misread, ignored, revised, rebutted, and ultimately digested and then surpassed in the initial making of the modern mathematical physics that only later came to bear his name. Robert Schofield correctly captured the program of this book in 1970 when he wrote that "the reasoning which underlay eighteenth-century . . . science cannot be found by determining what Newton really meant in the *Principia* or the *Opticks*. Indeed, for such a study, Newton's real meaning is essentially irrelevant. What is needed is an investigation of what eighteenth-century [savants] thought he meant and how and why their thinking changed through the century."[25] This is the historicist genealogical program that is pursued in this study of Newton's initial French reception.[26]

The Archaeology of Early Modern Mathematics

In building my arguments, I will accordingly avoid the retrospective conceptual teleologies of the Newtonian Revolution historiography, both old and new. But this is not to say that the contemporary historiography on Newtonian mathematics and mechanics will be eschewed altogether. Quite the contrary, a second and equally vital stream of aggressively historicist scholarship undergirds the arguments of this book. This other corpus of historical writing might be distinguished from the Newtonian Revolution literature by calling it the work of the "mathematical archaeologists." At the beginning of this historiography stands a single figure, Clifford A. Truesdell, and a single journal that Truesdell helped to found in 1960: *Archive for History of Exact Sciences*.

In the very first issue of *Archive*, Truesdell laid out his pioneering agenda, calling it the "Program to Recover the Rational Mechanics of the Age of Reason." Noting the tightly wound hagiographic knot that still tied together Pope's epitaph for Newton, Mach's foundational history of his achievement in mechanics, and the commonplace scholarship on eighteenth-century mechanics around 1960 (the heyday of Floris Cohen's "Great Tradition" literature), Truesdell charted a different historiographical path, one that began by taking Newton's *Principia* down off of its exalted pedestal. "Since the *Principia* is one of those works that everyone talks of but no one reads," Truesdell explained, "anything said about it other than the usual honey-sauced eulogy must stand up against righteous indignation from all sides. But it is a work of science, not a bible. It should be studied and weighed—admired, indeed, but not sworn upon. It has its novelties and its repetitions, its elegant perfections and its errors."[27] Viewed from this disenchanted perspective, the historical project that Truesdell imagined was not a continuation of the triumphalist storytelling about how eighteenth-century mechanics was unfolded out of Newton's *Principia*, but a new historical excavation of eighteenth-century science that would use period sources to reveal how the science we now call classical Newtonian mechanics came to be assembled.

As a further prerequisite for his project, Truesdell also isolated another distorting spell, one separate from the charm of the supposed Newtonian origination of eighteenth-century mechanics. This was the equally magical influence of Lagrange's 1788 *Mécanique analytique* on scholarly understandings of the history of mathematical mechanics after Newton. Truesdell saw two ways that Lagrange had cut us off from the actual history of rational mechanics as it had developed in the century after the *Principia*. First, since many view Lagrange's treatise as the definitive articulation of the foundations of this modern science, and as such as "the final repository of all the mechanics that went before it," rationalist historians have not been inclined to recover its history because, from their progressive point of view, these earlier steps amounted to little more than moments of adolescent growth along a path toward mature scientific perfection. Lagrange also included his own histories of statics, dynamics, and fluid mechanics in the *Mécanique analytique*, deploying the Enlightenment understanding of the Scientific Revolution to frame these modern sciences as progressive outcomes produced by the advances launched by Galileo and completed by Newton. Lagrange's histories "have been accepted as final in outline if not in detail," Truesdell explained, and they therefore reinforced the historiographical idea that the passage from Galileo to Newton, and then from Newton to Lagrange, was a natural progression. Viewed this way, it was a historiographical understanding that warranted no revision.[28]

Truesdell's "program of rediscovery" began with his critical detachment from these musty, ready-made Enlightenment just-so stories. The program then proceeded through a new inquiry into the actual history that produced eighteenth-century mathematical physics. In this respect, those that have answered his call are "mathematical archaeologists" seeking to unearth the historical steps that produced rational mechanics while not privileging any particular origin or outcome for these changes. Much indispensable work has been done in this vein, and the volumes of *Archive for History of Exact Sciences* offer an important reservoir of technical historical scholarship that is crucial to the arguments of this book. However, since the mathematical archaeologists overall have not broken as fully with the Newtonian Revolution framework as their program might suggest, there is still a need for more archaeological work.

Truesdell is a case in point. He rejects the idea that the "primitive mechanics" found in the *Principia* was a direct historical determinant of modern physical science, but he does see the book as a crucial turning point, and a kind of formative template, one that "led ultimately to mechanics as we know it today."[29] Seeking to sustain a singular heroic founding for rational mechanics, Truesdell also introduced an analytical bridge linking Newton with the modern mechanics associated with his name. This bridge has become canonical as well, problematically so this book contends, in the technical historical literature on eighteenth-century mechanics. It also needs to be exposed and scrutinized. For Truesdell, Newton's "primitive mechanics" of the *Principia* was not the direct source for the "most successful, the most thoroughly proved and understood, and the most perfect of the sciences of nature—the prototype of and paradigm of a mathematical theory for all physical phenomena," namely rational mechanics. Instead, he contends, in a historiographical move that exemplifies his break with the "Newtonian Revolutionaries" and anchors half a century of scholarship in the history of the exact sciences, the key figure who in fact pointed the way to these world-changing advances was Leonhard Euler.

Viewed as Truesdell sees it, Newton's *Principia* starts a process that eventually leads to the perfection of mathematical mechanics in the eighteenth century. But while Newton got things going, the actual accomplishment did not occur before "most of the life work of Leonhard Euler" was spent "clarifying and developing the Newtonian concepts, supplementing them by equally important new ideas, and demonstrating how real problems could be solved."[30] Situating Euler, along with his hometown colleagues in Basel, the Bernoulli family, as the actual founders of rational mechanics has now become a commonplace in the technical scholarship on the history of mathematical physics. While this has displaced Newton's singular heroic role in this history, his importance has been retained through the creation of another historiographic framework that this book will contest. This is the idea of a Continental translation of Newtonian mechanics after 1690, which is said to make possible the work of Euler and the Bernoullis in the making of modern mechanics after 1730.

In this understanding, which Truesdell helped to initiate and is today a fixture of the scholarship on eighteenth-century mathematical physics, Newton is the initiator of a newly mathematicized understanding of physics in the *Principia*, and the originator of a new set of mechanical principles that will ever after characterize this science. His example and influence is also seen as surpassing all others, making him the guiding thread in these later developments. But historically speaking, Newton's science is not yet modern mechanics, and it only becomes fully developed after the science of the *Principia* is translated into the Continental mathematical idioms of the Leibnizian calculus. Once it is made "analytical" through articulation via the Leibnizian calculus, Newton's mechanics is then developed into the modern analysis-based mathematical physics of Euler, the Bernoullis, and finally the French line of mathematicians that runs from d'Alembert and Clairaut at midcentury to Laplace and Lagrange around 1800. Modern mechanics when traced this way is at once the direct result of a mathematicizing revolution initiated by the *Principia* and at the same time the historical result of the technical mathematical work accomplished by Continental Europeans after 1730.

This particular historiographical understanding has produced a range of different outcomes in the literature. One thread simply starts with Newtonian mechanics after the completion of the so-called Continental translation and then traces its subsequent development after 1730.[31] Sometimes the term "Newtonian" is retained in this literature as the label best describing the object under study. But more often the history of this science before Euler is simply taken for granted and left unexplored. In other formulations, Newton is invoked as the distant source for this or that concept or idea, while Euler and his successors are treated as the mathematicians who actually worked to achieve the pioneering results suggested by, but not fully realized in, the *Principia*. Another strand of the literature simply treats the period between Newton and Euler (i.e., 1690–1730) as an analytical "black box," noting these decades as the time when the Newtonian foundations articulated in the *Principia* were transformed into modern mathematical physics, but also leaving the question of how this crucial historical transformation occurred completely unexplored.

Ivor Grattan-Guinness's multiauthored *From the Calculus to Set Theory, 1630–1910* illustrates well the historical narratives that these assumptions produce. Chapter 1 of his book massages all of the varied and contested seventeenth-century developments in algebra and infinitesimal mathematics into a progressive account of the development of the differential calculus by Newton and Leibniz around 1680. Chapter 2 picks up the story from there, looking in detail at the complexities of the calculus as developed by Newton, Leibniz, and others in the 1690s. Then, without pause or transition, the book shifts to Euler's work after 1740. To explain the jump in time, the section opens by declaring that "in the (about) 50 years after the first articles on the calculus appeared, the Leibnizian calculus developed from a loose collection of methods for problems about curves into a coherent mathematical discipline: Analysis." No account of the historical making of analysis is offered. Instead, discussion jumps to the 1740s, claiming that while "many mathematicians, such as Jean le Rond d'Alembert, Alexis Clairaut, and the younger generation of the Bernoullis" contributed to this development, it was in large measure the work of one man: Leonhard Euler.[32]

The historical effacements produced by these compressed and hero-burdened narratives are also illustrated by the historical presentation offered in a recent physics textbook, one that attempts to use history to introduce students to *The Elements of Newtonian Mechanics*.[33] In chapter 1, "The Foundation of Classical Mechanics," we learn that Newton's *Principia* was "the masterpiece" of modern physics, for in this book "we find a masterly synthesis of the concepts of motion and force" and a "formulation of the laws of motion [that] has, with superior strength and vitality, survived for more than 300 years." A summary of these concepts and laws then follows, one that translates the actual contents of Newton's treatise into contemporary mathematical nomenclature. Newton's second law of motion relating to acceleration, for example, is written using the algebraic equation F = ma even though no such equation appears in the *Principia*. Nor could it have appeared, because Newton resolutely avoided an algebraic presentation of his ideas, depending instead on an idiosyncratic form of geometry that he preferred on rigorously epistemological grounds.[34] Likewise, in chapter 2, when the textbook begins to develop the mathematical formulas and rules of operation essential for making Newtonian mechanics a predictive mathematical science, the authors shift seamlessly and without comment into the Leibnizian calculus, as if this is how Newton developed and presented his ideas. In fact, as we will explore in detail in the chapters to come, Newton did not use the Leibnizian calculus in developing his own mechanics, nor even his own fluxional alternative to it. The mathematics he did use, which he vigorously defended as superior to the calculus, was also at odds with the analysis that these contemporary physicists rightly teach as the very foundation of mathematical physics today.

Pulling these strands together, what the contemporary literature on eighteenth-century Newtonian mechanics generally offers us are two discrete histories joined tenuously together by a rickety and gap-ridden narrative bridge. On the one hand are the meticulous histories of Newtonian science as it was revealed in Newton's published treatise and the voluminous manuscript archive that he left behind. Much historical scholarship in this vein is pursued with the goal of isolating the essential features of the Newtonian scientific achievement. In the best examples of this recent scholarship, the scientist who emerges is an innovative experimentalist who channeled intense worries about epistemological rigor and the unruly character of sensate experience into innovative methods for reconciling divergent data and establishing reliable theories that explained observed phenomenon. As a mathematician, Newton was similarly complex, combining an intense attachment to what he believed were the epistemologically superior foundations of antique geometry, while sustaining an approach to physico-mathematics that made systematically quantified experiential and experimental reasoning a central tool for revealing the lawful regularities of the cosmos.[35]

This is a compelling and persuasive historical picture, but alongside it is another understanding, one that is crucial to the literature on the eighteenth-century making of modern mathematical physics. This other picture presents Newton as a mathematicizing analyst who, to use I. B. Cohen's conception, freed mathematical analysis from causal physical explanation, and through this split charted the course that prepared the way for Euler, Lagrange, and the calculus-based mathematical physics of the modern era. Sitting between these two divergent historiographies, moreover, is the scientific terra incognita of the decades around 1700, the period when Newton's very differently oriented work in the *Principia*, at least as revealed by the most recent scholarship, was somehow transformed into (or is it more accurate to say "was reconfigured through the making of") the new analytical mathematics and mechanics that Euler and his later eighteenth-century successors built into what we today call "classical Newtonian mechanics."

How are these two historiographical understandings to be reconciled? By entering into the scholarly terra incognita of the years 1685–1715, and by exposing for historical analysis the contents of the black box into which the traditional scholarship has cast the actual, contingent history of mathematical mechanics during its formative years. This is the primary purpose of this book. To realize its agendas, we will return to Truesdell's original call for an archaeology of eighteenth-century mechanics in order to pursue his program with a new commitment to precise, historicist rigor. This will mean explicitly avoiding any and all preconceived expectations about what this history must reveal in terms of progressive, rational development. It will also entail a strict avoidance of retrospective, teleological conceptualizations. Finally, it will eschew all urges toward hero making and resist every narrative of canonization that celebrates certain actors for their special, or even worse, supernatural, roles in the realization of these influential yet exceedingly contingent historical outcomes.

The Historicist Project of *Before Voltaire*

Fortunately, the rigorous and careful historicism that guides this book is already found in the work of several distinguished historians of eighteenth-century mathematics, and their scholarship will be an essential source for all that follows. What distinguishes these historians is their embrace of a "history-for-its-own-sake" approach to the history of the so-called exact sciences, and an unapologetic comfort with a fully mortal, human, and period-historical (as opposed to world-historical) conception of eighteenth-century scientific developments. In short, this influential cluster of work is as strongly historicist and value neutral with respect to the contemporary relevance of the Newtonian achievement as the philosophical historiography is presentist and partisan in its interest in disclosing the supposed heroic advances launched by Newton and his science.

Emblematic of this more historicist approach—and a scholar whose influence on this book cannot be overstated—is Henk J. M. Bos. In the short article "Philosophical Challenges from the History of Mathematics," Bos conveniently summarizes the methodological differences that make the historicist scholarship on Newtonian science that I will deploy completely different from the literature in the history and philosophy of the exact sciences summarized already.[36] "All mathematics was, is, and always will be incompletely understood," Bos asserts. The implication is that any overly seamless and rationalist narrative about the teleological unfolding of mathematical truth is historically untenable.[37] Also crucial for Bos is the need to avoid presentist, normative judgments about the contemporary value of past scientific work when writing the history of mathematics. It is essential to develop "a historical perception of mathematics through the actions of mathematicians," Bos contends, and not simply "through their mathematical results." This means avoiding contemporary standards of correctness when interpreting past mathematical work, and writing the history of mathematics as a contingent account of science in the making, not as a retrospective evaluation of past mathematical achievement. This also means letting the past mathematical actors themselves set the terms by which we judge them. As Bos writes, "The actions of mathematicians are to be understood as performing self-imposed tasks according to self-created criteria of quality control." In the same vein, Bos adds, "the use of terms such as 'truth,' 'proof,' 'rigor,' 'exactness,' 'purity,' 'legitimacy,' etc. generally indicates a conviction that mathematical endeavor can produce knowledge of a particular absolute quality." As historians, we cannot take such terms for granted, or treat them as ahistorical, universal values. Instead, we must work to understand the historical reasons why and when absolutist epistemological claims such as these are deployed. To be rigorously historicist, in other words, "the historical analysis of mathematical developments should *not* be based on [the a priori acceptance of] these convictions."[38] Bos also insists that "all mathematical concepts are fluid concepts," which is to say that mathematics, like every other aspect of human knowledge, develops contingently in time and according to shifting flows of locally regulated practices and quality control.[39] Any rigorously historicist account of the development of mathematical science must, therefore, be aligned with, and flexible toward, this conceptual fluidity.

The genealogical approach adopted in this book, along with its focus on the reiterative moments of translation that intervened between the initial readings of Newton's *Principia* and the construction of the mathematical mechanics that came to be associated (much later) with his name, is an attempt to apply Bos's tenets to a particular moment in the history of mathematical physics. The method of *Before Voltaire* is also influenced strongly by the contemporary historicist understandings of mathematics and its development. Ian Hacking, for example, starts his 2014 book *Why Is There Philosophy of Mathematics at All?* by trying (unsuccessfully) to define mathematics as something stable and universal. Ultimately, he concludes, following Wittgenstein, mathematics is best understood as a motley, one that may be no less historical than any other object of human thought.[40] Also influential are the historicist strands of contemporary science studies, especially those found in the work of Bruno Latour and Hans-Jörg Rheinberger. Latour makes translation a fundamental analytical tool for understanding diachronic change in the sciences. *Before Voltaire* does the same. As Latour writes: "Instead of opposing words and the world, science studies, by its insistence on practice, has multiplied the intermediary terms that focus on the transformations so typical of the sciences. . . . [Translation] refers to all the displacements through other actors whose mediation is indispensable for any action to occur. In the place of a rigid opposition between content and context, chains of translation refer to the work through which actors modify, displace, and translate their various and contradictory interests."[41] Rheinberger is likewise attentive to the historical contingencies that a focus on translation helps to illuminate. "A historical philosophy of science should not universalize its standards," he writes. "Both the ideas and the ideals of scientific method in the course of history can change, and what is or is not science can actually vary. . . . In order to understand particular historical developments, there is no alternative but to pursue detailed investigations."[42] Others have also followed the historicist spirit of Bos, Hacking, Latour, Rheinberger, and translation studies more generally in approaching the history of the mathematical sciences in a deeply historicist way, and the work of Niccolò Guicciardini must be singled out in this context as especially paradigmatic.[43] Also important is the work of other historicist historians of mathematics and physics whose contributions will be noted where relevant in the pages that follow.

Beyond the "Internalist" versus "Externalist" Dichotomy

Before turning to the substance of the book at hand, one final methodological and historiographical aspect of it needs to be clarified in order to prevent misreading and misunderstanding. This involves the position of this book with respect to the historiographical division (often more imagined than real) separating the so-called internalist and externalist approaches to the history of science.

By situating this book within the narrow and technically intricate literature on the history of the mathematical sciences, and by emphasizing the importance of the deeply historicist strand of this literature that Bos and Guicciardini represent, I have self-consciously tried to displace any perception that *Before Voltaire* is an externalist cultural history of mathematics and physics that has as its goal the repudiation of more technical and internalist understandings. According to the conventional understanding of the internalist/externalist division, Bos and Guicciardini would be internalists, because they are scholars who devote themselves intensely and precisely to the technical scientific matters present in the archive. The philosophical historians discussed earlier would also be grouped with them as "technical" and internalist scholars of Newton, because they also focus intensely on the precise technical details of his science. By the same standard, *Before Voltaire* might at first blush appear to be an externalist work discordant with these other studies given its broad focus on the cultural and historical context of the technical mathematics under study.[44] No understanding would be more misleading, however.

No one can deny the continuing presence of the internalist/externalist divide within the scholarship, and this book is certainly a deeply contextualized cultural history that sits comfortably alongside other so-called externalist studies of the history of science. Yet this book is in no way motivated by a desire to mobilize this distinction for analytical purposes, and it is even less interested in provoking externalist versus internalist polemics regarding proper practice when writing the history of mechanics. Instead, I encourage readers to think of *Before Voltaire* as a historicist synthesis of internal and external perspectives, and to read the book as a study that seeks to work across the imagined internalist/externalist divide in the name of a more complete and integrated historical understanding.

In a capsule description of the goals of his scholarship offered to the general public, Guicciardini articulates wonderfully the historicist synthesis of technical/internalist and cultural/externalist perspectives that this book aspires to achieve. "In my books," Guicciardini states, "I describe the birth of . . . [Newton's] calculus and . . . mathematical theory of gravitation. At school, we have been taught that these theories are indisputable and universal (I mean independent from cultural determinations). In my books, I show that the birth of calculus and gravitation theory interacted with many aspects of the culture of Newton's times, including especially philosophy."[45] *Before Voltaire* is inspired by the same historiographical spirit. Informed by it, the book works to integrate precise technical scientific understanding with an appreciation for its full immersion in the contingencies of ordinary human history. This is why Bos's methodological program is so resonant with my own, and why this book pursues a broadly contextualized approach while also remaining attentive to the technical scientific developments that it seeks to illuminate. Indeed, even if my comprehensively contextualized approach to the history of Newton's French reception has a very different look and feel than the technically focused historical literature published on these topics, and even if the book might appear to be offering an alternative to such histories, *Before Voltaire* is fully derived from, and in step with, the technical literature in the history of the mathematical sciences. It is at the same time a book committed to widening the lenses of technical scientific interpretation in order to include social, political, cultural, and institutional factors as well.

No challenge to specialized, technical, internalist history of science is therefore intended or present in this book. Quite the contrary, I see *Before Voltaire* joining with the aggressively historicist strand of contemporary Newtonian studies in ways that build from these more technical and specialized treatments a broader and more fully contextualized understanding. As such, *Before Voltaire* is a book just as attuned to the historicity of the scientific technicalities under study as it is attentive to the contingent institutional and cultural developments that were integral to these scientific changes. Each side of this equation is equally important, and the intended purpose of *Before Voltaire* is to synergize these different perspectives in the creation of a historical account that is accurate and nuanced in both its technical scientific and its social analytic claims.

The cultural history (externalist) literature on eighteenth-century science, and the Newtonian legacy therein, is therefore essential to this book as well. Whether it's Larry Stewart's insights about the role of public spaces and status-laden dynamics of sociability in the circulation of eighteenth-century Newtonian science, or Mary Terrall's arguments about the role of ambition and public sociability in the practice of eighteenth-century French academic science, or Andrew Warwick's demonstration of the role played by institutional politics and discipline in the refinement of "pure" Newtonian theory, the understandings offered by social and cultural historians of Enlightenment science are central to this book's arguments.[46] Since broad contextualization often takes one far away from the narrow domains of technical science, in this book I pursue inquiries in an aggressively interdisciplinary way, drawing upon work in literary studies, art history, and early modern cultural studies whenever it is fruitful in framing my interpretations. The point to emphasize is that these steps away from the technical study of mathematical science narrowly construed are not made in contradistinction to the perspectives offered by technical and internalist historians; they rather derive from an effort to enrich our understanding of the historical development of technical science through broader contextual interpretation.

My approach ultimately favors broad cultural history, and this orientation is present as well in my avoidance of the practice, common in technical histories of the mathematical sciences, of reproducing the technical details of the mathematics itself in the text of the histories accounting for it. The technical historical literature I will draw upon often has as much mathematical language on its pages as explanatory prose, and often the mathematical symbols do as much argumentative work as the prose descriptions offered alongside them. In this book, by contrast, I have reduced the direct mathematical content to almost nothing, substituting for it prose descriptions that explain the precise mathematical content at issue. My justification for this approach is multiple. First, I am not a specialist historian of mathematics who aspires to reveal technical shifts within the practice of mathematicians themselves. To pretend that I am by including the apparatus of such technical historical analysis would therefore be dishonest. Yet even if the literal mathematics and physics at issue is not present on these pages, my understanding of it is based on a careful study of this mathematics in the light of the best recent scholarship. My goal is to develop technical interpretations of the mathematical work pursued around 1700 that fill historiographical gaps in the technical literature. But I also want to show how the technical changes characteristic of these years were embedded in, and shaped by, wider cultural dynamics. For this broad cultural entanglement to be visible to the broad readership that I hope to address, the technical mathematics needs to be translated into the discourse of mainstream cultural history. This is thwarted by any use of technical, mathematical symbolism to sustain the narrative itself. I have accordingly made a choice to use a methodological translation—the transformation of technical mathematics into narrative prose—to produce my integrated cultural history of the historical translations attendant to the French reception of Newton's *Principia*.

As with any translation, some things are gained, and others are lost, by adopting this approach. Technical historians of mathematics will likely be disappointed by the inattention given to the precise conceptual and symbolic shifts evident in the archive, shifts that are indeed the bedrock of this history. But my goal here is not to pursue technical history of mathematics, but rather history as a means of highlighting how these detailed, technical changes in the nature and practice of mathematics were integral to, and in many ways provoked by, wider cultural transformations. My discursive choices are motivated by this historiographical agenda, and if it has meant avoiding the genre conventions of technical history of mathematics, I will at least point these historians to the sources where this material can be found.

I also believe that this is a history that should be read by a wide audience, so if I aspired to reach a broader readership by adopting the idioms most familiar to cultural history, I also charted my course in the hope of using translation to make the best and most highly nuanced scholarship in the history of the mathematical sciences accessible to readers who otherwise would have no reason to pick it up. My ultimate goal is to provoke cross-disciplinary dialogue and interrogation, and for students of early modern culture I have tried to make otherwise obscure developments in the mathematical sciences accessible, while at the same time showing the relevance of this technical science to familiar social and cultural shifts. For technical historians of mathematics, I have tried to write a history that accurately deploys the conclusions of the best recent scholarship while also translating and integrating this material into the broader scholarship in cultural history. If the result does not contribute directly to our technical understanding, it at least demonstrates how technical scientific change participated within, and was accordingly integral to, important historical processes of the early modern period. Readers in different communities will ultimately have to determine whether my skills as a translator were up to the many challenges of cross-disciplinary negotiation and communication that this project presented.

Making Newtonian (?) Mechanics in France around 1700

If translation is central to this book both conceptually and methodologically, it is also important to it literally, at least in a historiographical sense. This is because the dominant frame still governing writing about Newton's initial French reception treats it in terms of an imagined Continental translation of the *Principia*'s Newtonian mechanics into the mathematical language of European science, namely the Leibnizian calculus. The emergence of this particular interpretive understanding within the historical scholarship has already been discussed, and the goal of this final section is to summarize from the outset the reinterpretation of this conventional understanding that this book proposes to offer.

At the center of the story is Pierre Varignon. In the 1690s, Varignon, a fortyish-year-old member of the Académie Royale des Sciences, began to develop what he came to call his *nouvelle théorie du mouvement*, or "new science of motion." His innovation involved integrating a mechanical understanding of the composed nature of forces in moving bodies with the extremely novel mathematical methods of the differential calculus, which had only been introduced into France a few years earlier. The new science that resulted was the first articulation of what we today call "analytical mechanics," the science from which eighteenth-century mathematical physics was built. This is the science that is also taught today to all introductory physics students under the name "classical Newtonian mechanics." To secure the Newtonian pedigree of analytical mechanics, historians have turned Varignon into a translator who is said to have transferred Newton's supposedly revolutionary work in the *Principia* into something acceptable and communicable to European savants.

Varignon certainly read Newton's treatise soon after it was published, and he may even have been the reviewer who evaluated the *Principia* in the leading Francophone learned periodical of the day, the *Journal des savants*, little more than a year after the book's publication.[47] Varignon also invoked Newton's *Principia* frequently as a reference when articulating his own scientific positions. This has led those inclined to view this history from a Newtonian Revolution perspective to describe Varignon as France's first Newtonian, and to view his science as the first step in the larger Continental dissemination of Newtonian mechanics. E. J. Aiton speaks from this vantage point when he writes that "Varignon was in effect a Newtonian. His real achievement was the interpretation of Newtonian planetary theory to Continental mathematicians more conversant with the language of the differential calculus than the geometrical style of the *Principia*."[48] The mathematical archaeologists see things differently. Rather than view Varignon as the midwife who delivered Newton's scientific child and then passed it to its European governess, they view him as the creative translator who refashioned Newton's primitive science in the *Principia* into the raw material from which Euler and Lagrange would build modern mathematical physics. Michel Blay, whose work on Varignon was built upon Henk Bos's seminal rethinking of the historical influence of the Leibnizian calculus, articulates this second view concisely when he writes that "by allowing the science of motion to benefit from the recent progress of [mathematical] analysis, Varignon . . . truly paved the way for the immense development of mathematical physics in the eighteenth and nineteenth centuries."[49] In particular, Blay asserts, Varignon's new science of motion showed "in an exemplary, and, at long last, inaugural way, that scientific work must aim above all at obtaining, and rigorously manipulating, rules and formulas. The field of mathematical physics was now entered into once and for all, and that of the old science of motion, with its ontological and geometric ambitions, was left behind."[50]

While very little scholarship about Varignon and his new science of motion exists, all of it is framed in terms of one or the other of these two points of view.[51] Either Varignon literally translates Newton's science into Continental mathematical idioms, or he uses Continental mathematics to creatively translate the science of the *Principia* into something suitable for further development.[52] Never is he simply a late seventeenth-century Frenchman pursuing his own intellectual agendas under the influence of all the intellectual and cultural currents he was swimming within. The pages that Niccolò Guicciardini devotes to Varignon's work "on central forces and resisted motion" in his *Reading the Principia* have gone the farthest in breaking this pattern and recovering the historical contingencies and complexities present in Varignon's work. But even Guicciardini still frames Varignon in terms of his "role in the process of translation of the *Principia* into the Leibnizian calculus."[53]

Before Voltaire takes as its starting point the rejection of this "translator" conceptualization of Varignon's work. Varignon's reading and translation of the *Principia* will remain central to all that follows. The history that this book proposes will further be built upon a close analysis of the translations that ensued when he and others read the *Principia* in France at the turn of the seventeenth century. But as an actor in this history, Varignon will also be understood as a free agent who pursued his work without any determination by larger overarching historical trends. He was never a slave to Newton's influence, nor to any other influence. He read the *Principia* and learned many things from it, but he did not find in the book a historical destiny that he set out to fulfill. He rather encountered it as an intriguing mansion full of brilliant scientific marvels, but also a house with a musty old architecture and an oddly antiquated décor that was full of many closets stuffed with strange and eccentric curios. He also read the work in light of what other mathematicians of the day were thinking and doing, especially those with whom he worked at the Académie Royale des Sciences. The science that Varignon ultimately made from his wide and discerning reading is therefore best understood as a contingent result of all of these influences, and not as something specially determined by any one influence.

The previous scholarship has also viewed Varignon as a crucial mediator who facilitated the progressive development of modern mathematical physics from Newton to Euler and then to Lagrange. *Before Voltaire* will take a different approach. It will eschew the teleological framework that makes the period 1690–1730 nothing but a prelude to Euler by returning directly to the local historical context that produced this particular French mathematician and his new science of mechanics. It will also present Varignon's science as the product of his particular work in this precise place and time. By presenting Varignon and his science in these deeply historicized terms, *Before Voltaire* will ultimately argue for an understanding of his importance in the history of science that is inseparable from these specific human and historical contingencies.

Tying the whole history together will be the story of Varignon's development of analytical mechanics in the decades around 1700 in the context of the French Académie Royale des Sciences. A biographical mode of narration will therefore be deployed throughout that will make analytical mechanics appear to emerge as the individual creation of a single inspired mathematician. The actual historical argument, however, will proceed by continually immersing Varignon and his work in the cultural and institutional contexts that informed his activities and sustained everything he accomplished. Stated another way, *Before Voltaire* tells the story of the birth, development, reception, contestation, and then institutionalization of analytical mechanics as an established French science by 1715, and it does so through a biographical account that emphasizes Varignon's role in producing this precise scientific outcome. But in following the historical genesis of analytical mechanics via Varignon's career from 1680 to 1720, the book will also proceed by complicating at every step Varignon's singular role in realizing these achievements. It will also emphasize the collective and impersonal historical contingencies that were just as crucial in making this science.

The book overall is organized to achieve this collectivized historical approach, and taken as a whole it works diachronically to tell the story of the origination, reception, and eventual establishment of analytical mechanics as a permanent fixture of French science by 1715. Although chronologically arranged, the book also works synchronically by situating each moment of change in the broadest possible context. The book will unfold in three parts tied up by a coda.

PART 1. THE INSTITUTIONAL SOURCES OF ANALYTICAL MECHANICS

Part 1 opens the story by surveying the field of French mathematical science in the late seventeenth century, the field out of which Varignon's new science of motion grew. The two chapters in this section also initiate an argument central to the book overall, namely that analytical mechanics was authored as much by the intellectual and institutional dynamics of the French Académie Royale des Sciences as it was by the individual work of any one royal academician.

Chapter 2 initiates this analysis by looking at the seventeenth-century understandings of "mechanics" that were in play when Varignon set to work. Still caught around 1670 in a cross fire that pitted older understandings of mechanics associated with physical labor and artisanal know-how against newer trends that made it a scientific discipline within the increasingly conjoined field of physico-mathematics, mechanics was a zone of intellectual and institutional tension as the discipline worked in the late seventeenth century to clarify its identity as an important scientific subfield. French analytical mechanics as it developed around 1700 played a crucial role in bringing about this precise disciplinary clarification, and to understand its particular genesis, chapter 2 surveys the terrain that nurtured this outcome.

One set of dynamics were social, involving the changing understandings of and social role for mathematics and the mathematician in seventeenth-century France. Also important were the changing epistemological canons and the shifting hierarchies of knowledge that elevated certain pursuits to the level of science while leaving others among the lower ranks of the arts and crafts. Mechanics as a field was caught in all of these pulls, and as the discipline began to find a home in the newly self-conscious scientific institutions founded in the seventeenth century, these institutions became arenas where these struggles manifested themselves. Chapter 2 illuminates this history in France by looking at the mathematicians who founded the Académie Royale des Sciences in 1666, and at the particular mathematical culture of the academy during its initial years. The chapter argues that while mathematics was moving in a variety of directions in Europe in 1666, the French Academy adopted an unusually classical, liberal, and gentlemanly approach to its practice early on, one, it is argued, that fit with the wider self-conception of the Academy, at least at first, as a courtly institution beholden to the political logics of Louis XIV's brand of royal absolutism.

The courtly tendencies of French academic mathematics remained powerful into the eighteenth century, and for that reason chapter 2 looks in detail at this particular mathematical culture and the intellectual tendencies it supported. But alongside the courtly Academy, another conception of royal academic science emerged, one that initiated more modernizing trends. I use the term "administrative science" to describe this countervailing tendency within the French Royal Academy even as I stress the harmony that often reigned between the two competing orientations in the making of French academic science. What differentiated administrative science from its courtly sibling was a greater comfort with applied and utilitarian approaches to knowledge making and a greater willingness to blur any firm line separating the manual and the artisanal from the mental and the liberal. Also central to it was an administrative understanding of social status as meritocratic rather than grounded in aristocratic understandings of hierarchy and honor, understandings that defined the royal court complex. Chapter 2 concludes by tracing the parallel urge to foster administratively oriented mathematics at the French Royal Academy during its initial years, ultimately situating late seventeenth-century academic mathematics at a crossroads between a waning set of pulls toward courtly science and service and an increasingly strong administrative urge to promote practical applicability and utility.

Chapter 3 continues this theme by looking explicitly at French academic mathematics in the final decade of the seventeenth century, the years when analytical mechanics was first initiated. It emphasizes the crucial role that the new Phélypeaux de Pontchartrain ministry played, after its installation in late 1691, in pushing the Royal Academy emphatically in an administrative direction. The dramatic changes that the Pontchartrains began to enact, culminating in a complete administrative reorganization of the Academy in 1699, a change that is the topic of its own chapter, did not eliminate the courtly Academy altogether so much as challenge it through a new clarity about the administrative expectations of academicians and their science. Among the changes initiated was a new focus on the cultivation of individual research programs by academicians, and new imperatives to publish scientific work so as to mark academicians out publicly as individual scientific authors. Also innovative were new internal protocols that treated academicians as individual savants pursuing disciplinary research in an academic setting newly governed by protoprofessional expectations regarding conduct and collegiality. In this setting, Varignon began to fashion himself as an innovative practitioner of a new kind of mathematical mechanics, and to project this identity both inside the Academy and through publications and other performances into the wider public sphere. Other mathematicians around him did the same, and chapter 3 surveys the outcome this produced by looking at the leading academic mathematicians in the Academy in the 1690s and the different identities they cultivated and projected. Some of these identities harmonized with Varignon's self-conception, and others created friction with it; the chapter concludes by introducing us to some of the key players in the mathematical debates to come in a way that situates analytical mechanics, and the controversies that it triggered, as a historical outcome produced, at least in part, by the new institutional politics of the post-1691 Académie Royale des Sciences.

PART 2. THE INTELLECTUAL ROOTS OF ANALYTICAL MECHANICS

Having begun the book by situating the genesis of analytical mechanics in France in the widest possible social and institutional setting, the chapters in part 2 narrow the focus by looking at three precise strands of intellectual influence that were crucial to the formation of Varignon's new science of motion. Chapter 4 takes on the question of the Newtonian origins of analytical mechanics directly, looking at the earliest reception of the *Principia* in France. These include the particular readings of Newton's treatise that were initially pervasive in France in the 1690s, and the ways in which Varignon's science did and did not stem directly from them. The chapter ultimately argues against any direct Newtonian origin for analytical mechanics even as it emphasizes all the ways that Newton's work was an important, if indirect, influence on Varignon's thought.

Among the differences that separated Newton from Varignon was their respective relationship to the new analytical mathematics of the late seventeenth century, especially the new infinitesimal calculus developed by Leibniz and Johann Bernoulli after 1685. Varignon was an early user of and advocate for this new mathematics, and analytical mechanics was arguably the most important outcome of this mathematical work in France. Chapter 5 traces how the calculus came to be available to Varignon and others in France after 1691, and explores the particular mathematical communities, both supportive and resistant, that formed in and around the French Academy through the adoption (or critique) of this new mathematics. Nothing was more innovative about analytical mechanics than its use of the differential calculus to articulate its most fundamental claims. Given the novelty of its mathematics, the debates provoked by Varignon's science were often animated more by its mathematical claims than by the physical principles that Varignon attempted to capture through mathematical analysis. The beginnings of these mathematical struggles in relation to the early reception of the calculus in France are therefore examined in chapter 5 so as to illuminate the particular scientific debates out of which analytical mechanics would emerge.

With its emphasis on Newton's *Principia* in chapter 4 and the Leibnizian calculus in chapter 5, the first two chapters of part 2 largely stay within the traditional frame of scholarship about Varignon and his new science of motion, even if these chapters add more historical and institutional context to this story. They also pay closer attention than has been done before to the historical contingencies that were essential to the development of Varignon's work. Chapter 6, however, has no counterpart in the extant literature on Varignon, analytical mechanics, or the wider development of mathematical physics in eighteenth-century France.[54] It argues for the crucial, indeed perhaps even decisive, influence of Nicolas Malebranche on the development of French mathematical science after 1690. It also argues for the existence of a "Malebranchian moment" in France in the decades around 1700, a period when Malebranche's intellectual influence was enormous and astonishingly widespread. Chapter 6 surveys this influence by looking at Malebranche's direct role in the introduction of the Leibnizian calculus into France, and at the influence that his philosophy, as articulated in his monumental *De la recherche de la verité*, which was published in several editions from 1674 to 1712, exerted in giving wider scientific meaning to this mathematics and its potential for innovative scientific work. It also explores Malebranche's connection to the wider currents of Cartesianism in France and to the institutions of the reform-minded Oratorian religious order, arguing that it was through the institutional networks provided by the Oratorian colleges that Malebranche's uniquely influential fusion of Cartesianism, advanced mathematics, and modern scientific philosophy was disseminated throughout France. The chapter also traces the reach of this Malebranchian intellectual complex into the heart of urbane society, noting its presence in the salon culture and worldly periodicals that were coming to play an influential role in shaping elite sensibilities. The widespread presence and influence of Malebranchianism throughout the French public sphere around 1700 is shown in chapter 6 to have been a fundamental, if so far unrecognized, influence shaping French intellectual life overall in this period, and an influence indispensable to the development of analytical mechanics during these precise years.

PART 3. MAKING ANALYTICAL MECHANICS IN THE NEW ACADÉMIE ROYALE DES SCIENCES, 1692–1715

Building on the specific intellectual and cultural contextualization of analytical mechanics offered in chapters 4, 5, and 6, the chapters in part 3 narrow the focus one last time by turning directly to the story of the initiation, reception, battles over, and then institutionalization of analytical mechanics as an established French science between 1692 and 1715. These chapters combine technical scientific analysis with institutional and cultural interpretation to create a narrative history that argues for the role of a broad array of actors in the production of this scientific/institutional outcome. Some of the actors portrayed are individuals pursuing detailed mathematical agendas, but many were constituents of the institutional habitus within which this mathematics was pursued (the Royal Academy, the state, the Republic of Letters, the wider public sphere, etc.). These latter figures often combined an interest in the mathematical debates at issue with desires and passions that were tangential to the precise intellectual questions at stake. Nevertheless, these nonmathematical actors shaped the outcome of these struggles in crucial ways. Other determining agents were completely impersonal and nonhuman, yet they manifest their influence as well in everything from professional imperatives and political expectations to media dynamics, rhetorical constraints, and the all-too-human play of ego and passion. All of these factors were important, and together they conspired in the decades around 1700 to push French science toward this of all outcomes.

In narrating this history, I let the story be its own explanation of how and why this particular scientific result was produced at this particular time in this particular place. Narrating in this way, I also argue for the decisive role that local context in all of its contingent complexity played in the production of this particular scientific outcome.

Chapter 7 initiates the narrative by examining the key confluence in the 1690s from which analytical mechanics would emerge: the debates about Leibniz's infinitesimal calculus in France and the professional reorganization of the Académie Royale des Sciences in 1699. The first is examined by looking at two episodes from the 1690s: the debates triggered by the Italian mathematician Vincenzo Viviani's "aenigma problem" and those associated with Johann Bernoulli's public challenge to mathematicians to find the curve of most rapid descent. What these episodes reveal is that by 1700 a debate about the infinitesimal calculus had begun to polarize mathematicians in the Royal Academy, and that a loose set of rival parties had begun to form pitting self-proclaimed "Moderns" who supported the new mathematics against a party of "Ancients" who opposed it because of its failure to adhere to traditional canons of rigor. This Ancients-versus-Moderns battle begun in the 1690s would erupt with even greater fury after 1700 in response to Varignon's use of the calculus in his analytical mechanics. The analysis in chapter 7 sets the stage for this later struggle by looking at the sources and character of this particular division in the French Royal Academy.

While mathematicians in the Académie Royale des Sciences were beginning to declare their allegiance for or against the infinitesimal calculus, political changes were also pushing them to declare and enact their scientific commitments in new ways. The changes in the institutional governance of the Academy enacted in 1699 played an integral role in shaping all academic science over the next two decades, and analytical mechanics was especially influenced by these transformations. This was because this science entered the world at exactly the moment when these new protocols of governance were first being initiated. Analytical mechanics was therefore shaped by these institutional structures in particularly strong ways.

These changes also brought to the center of academic life a new figure—Bernard le Bovier de Fontenelle—who exerted at once a titanic influence over all of academic science in France after 1699 and a singularly important influence upon analytical mechanics in particular. This was because he was not only a massively influential figure in the academy, and in France overall, due to his control over the new publicity organs that the Pontchartrains made central to academic science after 1699. He was also a close friend and colleague of Varignon (as young men they even shared a Parisian apartment) and a practitioner of, and ardent advocate for, the infinitesimal calculus. The new academy regulations required Varignon, along with his allies and his enemies, to declare publicly their scientific commitments, and then to advance them through publication inside the Academy and in the wider public sphere. As the Academy's perpetual secretary, Fontenelle was charged in 1699 with promoting this new public face of the Academy in ways supportive of the Pontchartrains' political agendas. Analytical mechanics was born at precisely this moment, and it was initially received through these newly activated public channels managed by Fontenelle. The initial debates about it were also conducted by academicians who were playing new roles in a new public setting that was being managed in new ways by a new kind of academic official, Fontenelle, in conjunction with new political supervisors. Chapter 7 introduces this institutional dynamic, which was fundamental to all that followed, by looking at the new institutions created in 1699 and the institutional logic for the wider reform.

Chapter 8 continues the story by looking at both Varignon's precise steps in initiating his new science of motion after 1698 and Fontenelle's initial work as academy secretary at the same moment. Analytical mechanics would become an established feature of French academic science over the next two decades as a result of the conjoined efforts of these two individuals within the precise space of the French Royal Academy and its publics.

Chapter 8 initiates this story by looking at the foundations that governed the work of each. The discussion of Varignon returns to the many intellectual influences percolating in his mind in 1698, ultimately offering a detailed account of the particular fusion of Newton, Leibniz, Malebranche, and other turn-of-the-seventeenth-century intellectual and cultural influences that informed his innovative work. Fontenelle's role in this history is examined by looking at his new role as perpetual secretary of the Académie Royale des Sciences, and at the talents that he was able to channel into his new work. Analytical mechanics would enter the public sphere as an original exemplar of the new public science launched in 1699, and its reception both inside and outside the institution would be shaped as much by Fontenelle's use of the new public organs of the Academy as it would by Varignon's use of the same on his own behalf. Eventually the two academicians would join together as publicly visible champions of the new science, and chapter 8 illuminates the nature of this alliance by showing the institutional dynamics that made it so powerful.

Having narrated in chapters 7 and 8 the history of the gestation and early labor pains of analytical mechanics in France, in chapter 9 I turn to the turbulence of its actual delivery into the world, focusing in particular on "*la querelle des infiniment petits*," a widely noticed public scandal that raged in the first years of the eighteenth century, and which was the most dramatic and influential consequence of the initial reception of analytical mechanics in France. At one level, the quarrel was a simple re-eruption of the controversies over the validity of the infinitesimal calculus that had already occurred in the 1690s. What made this debate different, however, were the new provocations offered by Varignon's use of this mathematics to make innovative scientific claims, and the new institutional dynamics and public setting for the debate.

When asked to do so by the new academy regulations, Varignon declared his new science of motion to be his personal research project, and at two of the first four public assemblies (another innovation of the 1699 reform), he was invited to present papers explaining his new science to the elegant, urbane public assembled for these sessions. The papers, which were reviewed in the periodical press, announced Varignon's new science, and Fontenelle also opened the new era of public academic science by delivering an oration that celebrated without mentioning it by name the value of Varignon's brand of mathematical work. Accordingly, when opponents of Varignon's new science arose (and they did with increasing vehemence), they were confronting more than just a single academician advocating a particular scientific position.

The 1699 reform had made academic science newly public, and as Varignon's new science became situated at the authoritative center of the new public academy, its opponents were forced to contest it in new ways.

Michel Rolle, who emerged in this context as the most strident and aggressive opponent of analytical mechanics, illustrates well the outcomes that this confluence produced. As chapter 9 recounts, Rolle began as a vigorous opponent of Varignon inside the Academy. When he was shunned in ways that Varignon was not by the new public administration of the institution, he turned to nonacademic organs of publicity to advance his cause. His battle with Varignon, which was waged for almost a decade and constituted the most important theater of *la querelle des infiniment petits*, bears all the traces of a bitter professional struggle, and it may even be the first case of a professional battle over institutionalized scientific turf in the history of science. This is because no institution anywhere in Europe had created anything like the disciplinized professional environment initiated at the French academy after 1699, and the battle that ensued appears to have been as much a battle between two rival conceptions of academic mathematics, and the institutional power of each, as it was a battle between two enflamed individuals.

Whatever the deeper motives driving the struggle, Rolle and Varignon pursued their debate in a newly public way, and their struggle also activated a broader array of participants than was the case with previous academic disputes. Especially important was the entry of the Society of Jesus into this struggle through the initiation of their new monthly journal in 1701, the *Journal de Trévoux*. At one level, the launch of this periodical reflects the characteristic urge of the early modern Jesuits to position themselves and their intellectual work at the cutting edge of the most advanced scientific trends. The *Journal de Trévoux* was also representative of the new learned periodicals that were exploding in number around 1700. Like other learned periodicals launched at the time, theirs became an important organ of eighteenth-century scientific discussion and debate. But given the society's long-standing hostility to infinitesimal mathematics of all stripes, the first issues released in 1701 also allowed its editors, a team of savants connected to the esteemed Parisian Jesuit college Louis-le-Grand, to participate directly in the public calculus wars that were then erupting. The first two volumes of the *Journal de Trévoux* did just that, featuring pointed critiques of Varignon's science that echoed Rolle's critique of it as well. Chapter 9 examines this intervention and its role in both intensifying and complicating the quarrel overall.

Chapter 10 brings the story to a close by looking at the resolution of these battles, and the way in which the outcome secured a place for calculus-based mechanics at the center of French academic science for the remainder of the century. This outcome, chapter 10 argues, was not a consequence of any rational clarification of the issues at stake, and it was even less a result of some consensual agreement about the appropriate practice of mathematical science. On the contrary, the outcome is best described as a political settlement produced by two developments: the failure of the opponents of the new mathematics to obtain the necessary political power to secure their position, and the success of the advocates of the new calculus-based science, especially Fontenelle, to persuade both royal authorities and the public at large to accept the new science. Also crucial was the way that divisive argument itself proved incompatible with the political agendas of the new public academy, triggering interventions by royal officials in the name of decorum along with disciplinary sanctions that brought an end to the debate in ways that ultimately favored the continuation of Varignon's program. The result, chapter 10 argues, was a politically determined *pax analytica* that secured the ongoing pursuit of calculus-based physical science inside the French academy while making vociferous opposition to it anathema to the legitimate conduct of royal academic science. Amid this peace, which was further secured by Fontenelle's authority as the unified voice of French academic science for the next forty years, a broad public understanding of the new science was established that was friendly to the understanding of it that Fontenelle and Varignon shared. Chapter 10 shows how this settlement was secured by looking in detail at the institutional maneuvers involved and the public presentation of analytical mechanics orchestrated by Fontenelle that secured this consensus.

CODA: NEWTON AND MATHEMATICAL PHYSICS IN FRANCE IN THE TWILIGHT OF THE SUN KING

From 1715 on, analytical mechanics became an increasingly important centerpiece of French academic science, laying the foundation for the great eighteenth-century architects of modern mathematical physics: Maupertuis, d'Alembert, Clairaut, Lagrange, and, yes, Euler. A coda concluding the book reflects on the outcomes produced by the tumult of the decades around 1700 and at the ironic relationship of this history to the old story of the revolutionary Newtonian making of modern mathematical physics.

Newton appears throughout this story, and to call him an insignificant player in it would be to grossly misrepresent what the book argues. Yet Newton was neither the singular influence determining scientific change in France in the decades around 1700, nor a completely unknown figure being neglected because of French ignorance or prejudice. To sum up the precise relation between Newton's work and the development of the French science that this book traces, the concluding coda returns to the Newtonian theme in this history by reflecting on Newton's legacy as it was imagined in France in 1715, two decades before Voltaire would help to initiate the actual Newtonian Revolution through his retrospective storytelling about the imagined Newtonian achievement. In 1715, when Voltaire was still a teenager, Fontenelle and others saw Newton not as a world-changing hero but as one of many influences in the making of analytical mechanics. Yet since these same observers also viewed Varignon's science as an innovation that moved beyond Newton's work in the *Principia*, the references to Newton's genius that they routinely offered were made to praise Newton, not to connect him to the French mathematical physics that Varignon and his academic colleagues had initiated. Never were such statements used to position Varignon as derivative from or dependent on the *Principia*. Meanwhile, as this consensus about the Newtonian relation to mathematical mechanics was solidified, new developments arose that began to realign the *Principia*, and the idea of Newtonian science more generally, with something other than the calculus-based mathematical mechanics that Varignon had pioneered.

The book's concluding coda summarizes the events that opened the door to the culture wars that were fundamental in the making of eighteenth-century Newtonianism and the French Enlightenment built upon it after 1730, and because this is what my previous book *The Newton Wars and the Beginning of the French Enlightenment* examined, it is with the transition to this new climate of debate at midcentury that *Before Voltaire* ends. In describing the end of the first period of Newton's French reception, the period stretching from 1685 to 1715, the concluding coda emphasizes the striking differences between Newton's image and legacy as it had developed by 1715 and the image and legacy that would begin to take hold after 1730. Staying with the methodological convictions that sustain this book, the concluding coda has no master explanation to offer about why this history unfolded as it did. One set of historical contingencies brought about one outcome, and as those circumstances changed, a different climate was introduced that brought about still other changes. Ultimately, I contend, it is the historical developments themselves, however unpredictable and accidental, that explain how and why mathematical physics developed as it did in eighteenth-century France. And so with one episode of the story told, the book ends by looking briefly at some of the ironic perspectives it offers when looking at the triumphant Newtonianism of the French mathematical Enlightenment at midcentury.

The history of mathematics begins to look more and more like a history of events no more inevitable than the history of France.

Ian Hacking, *Why Is There Philosophy of Mathematics at All?*

PART I

The Institutional Sources of Analytical Mechanics

Mathematics at the Académie Royale des Sciences in the Late Seventeenth Century

If we use the label "analytical mechanics" to denote the use of the infinitesimal calculus to elucidate through differential equations the physical motions of natural bodies, then this new science was first introduced in earnest by the French royal academician Pierre Varignon in the summer of 1698. Work on his new science began as early as 1692, four years after Varignon had joined the Académie Royale des Sciences, and the records of the company offer a partial window into the development of Varignon's new science during the six years of its formation. In particular, we know that Varignon read several papers to the Academy before 1698 that prepared the ground for his new mechanics, but since only a few of these papers were transcribed in the Academy registers, and even fewer published—a fact that is not surprising given the absence of any regular organ for publication at the Royal Academy until after 1699—the precise development of Varignon's work in the 1690s is hard to trace. The Academy also was transformed by the royal ministerial shift that began in the fall of 1691, and these institutional shifts were not irrelevant to the development of Varignon's science.

Whatever its precise genealogy, analytical mechanics first appeared in its entirety in the paper "General Rule for All Sorts of Movements of Whatever Speed Varied Freely" that Varignon read to the Royal Academy in July 1698.[1] Two months later, he produced a second related study, and in 1699 he produced two more papers focused on what he later came to call his "*nouvelle théorie du mouvement*."[2] By the start of the new century, Varignon's program was in place, and over the next two decades (he died in 1722) he channeled all of his royal academic labors into the creation of a corpus of mathematical work that laid the foundations for what would later be called analytical mechanics, a science that would become a fixture of royal academic science in France for the remainder of the century.

The project of this book is to offer a multidimensional account of the historical changes that brought about this precise scientific development, and the goal of the chapters in this first section is to introduce the institutional setting that made Varignon's work possible. They especially situate Varignon's new science of motion as an outcome shaped by his membership in the thirty-year-old Académie Royale des Sciences. Analytical mechanics when viewed from this perspective appears as an outgrowth of the peculiar way that mathematics was privileged and institutionalized within the French Royal Academy during its initial three decades of existence. This new science is also shown here to be a consequence of the particular epistemic virtues that this institutional habitus fostered and prized.

Chapter 2 focuses on the nature of the mathematical culture at the Royal Academy at its founding, and the way that mechanics as a particular mathematical subfield activated socio-epistemic tensions that proved influential well into the eighteenth century. Chapter 3 continues this examination by focusing on the cultural and institutional changes that the Royal Academy experienced in the final decades of the seventeenth century. Chapter 3 also introduces some of the key dramatis personae of the later analytical mechanics debates. Taken together, these two chapters situate analytical mechanics at the institutional crossroads of an Academy undergoing transition. They also present analytical mechanics as a science produced by a peculiar seventeenth-century French intellectual and institutional environment. Together they set the stage for understanding the battles over this new science that erupted after 1699 as disputes driven, at least in part, by the institutional and professional pressures characteristic of this distinctive institutional environment.

CHAPTER 2

Academic Mathematics in France before 1699
The Initial Founding of the Academy and Its Legacies

Founded in 1666 through the support of the vigorous young sovereign Louis XIV and his ambitious chief minister, Jean-Baptiste Colbert, the Académie Royale des Sciences was at one level a concretization of the general trends joining sovereign state power with the new sciences born of the Renaissance. From this perspective, the company should be compared with others, such as the Florentine Accademia del Cimento (founded 1657) and the Royal Society of London (founded 1661), as an emblematic institution of the age. But if the Académie Royale des Sciences was a representative institutional embodiment of the European Scientific Revolution writ large, it was also a peculiarly French institution with its own exceptional culture. Especially distinctive was its particular attention to and cultivation of mathematics. The origination and development of analytical mechanics in the 1690s stemmed from this peculiarity, and to understand the specific history of this new science, the entanglement between it and the general academic environment that produced it, namely that of the Académie Royale des Sciences, must be explored.

The goal of this chapter and the next is to initiate this institutional thread of analysis, and the presentation will move in two stages. First, the category of mechanics will be used to show the competing views of mathematics that were institutionalized in the initial founding of the French Royal Academy in 1666. Stated simply, two distinct, if not necessarily incompatible, pulls dominated academic mathematics in the first decades of the institution's existence. One, which I'll call "humanist" and "gentlemanly" mathematics, pulled toward a more liberal and courtly understanding of the mathematician and his labors. The second, which I will call "administrative" mathematics, pulled toward a new mechanical conception that conceived of mathematics as a materially attached, applied, and instrumentally utilitarian endeavor. Each of these conceptions of the mathematician and his discipline was further reinforced by a corresponding notion of the Royal Academy as an organ of state. In the first, courtly conception, the company was conceived as an assembly of gentlemanly savants who brought renown to the sovereign through the acclaim of their work within the European Republic of Letters. In the second, administrative conception, the same aspiration toward estimable reputation was joined with a desire for academicians to produce instrumentally useful work that served strategic calculations of state. Viewed as a new kind of liberal and gentlemanly mechanic, the ideal academician in this second conception used mathematics to demonstrate his quality as a savant just like his courtly counterpart. But he also used it to serve the state by solving concrete, publicly relevant problems. French academic mathematics, it will be argued, developed in the final third of the seventeenth century through the competing pulls of these two distinct, though never irreconcilable, tendencies. Analytical mechanics, it is suggested, is best seen as a particular outcome produced by this characteristically French academic historico-epistemological dynamic.

The next chapter will continue the discussion by describing a set of political changes that began to push the Academy more fully toward the applied, administrative, and utilitarian approach to mathematics after 1691, a change that recalibrated the balance between the two tendencies established in 1666. Rather than fully eliminating the more courtly conception of mathematics initiated in 1666, these new political pulls in the 1690s, I argue, worked to activate in new ways the tensions between them that were always present, making the moment around 1700 one rife with socio-epistemic tensions. The changes also brought a new cast of characters into the Academy, including Varignon himself and a group of mathematician savants who became the key players in the development of, and debates about, analytical mechanics after 1698. In order to illustrate the changing culture of the Academy around 1690, and to introduce the important dramatis personae of the story to come, I conclude chapter 3 by offering brief biographies of the "second generation" of mathematicians who entered the Academy in the later seventeenth century. The goal of these brief biographies is to show how the particular intellectual allegiances of these academicians both reinforced and challenged the prevailing institutional fault lines of French academic mathematics as a whole. Varignon's early life and work life will also be introduced as part of this prosopography so as to show how his science grew out of these same institutional dynamics.

"Mechanics" and the Social Field of Seventeenth-Century Mathematics

A flurry of new French dictionaries appeared in the 1690s documenting the rapidly changing semantic field of the French language in the seventeenth century.[1] Especially revealing in these volumes is the complex and still unsettled meaning of the term "mechanics" as used by seventeenth-century French speakers. The new mechanics initiated by Varignon came to life out of this unsettled terrain, and the dictionaries of the 1690s offer an insightful point of entry into the socio-intellectual field from which the new science grew.

Two distinct yet historically entangled meanings were conveyed around 1690 by the word *méchanique*. The older of the two meanings associated the word with a lowly social station. As the dictionary of the Académie française explained, a mechanic is a person who is "sordid, petty. . . . He is base." The French Academy also joined with other contemporary lexicographers in connecting this lowly condition with the pursuit of bodily labor and the need to work with one's hands. In the French Academy's formulation, a "mechanic" is a manual artisan, and the practice of "mechanics" referred to the pursuit of those activities that require handiwork.[2] This notion of the "mechanic" also supported the traditional adjectival use of the word to describe those arts that were distinct from the thoroughly nonmanual arts of the free, or "liberal," person. Hierarchical assumptions that raised the liberal arts above those practiced by mechanical artisans, and the liberal theorist above those who worked with their hands, remained foundational in seventeenth-century France, and these dictionaries clearly marked out this persistent social division in their definition of mechanics.

Yet the dictionaries also included a newer meaning of the word, one that in its most forward-looking formulation, called *méchanique*, a discipline among the mathematical sciences. The French Academy listed this newer meaning first, recognizing it as the most common meaning, but it offered little elaboration. "Mechanics is that part of mathematics which has machines as its object," the dictionary stated.[3] Thomas de Corneille, who was charged by his fellow academicians with creating a supplement to the Academy's dictionary offering more developed definitions of technical words relevant to the arts and sciences, also felt no need to elaborate further about the meaning of *méchanique*.[4] Together, these conservative entries illustrate the persistent incuriosity among the lettered minds of the French Academy toward the new mathematical sciences of the seventeenth century. They also illustrate the way that the new mathematical mechanics of the seventeenth century came to life entangled with a set of traditional social assumptions that placed the liberal gentleman above the laboring artisan, and disembodied theory above the applied know-how of the skilled technician.

The Académie française claimed that its dictionary offered the supremely authoritative compilation of proper French usage, but the appearance of other, rival dictionaries in the same years showed that the *Dictionnaire de l'Académie française* was better described as a bastion of traditional assumptions out of step with the new linguistic trends of the time. Antoine Furetière's *Dictionnaire universel* stood at the opposite end of the spectrum, serving in many respects as the modernist alternative to the French Academy's traditionalism. In his entry on "mechanics," Furetière devoted five separate entries to the word, along with a separate entry for the adverb *méchaniquement*. His presentation reveals the new scientific conception of *méchanique* that had begun to challenge the traditional understandings, along with the contests that were afoot around 1700, realigning the relations between traditional and modern understandings.[5]

Furetière first clarified that *méchanique* properly used should be stated in the plural, and he then defined it as "a science that is part of mathematics, which examines [*enseigne*] the nature of moving forces, the art of making the design of all sorts of machines, and the raising of all sorts of weights by the means of levers, wedges, pulleys, etc."[6] He then offered a brief list of some of the "authors" of this mathematical science, noting ancients such as Aristotle and Hero, and moderns such as Jacques Besson, Guidobaldo al Monte, Simon Stevin, and René Descartes. In a separate entry, Furetière added a second scientific understanding of *méchanique*, defining it as "the manner of explaining the springs [*ressorts*] of machines, and the natural causes of the actions of animate and inanimate bodies." He illustrated this definition by citing the work of the *physicien* and founding member of the Académie Royale des Sciences, Claude Perrault, on "*la méchanique des animaux*," or the mechanisms of animals.[7] In a third entry, Furetière added yet another scientific understanding, describing *méchanique* as an adjective, to be contrasted with *géométrique*. This usage connoted "a mathematical construction, or a proof of a problem, that is not done geometrically, but through trial and error or with the aid of instruments."[8] This definition marked out a precise epistemological meaning of "mechanical" within the practice of geometry, one that had emerged in seventeenth-century geometry so as to distinguish lowly mathematical handiwork from the high theory of rigorous deductive demonstration.

We will have occasion to return to this particular early modern epistemological understanding of the mechanical very often in the chapters to come.

Yet even with these three distinct scientific understandings of *méchanique* denoted, Furetière was not finished. He defined the "mechanical arts" in a fourth entry, noting their distinction from the mathematical science of mechanics. The mechanical arts are "the servile arts, opposed to the liberal arts, that are practiced by laborers [*ouvriers*] who not only work on the construction of machines, but in manufacturing more generally and in the production of the necessities and commodities of life."[9] This definition distinguished a scientific interest in the design and construction of machines from the mere artisanal use of mechanized tools in ordinary labor. With scientific mechanics distinguished from the mere practice of mechanical artisanry, Furetière completed his entry by introducing the pejorative connotation of mechanical as "sordid, vile, and petty." Framed this way, Furetière's definitions worked to acknowledge the base connotation of *méchanique* as a lowly and manual state while distancing this understanding from the scientific understanding of the term that was semantically progressive and positive in its connotations.

Reading Furetière's extended definition of *méchanique* alongside the brief and traditional rendering offered by the Académie française reveals the unstable and contested terrain that was the field of mechanics, and mathematics writ large, in late seventeenth-century France. One dynamic fault line involved the ongoing negotiation between mathematics as an art versus mathematics as a science. As recently as a century earlier, the claim that mathematics was a field of the sciences would have been hard to sustain at all. As a discipline that involved the relations between nonmaterial objects (numbers, figures, etc.), mathematics had no place within the traditional understanding of *scientia*, which involved causal, natural philosophical inquiries into the transformations of material reality. At the same time, since the same disciplinary system also located the mathematical disciplines firmly within the arts, mathematics was doubly detached from science in this traditional hierarchy of knowledge.[10]

Geometry, arithmetic, music, and astronomy had been enshrined since antiquity in the quadrivium central to the liberal arts curriculum, and when institutionalized in the medieval universities after 1100 this made mathematics a liberal pursuit, but one that was perceived as preparatory to, not constitutive of, the higher sciences of natural philosophy and theology.[11] Also competing with the liberal mathematical arts were the mechanical mathematical arts, which ranged from "cossist" accounting and bookkeeping to architecture, engineering, and mechanics understood exclusively as work with machines. What separated liberal mathematics from mechanical mathematics within this traditional hierarchy was the possession by the former of a theoretical standard of rigor. The certainty attainable through geometric demonstrations made liberal mathematics distinct from the practical know-how of artisan mathematicians. Euclidean rigor also made the liberal art of geometry central to the practice of science, since geometry provided the model of certain demonstration that anchored scientific demonstration overall. As such, the study of geometry was crucial to scientific training, since it taught aspiring natural philosophers the appropriate manner of reasoning when inquiring into the nature of things. For this reason, the medieval university made mastery of the quadrivium, and especially Euclidean geometry, a foundational step in the progress toward mastery of the natural sciences. Mechanical mathematicians, by contrast, could make no comparable epistemological claim to certain knowledge, and they were accordingly excluded from the university and treated as simple manual workers with base mathematical tools.

In medieval and Renaissance Europe, terminological distinctions such as these marked a clear division between mechanical, artisanal mathematics and the pursuit of the liberal mathematical arts such as geometry. This same hierarchy also made any claim to something like "mathematical science" the equivalent of a category error. This premodern epistemological order continued to exert a strong influence in the seventeenth century, but it was also undergoing a massive reconfiguration as a result of the complex changes associated with the so-called Scientific Revolution in Europe. From as early as the fourteenth century, pressures ranging from the rise of mercantile capitalism to the reconfiguration of the political structures of European states began to disrupt the ancient hierarchies of knowledge. Mathematics was especially reconfigured as part of these transformations. The result was the creation by 1650 of a new set of "mathematical sciences" (or at least mathematical practices claiming scientific legitimacy), and a new status for mathematicians as savants worthy of liberal esteem. The favor that sovereign states showed these new mathematical sciences played a key role in providing social uplift for mathematicians, and nowhere was this better illustrated than in Louis XIV's France.

The Founding of the Académie Royale des Sciences in 1666

The late Roger Hahn's still-unsurpassed history of the Old Regime French Academy of Sciences describes its characteristic institutional dynamic in terms of a double allegiance.[12] On the one hand, Hahn argues, the Academy was guided by the agendas of the new sciences, but on the other, it was also beholden to the imperatives of the French crown. Stated simply, the Academy was indeed an institution pulled by the different forces of science and state, but to fully historicize this dialectic, each of these pulls needs to be complicated. Science, for example, was plural and diverse in seventeenth-century Europe, and while Hahn often uses the phrase the "Republic of Science" to describe the broader learned community that pulled the Academy scientifically, we would do well to remember that there was no singular "scientific community" in seventeenth-century Europe, and no "Republic of Science" in the specialized or professional sense that the term suggests today. All that existed were practitioners of the many and varied seventeenth-century European sciences, including those new ones, such as mathematical mechanics, competing for scientific status. There were also the negotiations of these practitioners with the still evolving and unspecialized constituencies of the Republic of Letters. There was likewise no monolithic state in Louis XIV's France, only competing constituencies negotiating for position and status with a newly assertive monarch. The royal court at Versailles was at once a new theater for marking out traditional dynastic hierarchies and a site where protomodern bureaucratic government and technocratic administration were pioneered.[13] Out of these plural and varied negotiations, what historians call "French absolutism" emerged. The Académie Royale des Sciences, founded in 1666, was a similarly multifaceted outcome of these same "absolutist" trends. It was at once a traditional courtly institution that mirrored the merging of science with court monarchy begun in the Renaissance even as it was also an incubator for a new relationship between science and state, one that pointed toward modern techno-bureaucratic scientific administration.[14]

The rationales leading to the founding of the new Academy illustrate well these competing historical tendencies, as does the particular role of mathematics within them.[15] Colbert considered two different proposals for the new company, and each illustrates the different conceptions of royal science that were present in France in the 1660s. The first proposal imagined a utilitarian Compagnie des Sciences et des Arts comparable to Bacon's conception of a "House of Salomon" that would channel work in the sciences toward the interests of state improvement administratively conceived.[16] In this imagined Compagnie, the sciences were to be valued according to a utilitarian calculus that measured their worth in terms of their contributions to instrumental goals in fiscal, industrial, military, and other strategic policy areas. We will consider Colbert's Compagnie proposal and its legacy in a moment. But alongside it, Colbert also considered an alternative proposal offered by the poet, polymath, and member of the older and more literary Académie française, Charles Perrault. Perrault proposed the establishment of a General Academy with a far more bookish and erudite bent. It would explicitly exclude the less noble mechanical arts from the company, opting instead for a liberal approach to the sciences that emphasized the superiority of mind over handiwork. Perrault also called for sections in belles lettres, history, and philosophy not contained in the Compagnie proposal, and whereas the latter would have been composed of academic specialists chosen for their expertise in given disciplines, Perrault expected his academicians to be wide-ranging polymaths versed in a wide variety of subjects.

A 1698 engraving by Sébastien Le Clerc captures well the spirit of Perrault's vision of royal academic science (fig. 1). Le Clerc had risen within Colbert's academic system to a position of great acclaim as an engraver, and when he set to work in 1698 on this image of the arts and sciences under Louis XIV,

FIGURE 1. *Sébastien Le Clerc (1637–1714),* L'Académie des Sciences et des Beaux Arts, *c. 1706. Etching and engraving; 9 3/4 × 15 5/16 in. (plate); 10 × 15 5/16 in. (sheet). Minneapolis Institute of Art, gift of the Estate of Kemper Kirkpatrick, P.92.8.80. Photo: Minneapolis Institute of Art. The image can be viewed online at https://www.metmuseum.org/art/collection/search/387878.*

he was a member of the Royal Academy of Painting and Sculpture, where he taught the mathematics of perspective, among other things. Le Clerc's unified vision of the French academies was likewise the vision of many who occupied places of prominence in Colbert's system.[17]

As a nineteenth-century historian noted with respect to Le Clerc and his widely esteemed polymathy: "No science escaped him. Geometry, physics, perspective, architecture, fortifications, he studied them all and made himself capable in each."[18] In this engraving, an equally universal representation of all scientific learning is offered, one that conformed to the common seventeenth-century understanding of the sciences (always plural) as rational knowledge in general, and not specialized disciplinary expertise. Bookish *érudits* mingle in Le Clerc's academic esplanade with mechanical artisans, astronomers, and musicians, while skeletons compete for attention with mathematical instruments, globes, paintings, machines, maps, and folio treatises. Prominent in the background is a library named "Theologia," acknowledging the continuing link, however distant and marginal, between the sciences and sacred knowledge. Overall, Le Clerc's image presents the sciences as all-encompassing universal knowledge, and Perrault's plan for the new Royal Academy of Sciences proposed the institutionalization of this idea through the creation of a single, grand academy of all scientific learning.

Although often held to be a paragon of the administrative technocrat, Colbert actually favored Perrault's plan at first, a fact that is not surprising when one takes seriously the minister's appreciation for the realities of Baroque-era statecraft, and his commitment to cultivating royal power in these terms.[19] As Bernard le Bovier de Fontenelle rightly characterized the minister's thinking at the time: "In 1666 M. Colbert knew how much the glory of scholarship contributed to the splendor of the state."[20] Yet even though Colbert, Perrault, and others saw the Grand Academy as the fullest realization of their ambitions, their dream was never realized. It was ultimately killed by a classic Old Regime corporate struggle.[21] One opponent was the Académie française, which found the idea of including belles lettres and history in the new institution challenging to its title as the royal institution of letters in the kingdom. Similar challenges came from the Sorbonne regarding natural philosophy and the Faculty of Medicine concerning medical science. Even the Parisian artisan guilds voiced opposition to the new institution despite its "liberal," erudite bent. Technical secrets were monopoly privileges within the corporate culture of Old Regime France, and an Academy of Sciences devoted to mechanics, no matter what its orientation, constituted a royal incursion upon traditional artisanal privilege.

In this respect, the ambition to found a unified academy of all the sciences in 1666 foundered on the same rocks that had constrained the establishment of the Collège Royale in the 1530s and the Académie française in the 1630s. It also mirrored the struggles that Theophraste Renaudot faced when he attempted, with the support of Louis XIII and Cardinal Richelieu, to sustain a royally supported Bureau d'addresse devoted to, among other things, the promotion and circulation of medical knowledge. In Renaudot's case, the Faculty of Medicine argued vigorously that his bureau was a challenge to the Paris medical faculty's right to control medical knowledge in the city, and this despite Renaudot's status as a titled physician. Using the courts to challenge the institution on legal grounds and the public sphere to discredit Renaudot and his work, the French medical establishment eventually succeeded in stopping his efforts.[22] Conflicts of this sort were inevitable whenever the monarchy attempted to establish a new and specifically royal institution within jurisdictions that were historically the purview of preexisting corporations, and 1666 was no different.

In the face of this resistance, the monarchy responded as it most often did by scaling back its proposal and by accepting a more limited outcome. An academy was established in 1666, but it was a decidedly less grandiose institution than many had envisioned. Colbert in effect built the institution man by man, quietly offering positions to those individuals who fit his conception of a royal academician while trying to restrict the profile of the company overall. Eventually the Academy would include twenty-two members, and by 1668 the group was meeting regularly in the Bibliothèque du Roi in Paris. It possessed a de facto leader—Christiaan Huygens—chosen because of his esteem within European learned circles. Yet no formal title was given to Huygens, or to any of his fellow academic colleagues, for at first "*La Compagnie*," as it was informally called, conducted its business through a strictly informal set of agreements and arrangements. The new Academy, in fact, received formal letters-patent registered by the Parlement of Paris only in 1713, which shows how misleading it is to view the institution as a royally supported leviathan ruling absolutely, from its inception, over seventeenth-century French science. The term "academy" was in fact widely used at the time to describe a wide variety of intellectual and sociable congregations, and at its founding the only thing that distinguished the Royal Academy from other academies in France was the status of the patron who had brought its members together.[23]

Royal support nevertheless offered a decisive difference, and while the new Royal Academy pursued its scientific activities in a manner wholly consonant with the wider sociable and intellectual norms of the period, its founding did inject a new set of centralizing impulses into the social dynamics of French science. The wider field of the mathematical sciences in France was shaped in several ways by the manner in which the Royal Academy came to life. Most important was the line that the institutionalized academy ultimately, if unintentionally, drew between academic mathematics and mathematics practiced elsewhere. Colbert, it will be remembered, considered another proposal for the Academy, one that would have included both liberal scientific savants and specialist experts in the various technical mathematical disciplines. By opting for a scaled-back version of Perrault's Grand Academy plan instead, the actual Royal Academy created in 1666 became a much more liberal enclave than Colbert had envisioned. It was also a company that included only a very small subset of the practices and practitioners constitutive of seventeenth-century mathematics as a whole.

Courtly Mathematics in the Founding of the Académie Royale des Sciences

Huygens personified the desired academic ideal in his combination of liberal mathematical theory with mechanical and technical know-how. The son of a distinguished Dutch statesman and poet, Huygens was born an elite gentleman with an unimpeachable claim to liberal *honnêteté*. Educated according to the model of Castiglione's courtier, he was also a skilled draughtsman and performer with the lute, and his fluency in French and Latin allowed him to write poetry and elegant prose in each language. Yet he was also a gifted mathematician, working at precisely the moment when the mathematical disciplines became newly liberalized and scientized. Huygens came to personify the new trend through his combination of classical geometry and instrumental, mechanical practice, and as such he became an influential pioneer of the new scientific discipline of physico-mathematics, and a guiding model for French academic mathematics.

Most important in making his reputation was his demonstration in the middle of the seventeenth century that a cycloid pendulum was isochronous. This discovery was important in and of itself, but it also led him to develop a new clock mechanism that improved the instrumental measurement of time.[24] Overall, the clock, and the mathematical theory that made it possible, exemplified Huygens's identity as the perfect combination of the liberal theorist with the mechanical practitioner. His invitation to lead the Sun King's first Royal Academy of Sciences attests to the status that this fusion gained for him, as did the lucrative six thousand livres per annum pension that he was awarded for his service. But if Huygens exemplified the ideal fusion desired in a French academic mathematician, he was also the exception that proved other rules, since the other founding mathematicians were less multidimensional than he was. His father also laughed derisively at Colbert when he described his son as a mathematician, a sign that the progressive aspects of Huygens's new mathematical identity look far more clear to us in retrospect.

Taken as a group, there was in fact a strong bias toward liberal, bookish mathematics among the founding class of French academic mathematicians.[25] The term "humanist mathematician" captures well the character of most of the mathematicians appointed to the Academy in the 1660s. By humanism, I refer loosely to the general early modern trend that turned lay elites toward the cultivation of intellectual identities as scholars through the pursuit of bookish learning. As Paul Rose has discussed most fully, humanism of this sort also contained a mathematical dimension centered on the recovery and renewed study of ancient mathematical learning.[26] Projects of mathematical book collecting, translation, and textual scholarship were one impetus, but geometric problem solving according to the ancient model was also important. The career of Federico Commandino illustrates well the character of the early modern humanist mathematician.

Born in Urbino in 1515 to a family with ties to the great Renaissance court of Federico da Montefeltro, Commandino turned a general Renaissance interest in ancient books and learned scholarship into a pioneering program of locating, translating, and publishing the works of the great mathematicians of antiquity. In the second half of the sixteenth century, his editions of Archimedes, Aristarchus, Euclid, Hero, and Pappus appeared, and his example provoked others to recover, translate, and republish the works of Apollonius, Diophantus, and other ancient mathematicians. Others found in Commandino's editions a spark for a Renaissance of mathematics, as Rose calls it, which entailed a new urge to recover and study the work of the ancients, an impetus to comment on the strengths and weaknesses of ancient mathematical work, and a desire to pursue new mathematical research guided by antique models. Commandino's influence was astonishingly strong and widespread, and it helped to establish a new persona of the mathematician as a humanist scholar, an identity that distinguished this sort of mathematician from both the university professors of the quadrivium and the lowly mathematical mechanic.

By 1600, mathematics practiced in this humanistic mode had become widely accepted as a worthy gentlemanly pursuit, and in this way the title "mathematician" started to isolate itself as one identity within the Republic of Letters. For gentlemanly mathematicians of this sort, textual translation and the production of modern editions of ancient works remained a central preoccupation, but other, more innovatory practices also emerged. Especially important was the urge born of this milieu to recover through rational reconstruction the lost mathematical knowledge of the ancients. The rational reconstruction of ancient learning often involved the conjectural deployment of mathematical reasoning beyond the extant textual record, and in this way mathematical research in this humanist mode became a prompt for a new interest in mathematical innovation and discovery. The widely accepted belief at this time that ancient mathematics was superior to its modern counterpart meant that any celebration of novelty or innovation for its own sake was next to impossible. By reconstructing ancient mathematics, however, mathematicians often resurrected old problems in ways that allowed them to be pursued in new and ingenious ways, and the new understanding of old material also led to innovation and expansion of the extant domains of mathematical knowledge. Epistemological space was thus created for humanist mathematicians to assert themselves as mathematical thinkers on a par with their ancient predecessors, and as authoritative innovators in their own right. Out of such thinking the idea of claiming modern mathematics as superior to that of the ancients started to become thinkable.

Such claims were also supported by the gentlemanly values that were always central to humanistic learning within the Republic of Letters. To be perceived as a mathematician who rivaled the ancients was to be recognized as a person of exceptional virtues who possessed profound intellectual gifts. These qualities in turn contributed to the mathematician being marked out as a distinguished person, one worthy of elite status. For this elevation to occur, however, the aspiring mathematician needed to displace any and all associations with the lowly mathematical arts. Accordingly, an intense preoccupation with standards of mathematical rigor also emerged as a core epistemic virtue among humanist mathematicians. Furetière pointed to one dimension of this gentlemanly epistemological canon when he distinguished between "geometrical" and "mechanical" modes of mathematical reasoning. Since the goal of the gentlemanly mathematician was to show his qualities of mind while displacing any attachment to the labors of the body, humanist mathematicians were particularly averse to the use of instruments or other mechanical means in the pursuit of mathematical knowledge. They also revered ancient canons of geometric rigor in an especially intense way, seeing in such venerable standards an ancient authority that embodied the purest form of truth.[27] Mathematicians who adhered less strictly to these ancient standards, or pushed at them in the name of innovation and novelty, were consequently viewed with suspicion. Very often the specter of "artisan labor" was invoked in this context through charges that aggressively innovative work, particularly if it involved trial-and-error reasoning or the use of instruments, amounted to nothing more than lowly mechanical tinkering.

In this way, the term *méchaniquement*, as Furetière recorded, emerged in early modern Europe as an epithet disparaging geometry practiced with insufficient attention to rigor. Humanist mathematicians in particular insisted on a dichotomy separating rigorous liberal reasoning from lowly mechanical reckoning. On the whole, they were also steadfast in defending the ancient canons of mathematical practice and rigor. To achieve results in this disciplined way was to demonstrate intellectual talent and ethical comportment at the same time. In short, it made the practitioner a distinguished mathematician and an *honnête homme* simultaneously. Mathematical work that was judged to be clever, elegant, or ingenious was not a problem within this frame since it simply added to the luster of the distinguished mind evident in the work. But for the same reason, whenever a solution seemed to sidestep the ancient epistemological canons, or appeared to play too aggressively with the standards of rigor that regulated this mathematical community, vigorous criticism often ensued, criticism that sometimes challenged the ethical character of the mathematician under consideration.

Two of the most senior French academic mathematicians appointed in 1666, Gilles Personne de Roberval and Bernard Frénicle de Bessy, epitomized humanist mathematics of this sort, and their academic appointments were representative of the general trend in this direction characteristic of the early founding. Each was an active member of the Mersenne circle, and together they personified the now-graying generation of humanist mathematicians that came of age after 1600 through the practice and scholarly exchange of liberal geometric problem solving.[28] A third academic mathematician, Pierre de Carcavi, was cast from the same mold, and he pursued a similar kind of mathematics when he was not pursuing his primary occupation as the royal librarian and intimate servant of the minister Colbert.[29]

The links that tied Carcavi to bookish erudition and mathematics were also in evidence in the two mathematicians chosen to serve as the secretary of the new Academy. Jean-Baptiste du Hamel was selected to be the primary secretary, but when diplomatic assignments drew him away from Paris, the abbé Jean Gallois was appointed in 1668 to fill in for him. Each was a practitioner of bookish mathematics. Du Hamel acquired his mathematical training from the Oratorians, a link that we will encounter again among French academic mathematicians. He authored several Latin treatises on ancient mathematics and philosophy. Each was a dialogue where the virtues of ancient and modern philosophy, especially Cartesianism, were discussed. Since du Hamel was an outspoken opponent of the new trend toward vernacular scholarship, the Latinity of his works was as important to him as the scientific content they contained. In his treatises, he adopted a middle position that was skeptical of radicalism at either extreme, and in this way du Hamel manifested the judicious moderation that was also characteristic of the humanist mathematician in this period.[30] Gallois established his equally *honnête* credentials when he joined with Denis de Sallo as the founding editor of the new *Journal des savants*, a pioneering learned periodical supported by both Colbert and Carcavi. When de Sallo proved to be too antagonistic in his editorial practices, Gallois was given sole control of the journal. Under his editorship the journal became exemplary of the wide-ranging polymathy and judicious decorum prized by Republicans of Letters.[31] Gallois's work as an academic mathematician was similarly moderated, lettered, and catholic in its interests. He also revealed his wider humanist orientation by holding a chair in both mathematics and ancient Greek at the Collège Royale.[32]

Of the eleven founding academic mathematicians in France in 1666, well over half exemplified the model of the humanist mathematician as it had developed into the seventeenth century. The original working practices of the Academy also reinforced this conception of academic mathematics. Especially influential was the early protocol stressing collective labor and a bias against individualized, specialist research.[33] The Academy was originally conceived as comprising two classes of savants, *mathématiciens* and *physiciens*, and the twice-weekly meetings were divided such that mathematical questions were pursued on Wednesdays and questions of *physique*, which included sciences such as chemistry, botany, anatomy, and medicine, on Saturdays. Despite this division, every academician was expected to participate in every meeting, so the split should not be construed as disciplinarily decisive. The typical working method of the early Academy was for all members to collaborate on a common project, and in the early decades, botany and natural history were the dominant sciences pursued.

An engraving of the early Academy at work conveys vividly the nature of their early endeavors (fig. 2). The setting is the Parisian royal library on the rue Vivienne, the Academy's first institutional home. A variety of activities are represented. Several individuals engage in solitary reading and other bookish pursuits, an indicator of the commitment to text-based scientific inquiry in the tradition of humanistic scholarship, which remained an important feature of early academic science. Also evident is the collective nature of the working

FIGURE 2. *From Guy Tachard,* Voyage de Siam des Pères Jésuites envoyez par le Royaux Indes & à la Chine *(Paris: Seneuze & Horthemels, 1686). Courtesy of the James Ford Bell Library, University of Minnesota.*

practices. Except for a few lone scholars isolated among the books, the academicians are presented as a gregarious ensemble collaborating in teams on a variety of projects. This representation captures well the emphasis on collaborative labor, sociability, and authorial anonymity that all scholars agree was central to the institution in its early years.

When accounting for the collectivist nature of the early Academy, the influence of Francis Bacon is often offered. His warning about the need for disciplined, collective scientific practice as a check on philosophical vanity and the delusions of the imagination was indeed influential in France. Accepting this line of reasoning, it is argued, the early Academy attempted to institute a Baconian antidote for such ills: an academic ideal rooted in subsuming philosophical egos within a collectivist and anonymous approach to learning.[34] Bacon was certainly an influence on the early Academy, but the engraving tells another story as well. In the back of the room a portrait of the king hangs above the empty throne of the sovereign. This was a material reminder that the Academy was not working only in the royal library, but also in the king's name and for his image. What is too rarely emphasized is how this courtly, absolutist conception of academic science also accounts for the collectivist, anonymous approach to science that the early Academy adopted.

Academic mathematics was especially shaped by this courtly conception of royal science. Mathematical questions were raised and discussed at the meetings, and a variety of mathematical topics were explored. But overall, very little mathematics was actually pursued in a systematic way in the first decades of royal academic science. The mathematics that was done, moreover, was of a fairly traditional sort, with geometric problem solving being especially prominent. Some physico-mathematics was pursued, especially the topics initiated by Huygens, but as will be discussed later, the initiatives in this direction did not come from the founding mathematicians of the company. Huygens was also supportive of the liberal, collaborative understanding of science dominant in the early Academy, and he encouraged rather than challenged the focus on natural history and anatomy, along with the collectivist orientation of the company.[35] His stance contributed to the narrowing of academic mathematics at first to questions of traditional geometry.

Mechanical Tendencies in Early French Academic Mathematics

In all these ways, mathematics at the early French Royal Academy was neither very active nor very representative of the wider currents of seventeenth-century European mathematics as a whole. A narrowly liberal and traditional orientation was dominant at first, but pressures pulling the Academy in other directions were also present. There were, for example, notable exceptions to the norm that made the label "French academic mathematician" largely synonymous with that of a liberal humanist mathematical practitioner.

Jacques Buot is a case in point. He served in the Academy from its founding until his death in 1678, and he exemplified the newly liberalized mechanic turned modern scientific mathematician that is one of the hallmarks of the seventeenth century.[36] One marker of Buot's mechanical origins is the relative absence of documentary traces accounting for who he was. One of the few extant documents is a contract between Buot, who describes himself simply as "*un mathématicien*" in the text, and the Parisian bookseller Charles Mondière. The contract arranged for the publication of Buot's *The Use of the Proportion Wheel, by which the Rules of Arithmetic Are Practiced Quickly and Easily*, which was published in 1647 and described a calculation machine of Buot's invention.[37] This book, and the contract for it, reveals Buot to have been a Parisian mathematical teacher and instrument maker with ties to the artisan mathematical world of bookkeeping and mechanics. A second document, dated three decades later, after Buot was already in the Academy, further confirms this picture. It is a receipt confirming a quarterly payment of thirty livres for his work as "mathematical teacher to the pages of His Majesty."

Taken together, these documents position Buot as someone who "excelled in mechanics," to quote Jean-Dominique Cassini, but in the traditional artisanal sense of the term.[38]

Yet if Buot began as a mechanical artisan, his career reveals the new social opportunities available for liberalized practitioners of these mathematical arts in the courtly spaces of Louis XIV's France. Buot is also the exception that proves other rules, for he was the only pensioned academician in the early Academy who possessed anything like these mechanical associations. He further achieved his elevated stature by distancing himself from his mechanical roots. Overall, the academic mathematicians as a group, even with their liberal character, also paled in influence when compared to the *physiciens*. Huygens was again the exception, but his comfort with the nonmathematical agendas set by the Academy's other members, especially the naturalists and anatomists, reveals the character of that exception. Other *physiciens* also complicated this orientation by contributing to physico-mathematics. But they came to mathematical theory from the empirical sciences rather than the other way around, and while their work could be and was classified as mathematics, it was largely marginal to the strand of academic mathematical work that would engender analytical mechanics three decades later.

Especially indicative in this regard was the work of Edme Mariotte, an academician appointed in 1668 as a *médecin*, but one who made important contributions to the new scientific field of mechanics.[39] Mariotte's contributions to the development of this science are well surveyed elsewhere, and for this discussion a few highlights need only be mentioned. One is his connection to Huygens, and the way that together they came to exemplify a particular French academic strength in this of all disciplines even if they were rather solitary exemplars of this discipline at first. Second is Mariotte's particular epistemological contribution to this general development, especially his emphasis upon experimental methods and empirical and inductive reasoning. As we will see, Varignon's analytical mechanics developed out of the liberal and theoretically inclined tradition of French academic mathematics, and it had little connection to the experimental and empirical orientation exemplified by Mariotte.[40] Third is Mariotte's idiosyncratic working method with respect to academic norms. Mariotte pursued individual questions of research and published singly authored books in contradistinction to the collaborative and collectivist ethos of the early Academy. These last two points converge in situating Mariotte as both a path-breaking French mathematician and one out of step with the Academy's dominant institutional currents before 1690.

When the Academy began to turn toward a more individualized conception of research later in the century, and toward more applied and experimental approaches to physico-mathematics as well, Mariotte's example became available as a model and an influence. But in the 1670s and '80s, he was more of a marginal figure whose "experiments in physics and mathematics," to use Cassini's description, remained isolated from the mainstream of French academic mathematics overall.[41] Buot and Mariotte, each in different ways, illustrate the bias against empirical physico-mathematics, and especially applied mechanics, present in the early Academy. Yet their presence at all also shows the distortion of any exaggerated over emphasis of this distinction.

Overall, the French Academy, like many other courtly scientific institutions of the time, was a site of negotiation among scientific intellectuals of a variety of sorts. If the Academy was over-represented at first by the more liberal and theoretical practitioners of seventeenth-century mathematics, it also possessed urges toward more practical and utilitarian understandings of the same thing. One further aspect of the original academic foundation shows clearly the way that the tension between the liberal and the mechanical tendencies in seventeenth-century mathematics were negotiated in the early Academy, namely the classification of "*élève*" created by the original founding.[42] The first members of the Royal Academy were by and large peers, with no official hierarchy distinguishing academicians by rank or disciplinary class. Even the distinction between *physiciens* and *mathématiciens* was more apparent than real, since every academician was expected to attend every meeting, and academic work was at first pursued collectively. Yet as egalitarian as the early Academy was, there was one classification that showed the institution's immersion in the hierarchical social values of the time.

Among the original founders were five academicians classified as *élèves*. Little is known about this classification and its original meaning, and two of the original *élèves*—La Voye-Mignot and Pivert—are so obscure as to be almost nonexistent in the historical record save for an occasional mention in the academic records. What did it mean to be a royal academic *élève* in France in 1666? The label suggests a student or apprentice role, and since two of the other three *élèves* were in their midtwenties when appointed, and the third still not yet forty, the *élèves* may have been imagined as a class of protégés destined to succeed the regular academicians when they got older. Charles Perrault, however, alluded to no such developmental conception when he defined the function of the *élèves* as those who "execute the decisions of the company, and especially conduct any observations which it required."[43] To follow Perrault's description is to see the *élèves* as the Academy's "mechanics," charged with performing the hands-on labor necessary for the liberal academic science of the other members. Much evidence supports this understanding. *Élèves* were admitted, for example, not as independent academicians but as the assistants (clients?) of other academicians. Antoine Couplet was made Buot's *élève* in 1666, and it is likely that this arrangement simply transferred into the new Academy a relationship that had already been formed inside Buot's Parisian workshop. Two other *élèves*, La Voye-Mignot and Jean Richer, were also attached to the academic astronomers Adrien Auzout and Jean Picard. We will return to royal astronomy and its relation to mechanical mathematics in the Royal Academy in a moment, but for this discussion the point to note is the highly technical and instrumental nature of seventeenth-century astronomy and the way that these academic *élèves* likely served as technical assistants for royal astronomical work. Antoine de Niquet also worked in the orbit of Auzout and Picard, and while he appears to have been more autonomous, he was most likely a technician in the service of each.

A sampling of some of the work done by these *élèves* in the first years of the Academy illustrates well their character. A 1666 letter from Auzout to Henry Oldenburg, secretary of the Royal Society of London, mentions the soundings of the river Seine taken by La Voye-Mignot at Rouen. The letter also refers to articles in the *Journal des savants* authored by the same on the worms to be found in oysters and on the methods for constructing large buildings. La Voye-Mignot is also noted as an assistant to Auzout and Picard in their efforts to establish a standard base measure of the *toise* for use in precision quantitative calculation.[44] The Academy minutes for 1667 and 1668 further report that La Voye-Mignot's traveled on Mediterranean warships to study whether Huygens's pendulum clock could be used to keep time at sea.[45] Niquet worked with Auzout and La Voye-Mignot on the production of reliable logarithmic tables for use in calculation, and he was also the academician responsible for the receipt and evaluation of new machines submitted to the Academy by outsiders.[46] Couplet was also connected to mechanics of this sort, serving as a technician in charge of maintaining the instruments at the Royal Observatory, and as the Academy's treasurer, which is to say its bookkeeper.[47]

Note that all of the *élèves* were mathematicians, and that their work was largely mechanico-mathematical in nature. La Voye-Mignot appears in the Academy registers reporting on anatomical findings, and his article on oysters has already been noted. Yet most of his extant academic work was mathematical, and that of the other *élèves* even more so. Pulling this evidence together, it seems warranted to situate the French academic *élèves* among the mechanical mathematical practitioners of the seventeenth century, and to see them as liberalized mechanics providing instrumental knowledge to the royal company. They also illustrate the continuation of social hierarchies that treated instrumental mathematical science as a lowly pursuit with respect to liberal science. Singled out as *élèves* in academic documents, paid lower pensions than their colleagues,[48] and often found working in service to more senior academicians, their presence within the early Academy marks the continuation of the traditional hierarchies that placed liberal, theoretical learning above mechanical, instrumental-knowledge work.

Nevertheless, the *élèves* were also recognized in other ways as full-fledged members of the Royal Academy, and since their subservient status was never formalized into an actual patron-client relationship, it is just as accurate to view the *élèves* as examples of how Old Regime institutions such as the Royal Academy worked to modernize the social status of previously lowly mathematico-mechanical practitioners. The *élèves* were in fact full-fledged royal academicians, and if it is right to see them as the mechanics providing instrumental, empirical, and quantitative assistance for the liberal scientific work of others, it is also right to see them as physico-mechanical contributors to the new academic science that was starting to imagine itself as both mechanical and liberal, empirical and theoretical at the same time.

Yet even if the presence of the *élèves* within the Academy shows the accommodation of new intellectual and social configurations within the space of liberal academic science, the Royal Academy remained a bastion of traditional gentlemanly values in its early years. A key reason for this was the Old Regime courtly ethos guiding the conception of the Royal Academy at its founding. Within this framework, the Royal Academy was imagined as a kind of a liberal scientific crown sitting atop the royal body of French learning. Having eschewed the idea of a utilitarian Compagnie des Sciences et des Arts that would have made the Academy a hub facilitating practical scientific work, the Royal Academy became instead, at least at first, an elite institution composed of gentlemanly savants serving as courtly embodiments of scientific learning. The scientific work of the Academy was emblematic of this self-conception, and academic mathematics was especially shaped by this ethos. Having internalized a rationale that made royal mathematics a practice designed to bring distinction to the name of the sovereign who patronized it, the early Academy tended to eschew modernizing innovation in the name of a classically conceived program of liberal scientific learning.

Whole domains of mainstream mathematical work in the seventeenth century were therefore alien to the practice of French academic mathematics.

That which was excluded ranged from military engineering and ballistics to hydraulics and the maritime sciences. None of these endeavors, nor any of the other more physico-mathematical endeavors, from machine and instrument making to quantitative calculation, had any place at first in the Royal Academy. The career trajectories of the *élèves* Niquet and Richer reveal the outcomes that this orientation produced. Confronted with the beginnings of what would become a half century of persistent warfare under Louis XIV, Colbert redirected both of these academicians to projects in military engineering in 1672. In their new roles, each savant continued to pursue the mathematical work that had brought him into the Academy in 1666, but each also became detached from the actual work of academic science as a result of his new appointment.[49] Chandra Mukerji's detailed history of the Canal du Midi, a massive public works project built largely during the 1670s and '80s, reveals a similar role—or, rather, a complete lack of one—for the Royal Academy in state-led public engineering.[50] The royal palace of Versailles, built largely in the 1670s and '80s, offers another example. It was famous for its state-of-the-art hydraulic engineering, yet only the *élève* Couplet played any role whatsoever in designing or building these aquatic systems.[51] Louis XIV in fact had a distinct, and comparably distinguished, corps of royal engineers, with expertise in everything from fortifications and naval science to hydraulics and political arithmetic.[52] Yet none of these applied mathematical savants were appointed to the Academy since no place for their utilitarian scientific work was created in the early founding. Royal academicians also played almost no role at all in any of the state-led projects in industrial science for which Louis XIV's reign is famous.[53]

Robin Briggs notes with surprise that an exhibition of models of innovative machines held in Paris in September 1683, and sponsored by Colbert's son, the Marquis de Blanville, "rather curiously . . . had no link at all with the Académie."[54] But this fact is only surprising if one views the absence of the seventeenth-century mathematical and mechanical arts within the royal company as a noteworthy deficit, and assumes that the Royal Academy was in fact created to be a central hub for all of seventeenth-century mathematical learning and practice under the Sun King. In fact, it was a peculiarly constituted institution devoted to a narrow set of largely liberal and theoretical concerns during its early years. And while the fusion of utilitarian mathematics with statecraft was indeed a hallmark of Louis XIV's reign, this kind of mathematics was pursued, at least at first, completely outside of the gentlemanly, courtly nexus of the Royal Academy.

Administrative Monarchy, Mathematics, and French Royal Astronomy

At first, the Académie Royale des Sciences was a liberal, courtly enclave isolated from the parallel but distinct pushes toward applied scientific state building under Louis XIV. The split was not absolute, however, and points of contact between the two streams of state science have already been noted. Yet the divide was real, and it is important to remember it whenever French academic mathematics is considered in relation to the wider practice of mathematical science in France and Europe. But if the presentation so far has been accurate in describing "two mathematical cultures" in France circa 1680, one liberal and academic and a second that was more utilitarian and extra-academic, the discussion has also been careful to stress the developmental pressures fighting against this division, and the complexities that brought these two spheres into constant contact and negotiation with one another. To complete this picture, one further layer of complexity needs to be introduced. This is the role that royal astronomy played in the Academy's initial founding. Taking the mathematical science of astronomy into consideration does not eliminate the "two cultures" picture of French mathematical science circa 1680, but it does reveal the vigor of the pressures pushing against this division in Louis XIV's France.

To return to the original moment of the Academy's founding, everywhere one looks one sees astronomy and astronomers as leading figures in the founding of the Royal Academy. It would be no exaggeration, in fact, to see astronomy as one of the central drive trains in the conceptualization and development of the new royal institution. Astronomers such as Adrien Auzout and Jean Picard were in Colbert's inner circle, and while Huygens's work with the pendulum clock was his greatest claim to fame in 1666, his discovery and analysis of the rings of Saturn in 1651 added enormously to his stature as someone worthy of leading the new royal company. Astronomy also played a key role in one famous story of the Academy's origination. A gathering arranged by Auzout in Colbert's apartments on the rue de Richelieu to watch the lunar eclipse on June 16, and then the solar eclipse on July 2, 1666, is often presented as a key moment in the creation of the new Royal Academy later that fall. In attendance at this astronomical soirée were Carcavi, Buot, Roberval, and Frénicle de Bessy, all core constituents of the early Academy. The group also reflected the connections between Colbert (Carcavi) and the leading lights of liberal mathematics (Roberval and Frénicle de Bessy), along with the central place of mechanics (Buot) and astronomy (Auzout) within it.[55]

This is not the only story of the genesis of the Royal Academy, but no matter how the seeds of the new company are sorted, astronomy played a seminal role in its birth. Auzout and Picard were among the most important founding members of the new Academy, and the first group of *élèves* was chosen especially because of their connections to these astronomers and their work. At the very first meeting of the new Academy in January 1667, discussion also centered on the ambition to build a new royal observatory, an edifice that in the minds of many was intended to serve as the physical and intellectual home for the new institution. A laboratory was quicker and easier to construct, and in its first few years the Academy's energies were focused instead on the rooms allocated for this purpose at the Bibliothèque du Roi.[56] Matters changed, however, with the appointment of the Bolognese astronomer Giovanni Domenico Cassini to the Academy in 1669. Cassini was given an extravagant nine thousand livres pension, along with promises from the crown that an elaborate observatory would be built to meet his needs. The Danish astronomer Olaus Roemer was added to the group in 1673, and with these appointments, and the construction of the new Royal Observatory, which was operational by the late 1670s and officially opened by the king in 1682 (Louis XIV's only visit to the Royal Academy), astronomy became a leading force within the new royal scientific institution.[57]

Cassini was a distinguished gentlemanly savant who, like Huygens, possessed impeccable credentials. He was both a courtly *honnête homme* who had served Popes and sovereign aristocrats and an instrumental mathematical practitioner who could aid in administrative calculations of state. The exceptional value of Huygens and Cassini, in fact, as manifest in their exceedingly large pensions, was found in precisely their capacity to join these two dimensions of French royal science together. Yet if the Academy, as I have argued, was a bastion of gentlemanly scientific values, the presence, and even dominance, of astronomy in the same institution reveals the power of utilitarian state calculations in its formation and development.

Alice Stroup's work on the Academy of Sciences's budget illustrates the point. Colbert was certainly fond of Perrault's Grand Academy vision of royal science, but despite his goal of surrounding Louis XIV with a glittering retinue of distinguished savants, Stroup shows that at the level of actual expenditures, utilitarian administrative considerations were a dominant if not exclusive motivation behind his management of the Royal Academy. Almost half of the 1.5 million livres devoted to the institution by Colbert was spent on the mathematical sciences. Astronomy was the clear beneficiary of this largesse. In addition to Cassini's exceptionally high pension, the Observatory, technically a division of the Academy in Colbert's budget, received 45 percent of the funds allocated. Another 25 percent went to fund practical projects in cartography and geodesy, which deployed astronomical instruments and methods. By contrast, support for research into general questions of natural philosophy accounted for only 4 percent of the total budget.[58] These numbers point to the clear connection between the Academy and Colbert's administrative orientation toward state science.

Astronomy was without question the most important mathematical science for princely states in the seventeenth century, and Colbert clearly understood its significance. No mathematical science held out more promise for present and tangible utilitarian returns than the study of the stars. One motivation remained the lingering belief that astrology connected the sovereign and his reign to the mystical power of the heavens,[59] but utilitarian calculations related to navigation, cartography, and other applied sciences had become after 1650 the dominant reason for sovereigns to cultivate astronomical knowledge. The strategic, administrative reasons for promoting royal astronomy were many. Every maritime state in the seventeenth century dreamed of possessing the power of a reliable method of determining longitude at sea, and the new "astral clocks" made available by the invention and perfection of telescopes had opened up the possibility of an astronomical solution to the longitude question.[60] Cassini's work on the ephemerides of the moons of Jupiter, in fact, held promise in this area, and this made him especially desirable to Colbert.[61] Huygens also presented himself as an administrative resource since his pendulum clock was both a mathematical marvel and a possible mechanical solution to the same problem through its potential to provide a reliable means for keeping standard time at sea.[62] One will remember the *élève* La Voye-Mignot reporting on tests with such clocks on warships in some of the first Academy meetings.

The link between astronomy and state power was indeed strong in the seventeenth century, and such considerations fundamentally shaped Colbert's approach to the Royal Academy and its mathematics. Two areas were of particular importance to the minister. The first was naval power. Colbert was one of the founding fathers of the French navy. Although created under Richelieu, the French fleet was still in its infancy when Colbert received his portfolio as naval secretary in 1661. Under his supervision, the French navy was to emerge as arguably the most powerful and technically sophisticated in the world.[63] Colbert served as both the secretary of the navy and as the Director of Royal

Buildings (the portfolio that included the royal academies and their patronage), and in this capacity he presided over the creation of the first publicly funded military-industrial complex in France. At Brest, Toulon, and Rochefort, most significantly, but also at other sites, Colbert created a royal arsenal of state-of-the-art naval science.[64] Determined to make French ships as fast, maneuverable, and reliable as their Dutch and English rivals, Colbert devoted large sums of money and copious administrative energy toward the creation of an equivalent French fleet. He was not averse to tactics of industrial espionage and subterfuge in this effort, and some of his success derived from his ability to lure Dutch and English experts to France. But he also worked to foster an indigenous culture of expertise that would make France the producer rather than the consumer of naval science.

By all accounts the program was successful. By 1715, the French navy was "for all intents and purposes the best in the world," and it was the English who were trying to pirate French naval designs while sea captains all over the Atlantic dreamed of capturing and piloting a French ship.[65] Mathematical knowledge of all sorts was crucial to this success, and while at first Colbert's naval administration and his Royal Academy of Sciences were discrete and isolated units, they were always linked conceptually in his mind. Astronomy especially brought the two together, since this mathematical science was crucial to the maritime sciences of navigation and hydrography, and mechanics, both as a science of machines and instruments and as a theoretical mathematical science of moving bodies, was also crucial to this nexus.

Colbert's third portfolio, that of Controller General of French finances, further brought his characteristic administrative approach to state building together with his promotion of the mathematical sciences. Having the best navigational instruments and the best maritime technologies was essential to building a successful state, and this meant developing the technological capacities of French industry. In the seventeenth century, this also meant developing and promoting mechanics, both that of skilled knowers and makers and also the fledgling science that sought to understand machines and their behaviors theoretically. Colbert was a strong advocate for mechanics in both senses. Consonant with his so-called mercantilist economic program, his ministry created monopoly industries in tin-plate and other metalwork, tapestries, gold braid, glass, crystal, cloth, china, and soap, to name only a few. To further this development, investments were made in domestic commercial infrastructure such as roads, bridges, and canals. Colbert also initiated a program of awarding patents to manufacturers of goods deemed inventive or innovative, and he oversaw state-sponsored study programs designed to explore techniques of manufacturing and to recommend improvements.[66] Each of these initiatives led to the royal empowerment of mechanics and mechanical mathematical knowledge.

Colbert also initiated a systematic effort inspired by such thinking to investigate the principles of the mechanical and industrial arts in the kingdom. In 1667, the minister purchased a print collection for the crown and directed its proprietor, the abbé Michel de Marolles, to organize the over twenty-three thousand prints representing the practice of the mechanical and industrial arts with an eye toward publication. Given its honorable, liberal orientation, the Royal Academy was not considered an appropriate place to locate such a project, and at first Colbert kept the two initiatives distinct. Nevertheless, academicians assisted Colbert in these efforts. In 1675, the king issued a decree formally shifting responsibility for the project to the Royal Academy. The decree also stipulated that the Academy should undertake the publication of a treatise on the mechanical arts "at once theoretical and practical and accessible to all." The death of Colbert in 1684 put a halt to these initiatives, and the project to publish official volumes devoted to the mechanical arts remained dormant until 1693, when it was revived in a manner to be examined in a later chapter. These efforts would continue with fits and starts throughout the eighteenth century, and they illustrate well the new alliance between scientific know-how and administrative monarchy initiated by Colbert's ministry in the 1660s.[67]

Colbert was without question a vigorous proponent of administratively oriented state science. Yet overall, as has been noted already, the Royal Academy developed largely in isolation from these utilitarian, state-oriented mathematico-mechanical agendas. The exception was astronomy, which was integral to all of these state programs, and also dependent on the instrumental, mechanical, and empirical approach to mathematics that was essential to each. The character of early modern astronomy encouraged this synergy. For those who saw it as a subfield within the mixed mathematics, seventeenth-century astronomy was rarely if ever concerned with cosmological theorizing. The question of the order of the cosmos was pursued within natural philosophy, a field isolated from mathematics and astronomy in the disciplinary taxonomy of the day. It was also an arena made particularly contentious by the Roman Catholic Church's condemnation of Galileo in 1633. Cosmology inevitably raised vexed theological questions, and astronomers rarely entertained them. Instead, astronomy, especially in France, was a largely empirical mathematical pursuit concerned with the accurate observation and predictive description of celestial phenomena.[68]

Mathematics was crucial to this enterprise in multiple ways. Astronomical predictions depended on geometrical models of the celestial system and rigorously accurate tools of instrumental measurement and calculation. Precisely calibrated instrumentation was also essential, as were sophisticated and often complex routines of calculation. These in turn depended on the rigorous development of quantitative standards and measures that were themselves intricately mathematical. Everything further depended on instruments such as telescopes and circle quadrants that linked astronomy with other mathematical disciplines such as conical geometry, optics, and mechanics, and with the artisanal world of instrument making.[69] What was rarely needed in all this complex mathematical work was demonstrative proofs in the manner of Euclidean geometry. This made astronomers key practitioners of empirical and inductive conceptions of mathematical science. Accordingly, the support that ministers such as Colbert lavished on astronomy produced a general impetus that advanced the more utilitarian mixed mathematical sciences and marginalized the liberal humanist conception of the same.

Under Colbert, France cultivated both courtly and administrative programs of scientific patronage simultaneously, and the first class of French academic astronomers illustrated well the particular outcomes that these dual urges produced. Auzout, Picard, and Huygens were all impeccable gentlemen who anchored the liberal orientation of the early Academy. Cassini and Roemer encouraged the same when they joined the company in the following decade. Yet each was also a committed empirical astronomer, and together with their team of *élèves* they also instituted the characteristic empirical and instrumental projects that Newton praised and put to world-changing use in his *Principia* when developing the quantitative empirical arguments in support of his theory of universal gravitation.[70] Overall, precise observation and measurement, along with comprehensive empirical calculation and prediction, became the hallmarks of French royal astronomy.

Testifying to the new fusion of utilitarian academic science, administrative monarchy, and royal grandeur at the center of seventeenth-century French science was the lavish Royal Observatory that Louis XIV and Colbert ordered constructed (fig. 3). The building was at once a separate institution and a branch of the Royal Academy, and it became the beneficiary of enormous state support while also serving as a publicly conspicuous site of official, academic science. Consecrated in 1682, the facility reflected this status. It was a lavish complex of buildings adorned with appropriate royal splendor.

The astronomers selected to staff this complex also received generous state pensions. The king further provided ample funds to buy the necessary

FIGURE 3. *Victor Jean Nicolle (1754–1826),* L'Observatoire Impériale, *c. 1810. Watercolor; 6.7 × 11.7 cm. Châteaux de Malmaison et Bois-Preau, Reuil-Malmaison, France, MM40.47.9043.44. Photo: Daniel Arnaudet. © RMN-Grand Palais / Art Resource, NY.*

technical instrumentation required for their work. Astronomy was without question the "big science" of the seventeenth century, and the price of a state-of-the-art *quart de cercle*, or circle quadrant, the foundational astronomical instrument of period, ran as high as 750 livres, the annual salary of a professor at the Collège Royale. Telescopic lenses and other instrumentation imposed further financial demands.[71] Maintaining the Royal Observatory as the leading center of astronomy in the world thus required enormous financial support, and Louis XIV committed the state to this enterprise in a typically grandiose way in the 1670s.

Under the direction first of Giovanni Domenico Cassini (d. 1712), who became Jean-Dominique when he naturalized as a French subject, and then his son and grandson Cassini II (d. 1757) and Cassini de Thury (d. 1784), the institution established itself as arguably the premier site of astronomical work in Europe. The fact that before the nineteenth century, Paris, not Greenwich, served as zero longitude for astronomers and mariners offers a good indicator of the prestige that the French Royal Observatory held. Newton's authoritative citation of the precision quantitative measurements of Cassini, Picard, Richer, Couplet, and other royal academicians in the *Principia* was another testimony. *Les Connaissances des temps*, the Royal Observatory's annual almanac of astronomical data, was similarly authoritative and widely used. Early modern astronomy was also by its nature an international enterprise, requiring the coordination of astronomical data from a variety of locales and the wide circulation of such data once acquired. The French Royal Observatory became a center of this international network of astronomical exchange. French Jesuits in particular played an important role in collecting and delivering astronomical data to Paris from around the world.[72] In this way, other astronomers around the world, including many with no allegiance to the French crown at all, came to see Paris as the ultimate destination for their scientific work.

The prestige of the French royal astronomers within the international networks of astronomy translated into international prestige for the French monarchy as a whole, while back at home it triggered widespread acclaim and public interest.[73] This prestige helped to mark the royal astronomers as a class apart. The actual scientific life that the astronomers led also reinforced their isolation. Needing to work in the evenings, when the stars and planets were most visible, astronomers lived a life at odds with the quotidian rhythms of their scientific colleagues and neighbors. To facilitate their work, the crown housed them at the Observatory in a residential setting, albeit one with lavish aristocratic appointments. Their daily material needs were also met in this communal way, and these living conditions bred a unique community esprit among the members. Perhaps for this reason, French astronomy also became a family business in Old Regime France. The Cassinis were the most famous astronomical family in France, but the de la Hire and Le Monnier families also produced multiple generations of savants connected to the Royal Observatory. Such family ties further cemented the cohesion of the astronomical community.

The actual scientific work of the astronomers supported this communal ethos. Much of an astronomer's nightly work amounted to making detailed records of stellar phenomena. This was tedious precision work. But since the acquisition of these observations, and the construction of astronomical theories from them, depended upon collective labor, the discipline was also intrinsically collaborative in a way that other disciplines were not. Astronomical work also involved traveling to distant and inaccessible areas with cumbersome and sensitive scientific equipment, and the successful realization of such scientific trips required astronomers to work together in teams. This further fostered an esprit de corps among them. These aspects of the astronomical life as it was lived in the seventeenth century also shaped the scientific assumptions of the astronomers. Their routine practice of detailed observation attached them to an empirical and inductive approach to science in a particularly strong way. To them, the senses were the royal road to science, and only theories supported by vast quantities of empirical data were tenable. The work of the entire Cassini family embodied this spirit, and they set the tone for the entire astronomical community in France.

As a result, astronomy in Old Regime France was strikingly empirical and descriptive in character. What theories astronomers offered were of a limited nature, rooted in establishing the precise geometric curves traced by planets and the exact patterns of celestial movements. They completely eschewed the great cosmological questions that preoccupied figures like Galileo, Newton, and Leibniz, and they also by and large avoided for the most part any exploration of causal, celestial mechanics. When Cassini I challenged Descartes's theory of comets in 1699 by arguing that comets followed regular, closed orbits, he did so by offering detailed astronomical observations that showed how actual comet behavior supported his position. In making his argument, he did not concern himself with the problems that this theory posed for the vortical system of celestial mechanics, leaving it to others to try to reconcile his closed orbits with Cartesian fluid mechanics.[74] Cassini's methods in this case were representative of Old Regime French astronomy as a whole.

There is good reason to think that this was precisely the way that French royal administrators wanted it. Central among the justifications for the lavish expenditures made on astronomical work was the promise of direct, utilitarian results. The longitude problem remained a central preoccupation. In 1716 the Regent Philippe II, Duc d'Orléans, revealed the depth of the state's interest in a solution by offering a prize of 100,000 livres to anyone, regardless of nationality, who could provide a reliable means of determining longitude at sea.[75] This was a staggering sum, and it reflects the importance of this question for the government. Strategic military and fiscal reasons also made scientific cartography a priority for royal officials. Since scientific map-making was closely linked to astronomical practice in this period, the Royal Observatory also became a center for cartographic work.

Yet while the emergence of the Royal Observatory as a new center of state scientific power undoubtedly changed the character of French academic mathematics, rivalry and contestation were not the primary result of this new situation. Astronomers practiced a very different kind of mathematical science than classical mathematicians such as Frénicle de Bessy and Roberval, and having precise quantitative and empirical mathematics at the center of the Royal Academy certainly changed the scientific community that gathered there each week. But the difference between the mathematical attitudes of the royal astronomers and the mathematical work of academicians such as Huygens, Mariotte, or Buot was not so great. The tension that existed between the empirical, hands-on practices of the astronomers and the more theoretical, mind-intensive practices of the classical geometers is also easily exaggerated. Virtuosi in each discipline rarely felt anything but admiration for talented individuals with different skills, and overall the divide that separated savants with a more liberal scientific bent from those with a more mechanical or utilitarian orientation was far less pronounced than the divide that separated each from the ecclesiastical scholastics of the universities or the artisanal practitioners of the thoroughly unliberal mechanical arts.

Similarly, the growing intellectual respectability that astronomers acquired over the seventeenth century was part and parcel of an overall elevation of the status of nonscholastic learned men more generally. Practitioners of the new physico-mathematical disciplines especially benefited from this general social uplift. Seventeenth-century sovereigns wanted men of learning at their courts who personified the kind of elite culture that they increasingly prized. But they also wanted practical, utilitarian problem solvers who could advance the agendas of royal government as well. Courtly savants in this context often found it easy, therefore, to negotiate between their liberal and mechanical identities, and as the elevation of astronomy and other mixed mathematical sciences during this period suggests, criteria of value were shifting in new directions, often in ways that were beneficial to mathematicians of every stripe.

Fashioning oneself as an "*honnête homme*" and a "citizen of the Republic of Letters" further worked in most contexts to resolve whatever particular intellectual or institutional differences might have divided savants. Moreover, since such identities could be fashioned equally by an empirically minded astronomer, an experimentally focused student of optics, or a classically oriented student of geometry, the particular disciplinary or scientific affiliation of the mathematician was normally less important than his commitment to the wider values of the learned community as a whole. In France, where the title "royal academician" could also be joined to any of these precise self-conceptions without creating any friction, the result was the formation of a powerful esprit de corps within the Academy that bound all the different mathematical constituencies together despite their intellectual or professional differences. The fact that Louis XIV's generous patronage also floated the fiduciary boat of almost every academician further made this collegiality easy to maintain.

The Royal Academy was without question an elite and courtly institution oriented toward a liberal, *honnête* conception of academic science. Yet it was also an organ of the administrative monarchy committed to the often very unliberal mission of using science to improve the strategic commercial, industrial, and military needs of the French crown. One way that these two missions harmonized was through structures that allowed savants to see themselves simultaneously as honorable savants and technical state servants. The shrewd division of labor that kept these two agendas linked, yet also isolated them through the labor of the *élèves*, illustrates perfectly the outcomes that monarchy under Colbert and Louis XIV produced. It also illustrates perfectly how the political culture of Louis XIV's France both modernized the mathematical sciences while also contributing to the persistence of traditional patterns. Emmanuel Le Roy Ladurie describes the political "double business" of the Sun King's reign by saying that, "[The king] submerged distinctions and privileges in a 'revolutionary' fashion before the state, while at the same time respecting and even reinforcing the 'Jacob's ladder' of social hierarchy."[76] Contradictory formulations such as these are the best that can be offered when trying to capture the complexities of royal government in seventeenth-century France. They also capture well the usually harmonious, but sometimes conflicting, impulses that shaped French academic mathematics in the same period as well.

CHAPTER 3 Academic Mathematics in France before 1699

The Administrative Turn at the Académie Royale des Sciences

As we will see in later chapters, the institutional dynamics set in motion by the founding of the Académie Royale des Sciences in 1666 continued to play a role in the development of academic mathematics into the eighteenth century. The new analytical mechanics was one of many outcomes shaped by this legacy. But in the 1690s, when Varignon began to innovate his new science of motion, the academic environment in Paris was also going through a decisive process of change. Varignon's position in the Academy was itself one sign of the shift, since his appointment in 1688 was one of several that marked a generational turn over as new members, many with new orientations, were added to the company to replace the original founders after their deaths or retirement. Ministerial shifts after the death of Colbert in 1684 also brought other realignments, and the changing nature of Louis XIV's monarchy during his exceptionally long reign (he died in 1715) also brought new understandings of the place of the Academy within the French state. In 1691, the Academy was still a collection of about two dozen royally pensioned gentleman-savants meeting twice a week in the royal library on the rue de Richelieu in Paris as it had been doing for twenty-five years. But in other ways it was also a completely different institution with a new and changing mathematical culture. Analytical mechanics was propelled into existence by the older and the newer institutional logics of the Royal Academy simultaneously, and to complete this survey of the institutional climate that produced this new science, let us look at the wider changes that were under way within French academic science in the 1690s.

The Academy and the New Political Climate of the 1690s

Although they started percolating earlier, the political changes that proved most influential began in 1691, a year that was also significant with respect to the history of the infinitesimal calculus in France as we will soon see. The source of the change was completely mundane: a ministerial shake-up triggered by the death in July 1691 of the Marquis de Louvois, Colbert's successor in 1684 as minister in charge of the Royal Academy. A strand in the historiography of the Royal Academy, which traces its origin all the way back to Fontenelle, the Academy's first historian, views the Louvois administration negatively. It claims that Louvois let the Academy languish during his watch, and that he did not understand, as his predecessor had, the value of a royal company of science. From the perspective of this historiography, 1691 marks the end of a period of decline and the turn toward renewed academic vigor.

Yet Louvois was not the ministerial villain that he is often portrayed to have been. A less personal assessment might describe 1691 not as a time of revival but rather as a moment of intensification of administrative trends already present under both Colbert and Louvois. It might also mark 1691 as a time when the old courtly understanding of the Academy started to actively move into eclipse. A military-minded realist who governed during a period of constant warfare, Louvois valued the Academy solely for its promotion of useful knowledge applicable to royal administrative needs. Colbert was certainly no enemy of this rationale for royal science, and if Louvois more clearly aligned the institution with these values and agendas in the 1680s, the new ministry which succeeded him in 1691, although often labeled "neo-Colbertian" by historians wanting to mark a change from Louvois, is perhaps best viewed as an even more emphatic champion of this conception of science as a tool for strategic administrative governance. Whatever the accurate genealogy of these developments, the Academy became after 1691 an ever more emphatically administrative institution than it had previously been, and it also began to conduct itself in ways radically different from the patterns and protocols instituted in 1666.

The key moment of change was the ministerial shake-up that followed Louvois's death in the summer of 1691. The Old Regime French monarchy worked through a web of patron-client networks that were always torn up and thrown into a tangle whenever a central minister of the royal government disappeared. The year 1691 was no different, and after the clientelist threads were disentangled and the knots retied in the weeks after Louvois's funeral, responsibility for the Académie Royale des Sciences had passed into the hands of the Phélypeaux de Pontchartrain family through a fairly routine restringing of the patronage cords that constituted the heart of the Old Regime French monarchy. The ascent of the Phélypeaux clan in 1691 was anything but surprising, for the family possessed historic ties to the French court, and the ascendant titularies were men of talent devoted to royal service.[1] Yet despite their obvious legitimacy, the appointments of Louis de Pontchartrain as controller general and director of the royal household in 1691, and then his son, Jérôme de Pontchartrain, as secretary of the navy in 1694, signaled a new preeminence for the Phélypeaux within the French monarchy. The family position was further solidified in 1699 when the king awarded Louis de Pontchartrain the chancellorship, an office that the crown had largely marginalized early in the reign through its new focus on governance through ministerial bureaucracy. The king also supported Pontchartrain's efforts to reclaim the historic importance of this office at the center of the French state. Concurrent with his father's appointment, Jérôme was elevated to his father's positions, effecting a ministerial unification of the key portfolios of fiscal *contrôle*, royal buildings, and the navy not seen since Colbert's ministry.[2] The Phélypeaux would sustain their ascendant ministerial role for decades after this initial climb, producing among others the Comte de Maurepas, a titan of eighteenth-century royal administration who began serving as a royal minister in 1718 at the age of seventeen and remained crucial to royal government through the reign of a regent and two monarchs until his death in 1781.[3]

Installed in these positions of power, first in 1691 and then more decisively in 1699, the Pontchartrains (as I will refer to the Phélypeaux de Pontchartrain clan) were without question the most important players in Louis XIV's monarchy during these years. From this ascendant position, they also initiated a series of reforms that transformed the Royal Academy into the protoprofessional and publicly oriented institution that would become its hallmark during the French Enlightenment.

The reform impulses that characterized the new ministry were motivated by a host of larger political and administrative concerns, and they were also sustained by the changing political situation within France in the 1690s.[4] Louis XIV ascended the throne in 1643 at the age of five, and historians often mark the real beginning of his reign in 1661, when he effected what has come to be called his administrative revolution by choosing to govern France himself through personally appointed ministers, and to eschew the ancient pattern of treating the court of his family and dynastic peers as the heart of the state. The construction of the royal palace at Versailles, which began soon after the administrative revolution, was part of this political program, for here the king installed the aristocratic constituents of the realm, including his new, personally selected royal ministers, in a new kind of courtly setting designed to promote a new, king-centered conception of royal governance.

As we saw in the previous chapter, the Académie Royale des Sciences was one of many entities instituted at this time as a result these new courtly political logics, and in its early years it bore many traces of its court-centered royal origins. But it was also shaped by its parallel administrative rationale as well, a rationale that mirrored the wider administrative conception of royal government that had informed Louis XIV's administrative break with traditional dynastic courtliness in 1661. The Versailles court system was constructed as a way to insert these new administrative tendencies into the traditional hierarchical scaffolding of Old Regime French society, and as the system reached its apogee and began to wane toward the end of the century, the tendencies pushing away from the old courtliness and toward the administrative systems of government began to increase. Using 1661 as the date when Louis XIV's monarchy began in earnest, the midpoint of his reign passed in 1688, the year of the king's fiftieth birthday. The ascent of the Pontchartrains three years later thus marks an important turning point, when the courtly absolutism of the Sun King began to become the more emphatic and assertive Enlightened absolutism of the eighteenth century.

The reforms initiated by the Pontchartrains at the Royal Academy should be viewed in light of these underlying changes, and one important context for them was the apparent disintegration around 1700 of Louis XIV's ability to use the Versailles complex to synergize modernizing administrative agendas with the traditional dynastic ties of the court aristocracy. The early success of the Sun King's reign derived in large part from the king's ability to use his court system to forge a unity out of precisely these contradictory pulls, but increasingly this unity disintegrated. Historians looking at the second half of Louis XIV's monarchy consequently present widely disparate pictures of the king and his priorities.

One view sees the aging monarch as an emphatic modernizer who provoked instability at the center of the Versailles court system by siding with the new power of the administrative state. François Bluche argues in this vein, writing that after 1690, "the king, anxious to reinvigorate national energies, introduced more and more meritocracy into an aristocratic system."[5] Among other innovations, argues Bluche, Louis created a new royal order in 1693, the Order of St. Louis, which did not make aristocratic rank a prerequisite for admission. Only the discretionary favor of the king was required for admission, and this allowed royal distinction and title to be conferred on men of talent irrespective of their birth or dynastic rank. Louis's appointment of the lowborn Catinat as a French Marshall in the same year is a further example of this modernizing tendency. The king also approved the introduction of the *capitation* in 1695, France's first graduated income tax, imposed on all French subjects regardless of rank. Vauban and Boisguilbert, two paragons of administrative thinking, had conceived of this tax, and together with its even more controversial twin, the *dîme royale*, an across-the-board 10 percent income tax, these fiscal instruments pointed to a new impetus coming from the king himself pushing toward administrative modernization and meritocracy.[6] As the arch reactionary the Duc de Saint-Simon lamented at the time: "Little by little the king and his ministers are putting this reign at the level of everyone."[7]

Another view locates the agency for these changes instead in the simple decline of the monarch's capacities as a vigorous sovereign and in his turn backward toward older political calculations. Jean-François Solnon, for example, charts a decline of the royal court at Versailles beginning in the final decades of Louis XIV's reign, and one essential break occurs in 1683 with Louis's secret marriage to Madame de Maintenon. Maintenon exerted an increasing influence on the king after this date, and by the 1690s Louis had made the practice of holding royal audiences in her presence commonplace at Versailles. He also increasingly used her bedroom as a retreat from the public duties of court. Trivial as this shift may seem, Louis's new interest in privacy is no insignificant development given the important public role of courtliness in his original system of absolutism.[8] Other factors also fostered a decline of court-based absolutism after 1690. First was the long and costly War of the League of Augsburg (1688–97). The financial burdens of the war forced the king to subdue court festivities and reduce their expense—maintaining the king's private retreat at Marly alone cost more than two thousand livres a day—and the protracted and bloody nature of the conflict also put a damper on court life.[9] "The heart was not in the diversions of court," Solnon writes. "The courtiers lived each day in anticipation of news from the front. The battles decimated families, and everyone noted the signs of a new climate."[10]

Other signs of change were also in the wind. A sort of fatigue set in with some veterans of court life. "Always the same pleasures," lamented Madame de la Fayette, "always at the same time and with the same people."[11] Similarly, a generational shift of sorts occurred, or at least the perception of one, as a new generation of courtiers came of age after 1690. Many newcomers found it hard to comprehend the rigorous ethic of their parents, and as these younger courtiers began to become adults within the court system, many began to take the practices of court for granted, or to scoff at its burdens. Changes such as these were not trivial. Many traditionalists found their conception of monarchy reinforced through the rigid hierarchies and protocols of court life. Accordingly, as the power and prestige of the institution waned, the court (which is to say royal government, given the political realities of the Old Regime state) found itself newly alienated from the monarch and his entourage. The king himself also encouraged these developments whether unwittingly or not. His increasing isolation from the public life of court and his private devotion to Madame de Maintenon were signs of a new climate. His perceived turn away from the pleasures of court sociability and toward the ascetic devotions of religion was another. As Jean de la Bruyère summed up what he perceived to be the decline of the grand old courtliness of the past: "The courtier used to have horses, spend his time hunting and dancing, carry a large gun, and practice libertinism. Now he wears a wig, a close-fitting jacket, matching trousers, and he is devout."[12]

Equally important, however, was a marked decline in the king's ability (or should one say desire?) to use the power of his royal gaze to make the court complex a concentrated site of royal authority. After his public duties were completed, the king increasingly retired to the bedroom of Maintenon with his ministers or his musicians, while his courtiers, taking their cue from the monarch, deserted the great halls and concentrated their festivities in smaller, more intimate surroundings. As a result, over time, small entertainments in private apartments gradually supplanted the grand spectacles in the halls of Versailles. Similarly, as Solnon observes through the subtle shifts of reporting found in the diaries of court elites, social calendars dominated by court-centered entertainment and by commitments to attend functions attended by the king himself gradually gave way to social time spent away from the monarch and in sites other than the royal palace.[13]

This political cultural shift also supported a demographic shift of equal importance. Court-based absolutism had always been based on the physical proximity of aristocratic elites to the body of the king, and from the moment that the provincial nobility was installed at Versailles in the late 1670s, Frenchmen seeking public status and authority established themselves in and around the royal palace so as to be present at the heart of the society upon which their ambitions depended. By 1700, however, a clear demographic trend in the opposite direction was occurring. Most important was an exodus of court elites from Versailles to Paris. In 1704, the Hôtel de Soubise was opened, a building shocking at the time for its royal pretensions and decorations. The architect Pierre-Alexis Delamair constructed a similarly palatial *hôtel* for the Duc de Rohan between 1705 and 1708. New Parisian neighborhoods also grew in size and notoriety as elites began to establish themselves in the city rather than in the environs of the royal palace. The Faubourg Saint-Germain in particular witnessed a population explosion beginning as early as 1697 as a new generation of Enlightenment architects began building new residences for elites in this part of Paris. On the other side of the Seine, the Faubourg Saint-Honoré became similarly populated by many of the same class of people for many of the same sorts reasons.[14]

Rival courts, largely suppressed in the early years of Louis' personal rule, also began to reappear after 1700. Urban sociability in places such as Paris also led to an expansion of noncourtly social networks. Salons such as those of Mademoiselle de Scudéry and Madame de la Sablière were fixtures of Parisian life in the 1650s, but during the heyday of Louis XIV's Versailles system they found their importance eclipsed by that of the royal court. By 1700, however, this hierarchy had begun to invert itself.[15] New periodicals and forms of print media, such as Donneau de Visé's *Mercure galant*, to be discussed in detail in a later chapter, offered virtual spaces where political and social hierarchies could be reconfigured.[16] This was the political and cultural environment that supported the "Malebranchian moment," to be discussed in the next section, and by the 1700s Madame de Lambert, a fixture of this new Paris-based sociability, could plausibly argue that her salon, frequented by Fontenelle and other members of the French academic intelligentsia, was as influential as the Versailles court in setting the cultural agendas of elites. By 1730, the argument was no longer necessary—the Parisian salons had clearly eclipsed the royal court at the heart of elite sociability.[17]

In 1715, when Louis XIV died, leaving his five-year-old great-grandson as heir to the throne, the regent Duc d'Orléans officially designated Paris as the seat of royal government. Historians often treat the Regency period (1715–23) as a great historical rupture, the moment when the death of the Sun King finally released elites from their fifty-four-year attachment to Versailles and court-based monarchy.[18] But 1715 marks far less of a break in the political history of France than is sometimes realized. Most of the crucial changes began much earlier, most notably in the 1690s, and overall the Regency period stands more as a culminating moment when a set of changes already under way were solidified.[19] The link with the Phélypeaux was sustained, for example, with the appointment of the teenage Maurepas to the royal council in 1718, and in 1723 when Louis XV reached maturity and ascended to the throne, he chose to continue and intensify the administrative turn begun in the 1690s and to allow the old court complex to continue to go into retreat.[20]

This recognition brings us back to the first Pontchartrain ministry and to the important changes that it helped to catalyze after 1691. Clearly, a number of new dynamics were under way making this decade a moment ripe for change. Among the sites that became newly important to the Pontchartrains as a locus for the needed reforms they built their ministry pursuing was the Académie Royale des Sciences. The institution accordingly became after 1691 a crucial arena for these new and transformative political developments. Analytical mechanics came into the world at precisely this moment, and the new science was shaped in crucial ways by the wider political cultural changes under way during these years. In order to appreciate this influence, let us complete this survey of the institutional dynamics supporting the genesis of analytical mechanics in France by considering the new changes at the Académie des Sciences initiated by the Pontchartrains.

The New Administration of the Académie Royale des Sciences after 1691

The first steps toward change occurred within weeks of the establishment of the Pontchartrains at the center of Louis XIV's monarchy. Most crucial was the creation of a new management structure for the Academy. Only a month after his appointment as Director of Royal Buildings, Louis de Pontchartrain appointed his twenty-nine-year-old nephew, the abbé Jean-Paul Bignon, to be the Academy's president, a new office that had never existed before. From this position, Bignon would go on to play a pivotal role in reshaping French academic science over the next four decades. He is, accordingly, a figure whom we will encounter frequently in this book and who needs to be introduced here.

In 1691, Bignon's primary title was royal librarian, an office that ran in his family. During these years, he was also beginning to translate his educational formation within the Oratorian religious order, another nexus we will encounter frequently, into an emerging career within royal administration and the Republic of Letters. In 1693, he was admitted to the Académie française, an appointment that was based as much on his political favor as on several sparkling orations he delivered in praise of Saint Louis. He was also a vigorous antiquarian scholar, with a special interest in numismatics, which brought him into the fold of the learned men who assembled each week at the sibling institution of the Académie Royale des Sciences, the Académie des Inscriptions et Belles Lettres. Bignon would eventually acquire a supervisory role over all of these French royal academies, and the new impetus pushing him in a managerial direction after 1691 was a result of his ties to the Pontchartrain family.

As the new Pontchartrain ministry solidified and became a formal royal administrative structure, Bignon came to play a sort of middle-manager role, presiding over the cultural wing of the administrative monarchy and navigating between it and the councils of state. This position gave the abbé tremendous real power in shaping the actual institutional development of French culture even if it did not give him the notoriety that accrued to those with whom he worked. As a result, Bignon remains an astonishingly understudied figure. Jean-Jacques Dortous de Mairan, the celebrated eighteenth-century academician and interim successor to Fontenelle as the Academy's perpetual secretary, called Bignon the "guardian angel of the sciences and of savants."[21] Françoise Bléchet, virtually the only modern scholar to have systematically considered Bignon's importance, describes him as "the Enlightened despot of culture" in the period.[22] Using the language of Old Regime patronage that was equally applicable to Bignon's work, his friend and academic ally René-Antoine de Réaumur called him simply "the great Maecenas."[23] These testimonies are not exaggerated, for from 1691 until his retirement in 1734 Bignon played a singularly decisive role in shaping French culture, not least the culture of the Académie Royale des Sciences.

Several features of his rich array of responsibilities are relevant here. Most important was his managerial role at the Royal Academy of Sciences. He ultimately came to have direct oversight responsibility for virtually all of the French royal academies, along with their related institutions such as royal publication, and his surviving correspondence reflects the wide range of his administrative duties.[24] At the Academy of Sciences, he served until 1699 as president, and then, when the Academy reform of that year, which will be discussed in detail in a future chapter, institutionalized this post and a parallel vice president position around an annual appointment schedule, Bignon served in one or the other of these offices in thirty-two of the thirty-six years that he was eligible.[25] He also played a very active role in the Academy's affairs, and it is clear that his authority was the most important even when he was not the titular head. His influence was especially important during moments of controversy, when the Academy often appealed to Bignon to effect resolution or restore order.[26]

Typical of Bignon's method of management was his use of allies within the Academy to shape affairs within it. His early correspondence, which would reveal these habits during the years considered in this book, has not survived, but by 1714 it is clear that René-Antoine de Réaumur had become Bignon's chief agent inside the Academy of Sciences. There is also ample evidence that after 1714 Bignon worked through Réaumur to shape outcomes within the Academy, and that he maintained a very close, day-to-day interest in its activities. Overall, he saw it as his job to ensure the Academy's success and to secure its favor in the eyes of his ministerial patrons. Soon after being named president of the Academy in 1691, he also began to actively pursue these goals. An early change was his decision to appoint two new academicians: Joseph Pitton de Tournefort and Guillaume Homberg.[27] The first was a botanist and the second a chemist, and since no one had either died or retired in advance of these appointments, all indications point to a decision, supported by Pontchartrain, to expand the size of the Academy by appointing two new members with close ties to Bignon. The institution also experienced another first in December 1691, when Father Thomas Gouye of the Society of Jesus, who was also the mathematical tutor of Jérôme de Pontchartrain, entered the Academy to present the latest astronomical work completed by his Jesuit brothers in Asia.[28] The Academy had been receiving reports about this work for years, but never before had a Jesuit entered the Academy to present the reports directly. In 1699, Gouye would be made an official member of the Royal Academy, the first and only Jesuit to earn that honor, and his appearance in 1691 was a sign of the political changes afoot.[29]

The Academy records also reveal a set of changes in academic practice attributable to Bignon's new and more assertive management style. The Academy had named a secretary at its initial founding and charged him with the maintenance of a record of the academic proceedings. It will be recalled that these duties fell initially to the distinguished Latinist Jean-Baptiste du Hamel, and then to the abbé Gallois, editor of the *Journal des savants*, when du Hamel's other duties kept him away from the Academy. In 1691 this arrangement was still in place, and by this date it had resulted in a set of comprehensive *registres de l'Académie Royale des sciences* that documented the work of the institution over its first twenty-five years. These registers offer a revealing window into what the Academy did and how it worked, but under du Hamel they were anything but a complete record of everything that transpired at academic meetings. Most entries simply note that the Academy met and talked about this or that topic, and when a single academician presented research over a series of sessions, the entries often amounted to little more than the statement that "Mr. X" continued to read his *mémoire* on A, B, or C. Sometimes the *mémoires* themselves were transcribed into the registers, but before 1691 this was by far the exception rather than the rule. In fact, the entries for particular meetings were often so short (a few sentences was the norm) that several could be included on a single page, and this despite du Hamel's large and flowing handwriting. The deliberations and discussions that accompanied a presentation were never entered into the record, and actual academic debates, which were no doubt spirited, are impossible to reconstruct from these records. Early on, more detail is present, but by the 1690s it had all but disappeared as the records became brief and formulaic. No systematic recording of attendance at the meetings is to be found in these early records either, and while occasional entries recording judgments about a machine or a new book submitted for evaluation sometimes appear, mundane academic activities such as these were rarely included.

After 1691, these practices changed, and the Academy records after Bignon's appointment suggest that the new president perceived the prevailing habits to be negligent and in need of reform. His efforts were guided by a number of related agendas simultaneously. One involved simple activity. Whether coming directly from Pontchartrain, or from Bignon with his patron's support, the message began to be delivered with increasing vehemence that the royal academicians needed to pick up the pace of their work. Bignon also wanted his Academy secretary to be far more vigorous and meticulous in recording the activities of the company. He especially wanted to see more papers read, and then, once they were presented, to see this academic work transcribed into the records. Finally, he wanted all of this new work to be made accessible to the public in a new way. Save for the few collective volumes published by the Academy in the 1670s and '80s, and the occasional publication by exceptional academicians—Edme Mariotte's books from this period are a good example—the labors of the Academy had not resulted in much publication.[30] This lack of publication fit with the Academy's early courtly orientation, but with the turn toward a more administrative conception of academic science, a new imperative to publish became the norm. Bignon made these agendas specific in a new set of expectations, which he articulated forcefully after his appointment in 1691. They asked academicians to become vigorous workers pursuing active personal research agendas that resulted in visible publication.

The Academy records from the 1690s record the precise changes that Bignon's assertive management provoked. Having already appointed Tournefort and Homberg to the company in late November, and having sent the Jesuit Father Gouye to the assembly on December 15, Bignon appeared at the assembly of December 19 to issue the following directive, which was duly recorded in the Academy minutes. The entry reads: "M. l'abbé Bignon said that it is the intention of M. de Pontchartrain that the company give a report [*mémoire*] to the public each month of everything that it is doing. To which the company concluded that in order to comply with this order each person at the assembly will present to the Academy by the following Saturday a project appropriate for the public."[31] At the next meeting, the Academy discussed what it would include, and the list of first responders included both Tournefort and Homberg, academicians who came to model Bignon's academic ideal in their frequent contribution of *mémoires* to academic meetings. Philippe de la Hire and Jean-Dominique Cassini also proposed work in astronomy for these monthly reports, and the celebrated public anatomist Joseph-Guichard Du Verney was quick out of the gates as well. Varignon also contributed a number of mathematical papers to this enterprise, and while the intellectual substance of this work will be considered later the point to emphasize now is the direct entanglement of Varignon's mathematical labors with the new administrative imperatives of the Academy provoked by the ascent of Bignon and the Pontchartrain ministry.

The monthly academic reports were published throughout 1692 and 1693, and in the first number, which included a paper by Varignon on fluid mechanics alongside others by Cassini, de la Hire, and the mathematician Michel Rolle,[32] an editorial *Avertissement* explained that these were short pieces, taken from the academic registers, that offered readers "little presentations drawn from the larger projects upon which the academicians are assiduously at work."[33] It is unclear who authored the *Avertissement*, but it was likely Gallois, who appears in the Academy registers as the point person charged with receiving academic *mémoires* destined for publication. But it might have been Bignon. Whoever it was, the monthly reports were clearly edited. Some entries offered direct transcriptions of the papers read to the Academy, and these were left in the academician's voice, as if the reader was listening to the paper as if it were delivered at the assembly. Many more, however, were presented as summaries written by an editor that explained to readers the significance of the academician's work. A disciplinary split also governed this editing, because technical mathematical papers were transcribed and published verbatim, sometimes with a narrative paragraph at the outset explaining their significance, while works in applied and empirical mathematics (astronomy, machine-based mechanics, physic-mathematics) and papers in botany, chemistry, and natural history were almost always turned into narrative summaries.

The Academy issued these new monthly reports without fail every four weeks between January 1692 and December 1693, and the Academy registers in this period reveal the changes to academic practice that this publishing program provoked. Ultimately the initiative died after two years because the Academy simply could not meet the burdens of monthly publication given its small size.[34] The problem is evident in the publications that were offered since more than 70 percent of the reports (sixty-one of the eighty-seven total) were authored by only five academicians: Cassini, de la Hire, Homberg, Varignon, and an obscure M. Sedileau connected to the Observatory who often published together with de la Hire. To meet the ministry's expectations for the Academy, the members would have to become more numerous and more productive.

Adding new members was one way to do this, so after the appointment of Homberg and Tournefort in 1692, Pontchartrain approved the appointment of sixteen new members between 1693 and 1699, the date when a comprehensive reform of the Royal Academy led to further expansion of its membership. Only one of these new additions can be described as a replacement for a deceased or retired academician.[35] This means that the Academy began to significantly grow in size under the Pontchartrain administration, a fact that was not without financial consequences for the crown, as Alice Stroup has shown.[36] Some new members came in as *élèves*, which may have been a cost-saving measure, and these included the sons of de la Hire and Cassini, who began to work as academicians under their fathers in 1694. Giacomo Filippo Maraldi, Cassini's nephew, also became an *astronome élève* in 1694. The Oratorian mathematician Louis Carré also joined as Varignon's *élève* in 1697, an appointment important to the history of analytical mechanics, as we will soon see. Varignon in fact welcomed a number of his future allies into the Academy during these years as a result of the expansion, including especially the Marquis de l'Hôpital, who was appointed in 1693 in time to contribute three *mémoires* to the monthly reports before they were terminated. Fontenelle, who was appointed to succeed du Hamel as Academy secretary in January 1697, was another of these additions. Thomas Fantet de Lagny and Joseph Saveur, each appointed in 1696, were also important figures in the history of analytical mechanics appointed during the expansion. We will return to these appointments and discuss their significance with respect to the development of analytical mechanics later.

If adding new members was one way to add vigor to the Royal Academy, another was to improve the working practices of the academicians once they were appointed. Bignon pursued both agendas in the 1690s, and his efforts at creating a more productive Academy focused in particular on making the existing members more active and expansive in their scientific reach. To stimulate work from within, Bignon especially pushed the secretary to become a more active manager and recorder of the week-by-week activities of the company. The production of the monthly reports had required a new diligence on the part of the Academy secretary regarding the receipt and transcription of academic *mémoires*, and even when the monthly reports stopped these new expectations continued. Bignon made it clear that he wanted the academic *mémoires* read at the assemblies to be transcribed in the registers, and as the academicians, together with the secretary, worked to comply, the registers ballooned in size as a result of the new texts being added.[37] To ensure that all of this was done to his satisfaction, Bignon also asked du Hamel to personally scrutinize these transcriptions and to apply his signature at the bottom of each entry confirming his approval. The registers were then passed to Bignon for his signature, and in some cases, especially at first and during moments of change, Pontchartrain signed the academic registers as well. This practice would continue into the eighteenth century, and it only ceased when the reforms initiated after 1699 were fully institutionalized.

The frequent presence of Bignon's signature in the Academy registers after 1691 marks his active presence as a reform-minded manager of the institution. The records also reveal other aspects of his work. Early on the Academy adopted the practice of taking a long holiday in the fall, usually stretching from September into November. The monthly reports of 1692–93 marked this practice by having no issue for September or October. "*Les vacances*," as it was called in the registers, served as a kind regular break for the Academy, serving to end one year and inaugurate the next. In the September session before *les vacances* of 1694 Bignon signed the registers, as was becoming customary, signaling his approval of the year's work so far. But then on December 9, the first session after the break, a new practice began: the recording of the names of the attendees present at each session.[38] From this date forward, every entry for every session began by recording the names of the members present. One month later, on January 8, 1695, du Hamel announced an explicit directive from Bignon asking the members to submit to him an extract, suitable for submission into the registers, of every word presented orally at the assemblies. Du Hamel also announced that Varignon had been assigned to help him with this work especially with respect to mathematical papers.[39] In March 1695, du Hamel read another directive from Bignon explaining to the members the practice of ministerial oversight of the academic registers. He also delivered a proposal to them from Bignon inquiring whether the Academy would like to choose "some book of physics which is thought to have some utility and to then examine its propositions together." The registers report that the Academy agreed by selecting "the book of M. Borelli on the movement of heavy bodies." No further record of this initiative is found in the records.[40] In 1687, Varignon had published his own work of mechanics directly engaging with the ideas of Giovanni Alfonso Borelli, so in this initiative too Varignon appears to have been close to the people in charge of the initiative.[41]

The changes in 1695 further expanded the size of the registers as ever more papers were now entered directly into the records. The entries also reveal the difficulties that these new practices posed, especially for the secretary. Bignon, it should be remembered, made it a habit of signing each entry in these years signaling his approval, and while some entries followed the old pattern, with only a few sentences entered summarizing each presentation, more elaborate entries, including complete transcriptions of academic papers, started to become commonplace. The registers also reveal the fits and starts of these new managerial efforts. On January 12, 1695, for example, just four days after Bignon had issued his decree asking academicians to submit transcripts of their academic presentations for the records, the anatomist Jean Mery offered a comparative anatomy presentation that examined the pulmonary artery in a human and a turtle. Rather than the usual sentence or two describing the topic and its key highlights in the manner of his earlier entries, du Hamel devoted several pages to a detailed summary of Mery's presentation. He also included an account of the questions and discussion that ensued afterward. He recorded, for example, that Varignon had some suggestions to offer Mery on this topic. He also left several blank sheets in the register after his summary, perhaps in anticipation of a transcription to be submitted.[42]

This more comprehensive approach to academic reporting was new for the registers, but it was also exceedingly short-lived, for in the next session du Hamel offered a brief report of a few sentences, and Bignon's signature appears at the bottom of each.[43] Nevertheless, the habit of leaving blank pages in the register after each session persisted, and even more pages were filled with ever more transcriptions of academic papers read at the assemblies after 1695. Yet the disciplinary stresses evident in the monthly reports also continued as well. Many of the academic papers entered into the registers are in du Hamel's hand, but Varignon's mathematical papers appear in the registers in his own hand, and on different-sized paper. They were also clearly prepared by him without du Hamel's intervention. Varignon's hand also appears in other mathematical papers entered into the registers, and the handwriting of other academicians is also present in the mathematical papers.[44] What this variety reveals is the difficulty of translating the new ministerial expectations regarding the work of the company into a clear set of protocols for academic work.

In all these ways and others not evident in the Academy records, Bignon worked after 1691 to prompt more and better activity from the Royal Academy. He also applied himself to expanding its scientific reach, and one area of particular interest to him was the mechanical arts. It will be remembered that while some, such as Colbert, imagined the Academy as a possible venue for this kind of applied and techno-utilitarian scientific work, the Royal Academy as it was founded did not include *méchaniciens* of this sort among its members. To be an academician meant being a liberal theoretician of one sort or another, and while all kinds of applied and utilitarian work was supported by the company, the Academy was not explicitly instituted to include the mechanical arts within its purview. Bignon, following the wider political agendas of the Pontchartrain ministry, did not like this situation, and soon after his appointment he began to create a better synergy between the mechanical and technological needs of royal administration and the Royal Academy.

Toward this end, and only weeks after his appointment, he formed a special group of advisors to consider problems of technology in the kingdom. The group did not consist of academicians, even though Bignon was already supervising the Academy by this time, but it was not composed of artisans either. Instead, he selected three individuals who personified the administrative compromise between technical service and gentlemanly honor so typical of French science in this period.

The first was Jacques Jaugeon, a descendant of minor nobility who invoked the aristocratic title of *écuyer* in all of his formal correspondence. He earned a modest income from his family estate, and this allowed him the leisure to pursue a career as a savant in Paris, where he participated actively in the many sites of scientific sociability in the city. Jaugeon's colleague, Gilles des Billettes, was likewise from the lower ranks of the French aristocracy, and he too lived a gentleman's life in Paris, working as an independent savant. The third member of the group, Father Sébastien Truchet, rounded out the trio by representing the First Estate. He was a Carmelite monk, with a gift for mechanical handiwork, who used the leisure that his Parisian monastic life gave him to produce elaborate automata, which were all the rage in elite Parisian society. These three men shared a devotion to the rational pursuit of the mechanical arts, and every Monday after September 1691 they convened with Bignon to discuss technical matters and the progress of the mechanical arts in the kingdom.[45]

Evidence of close ties between this group and the Royal Academy, at least in Bignon's mind, exists as a *procès-verbaux* kept by Bignon's *méchaniciens* between 1693 and 1696 is appended to the Academy's *registres* between the entries for 1693 and 1694.[46] Yet the two groups remained institutionally separate despite Bignon's centrality to each. No doubt, the ongoing problem of connecting the mechanical arts with the elite, liberal, and courtly conception of academic science sustained this division. In 1694, des Billettes composed a *mémoire* exploring the future prospects of the group. The addition of the *méchaniciens* to the Academy was one option, but more desirable in his mind was either the creation of a separate Académie des Arts or the attachment of the group to the Royal Ministry of Arts and Manufactures.[47] In 1699, Jérôme de Pontchartrain opted for the first scenario, adding all three members of Bignon's group to the Academy as *pensionnaires*, and joining them with a newly appointed *associé*, Jean-Mathieu de Chazelles, and two *élèves*, Antoine Parent and Michel de Senne, to create the Academy's first class of *méchaniciens*.

The addition of the formal title *méchanicien* to the newly defined and regulated disciplinary classes of the Royal Academy was one piece of the broader reorganization of 1699, which will be discussed later. But the early steps in this direction under Bignon show the nature of the socio-intellectual changes already under way by 1691. Here, for the first time, the administrative agendas in the mechanical mathematical arts, present in the discussions that led to the founding of the royal company, but excluded at its initial formation, were coming to be channeled squarely and unequivocally into the mission of the Académie Royale des Sciences. Bignon especially translated this new orientation into his management of the company. Pontchartrain also supported him in this effort, working on the ministerial front to intensify the role of the Academy as a resource in the solution of state problems. He encouraged those within the administrative monarchy to turn to the Academy for assistance with technical scientific needs, and he reemphasized the role of the Academy as a technical consultant in the solution of state problems.[48] Bignon pushed the institution in the same direction from the inside, and when the members of his separate Academy of the Arts were added to the Academy in 1699, they were immediately directed to resume the program, launched under Colbert, but later shelved by Louvois, to produce a set of official academic volumes on the mechanical arts. A manuscript version of one volume, dated 1704 and signed by Jaugeon, survives, but the work was never published. Truchet's papers similarly reveal work on printing technology and the manufacture of books. Collections of images designed to be included in these volumes also exist, and they indicate that the work began as early as 1693.[49] Bignon also used the new public orientation of the Academy to further this program. The new twice-annual academic public assemblies instituted in 1699 will be discussed later, but in the first one ever held in November 1699, much of the session was devoted to separate papers by Truchet and de la Hire on questions pertaining to the mechanical mathematical arts.[50] The published monthly reports from 1692 and 1693 are also indicative of this orientation in their inclusion of numerous papers in mixed mathematics with explicitly utilitarian concerns, along with the relative absence of purely theoretical mathematical papers. This trend would remain in evidence at the public assemblies after

1699, and Bignon appears to have taken an active role in pushing activities in this direction.

Yet even as he actively encouraged the technical agendas of the administrative monarchy inside the Royal Academy, he also sustained the *honnête* image of the Academy as a company of elite, liberal savants. Bignon exemplified the ideal of the *honnête homme*, and as his star ascended he became a widely esteemed citizen of the Republic of Letters with no hint of the mechanic in him. Especially important were his rhetorical gifts as manifest in his most important labor: epistolary commerce. Bignon's position as de facto French cultural minister meant that he was forever engaged in correspondence with savants seeking pensions, patronage, or favor with the French crown. He also represented the Academy in its relationship with other scientific bodies, such as the Royal Society of London, and maintained an active, if thoroughly formal, correspondence with the Royal Society's leaders during his lifetime, Henry Oldenburg and Hans Sloane. His royal duties similarly placed him at the center of the correspondence networks that constituted the bureaucratic core of the French state. The minister's success in this epistolary labor stemmed from his ability to translate the mundane task at hand into a letter that both achieved royal administrative agendas and demonstrated the author's own quality as a man of letters. Bignon accomplished this synthesis with ease, and while he published only one work of literature during his lifetime, an oriental novel published in 1717, his name and esteemed reputation were well known in France and throughout the communication networks of the Republic of Letters. This allowed Bignon to personify the *honnête* values essential to the image of French academic science, and to translate his cultural authority into political power for him and his ministerial patrons.[51] Working in this way, his influence in shaping the development of analytical mechanics was also significant, and it will be noted in detail in the coming chapters.

The Second Generation of French Academic Mathematicians

A generation is an imprecise classification, and the history of the Académie Royale des Sciences between its founding in 1666 and the 1699 reform does not offer any decisive breaks from which to mark out its old and new cohorts. Continuity across this period also exists, with founding members of the company still active in the eighteenth century. Some of these elders played a vital role in the analytical mechanics debates that erupted after 1699, bringing to the struggle their understanding of mathematics cultivated over four decades as royal academicians. Yet these senior academicians were also not the major players in this history. Much more important were a collection of young to middle-aged academicians who entered the Academy amid the transitions just described. The transformations brought by the Pontchartrains, especially the 1699 Academy reform, brought a flurry of new appointments, ultimately doubling the overall size of the company. Many of these additions, such as l'Hôpital's appointment in 1693 and Fontenelle's in 1697, were directly connected to the specific development of analytical mechanics, and will accordingly be treated in detail in later chapters. Varignon, however, entered the Academy in 1688 at precisely this moment of change. Many of the other dramatis personae in the history of the new science of motion were also second-generation French academic mathematicians such as Varignon. So to set the stage for the narrative to come, we will conclude this chapter by looking at some of the kay actors, and the careers they brought with them when they pursued French academic mathematics in the 1690s.

To set the context for their arrival, let us look briefly at the changes that led to the departure (by and large) of the founder's generation by the year 1691. Huygens's death in 1695 complicates any easy use of 1691 as the time of rupture between old and new, but since Huygens had already returned to Holland in 1684, for reasons that still remain mysterious, and since he ceased after this date participating in the French Royal Academy directly, we can use this date to mark one moment of change. To be sure, he remained an active correspondent with many of his older colleagues after 1684, contributing through his letters to their academic work. He also formed epistolary relations with younger academicians as well, exerting an indirect influence on the work of the academic mathematicians right up to the moment of his death. Yet by the time of Huygens's departure in 1684, many of the academicians who had founded French academic mathematics with him were gone, and even if it is arbitrary we will use this date as the moment separating the old Academy from the new.

Three original academic mathematicians died in 1675: Roberval, Frénicle de Bessy, and Buot. Picard passed away in 1682 and Mariotte in 1684. The astronomer Auzout left France permanently in 1668, moving to Italy in disgrace, and the arrival of Roemer in 1673 as his ostensible replacement proved short-lived as he was back in Denmark by 1678. The opening of the Dutch War in 1672 led to the transfer of Niquet and Richer away from Paris and into military service, and Colbert's effort to replenish the Academy with new mathematicians before his death in 1683 was not overwhelmingly successful. One appointed in 1678, Pierre de Lannion, was removed in disgrace in 1686 and sent to the Bastille for his crimes. A second, appointed in 1682, Jean Le Febvre, was dismissed in 1702 in a scandal involving his editorship of the academic almanac *Les Connaissances des temps*. A third, also appointed in 1682, Laurent Pothenot, was dismissed in 1696 because of his failure to attend meetings, a sign of the new administrative rigor instituted by Bignon and the Pontchartrains.[52] The opening of the new Royal Observatory in 1682 was the exception to this history of change; through its family dynasties it served as a site of continuity between the mathematical founders of the Royal Academy and the academic mathematicians of the eighteenth century.

Yet if the distinguished savants who initially established French academic mathematics did not leave behind a clearly defined group of successors, those that came after them were not all insignificant. Many of those appointed after 1684 went on to become leading members of the institution in the eighteenth century. Many of them were also instrumental in shaping French academic mathematics, not least analytical mechanics. To illustrate, we conclude this chapter with a brief biographical account of five academicians who were emblematic of what French academic mathematics became in the final decades of the seventeenth century.

NICOLAS-FRANÇOIS BLONDEL (1618–86)

Not all of Colbert's appointments to replace the original founders failed, and since one became exceptionally distinguished through work that illustrates the larger trajectory of change described above, I will begin with him as an early indicator of the new climate.[53]

Nicolas-François Blondel was born of middling stock into a family that illustrated well the opportunities for social elevation offered by Louis XIV's monarchy. He rose from youthful service in the king's armies to become first an acclaimed military engineer and then one of the most distinguished scientific savants in seventeenth-century France. The steps of his career progression also personify the broader changes under way within France and French academic mathematics. Blondel was given a liberal education by his recently ennobled father, and he entered military service in his teens with a strong grasp of ancient and modern mathematics. This led to his first steps toward fame as he began to acquire a reputation as a precocious student of the military sciences. By the 1640s, though still in his twenties, Blondel was advising the royal minister Mazarin on maritime defenses in the Mediterranean and French fortifications in Provence. He rose through these efforts to the rank of *maréchal de camp*, and in 1652 he left the army to become the tutor of the son of the reigning secretary of war, Henri-Auguste de Loménie, Comte de Brienne. Blondel was charged with taking the secretary's son on a sort of modern scientific Grand Tour, where instead of the great artworks of Europe he and his charge visited martial installations and consulted with experts in military engineering. Instead of Italy, Scandinavia, Holland, and Northern Europe were their pedagogical destinations, and when they returned, Blondel was rewarded with Pierre Gassendi's chair in mathematics at the Collège Royale, a replacement that reveals the changing hierarchies of scientific learning in Louis XIV's France. Colbert also drew Blondel into state service in the 1660s, charging him with advisory duties in the planning of French Atlantic naval defenses, and the planning of the new military port at Rochefort. He was also sent on a trip to Martinique and Saint-Domingue to consider the commercial maritime development of the islands. Upon his return from the Caribbean in 1669, Colbert appointed him to the Académie Royale des Sciences.

Blondel was accordingly the first royal academician to win academic appointment because of his work in utilitarian mathematics as it related to state building. As an academician, he also continued working in this mode, helping to move the Academy toward a more utilitarian and applied mode of mathematical work. He enlisted other academicians in the production of his *Art of Launching Bombs*, published in 1683, an influential work in ballistics that also demonstrated the state of the art in applying mathematics to military affairs.[54] He also contributed papers to the academic meetings on questions of mechanics. In 1671, Blondel was appointed to direct Colbert's new Academy of Architecture, and it was here that his impact was most strongly felt. His "Cours de l'Architecture," which he delivered as public lectures at the new architectural Academy, was at once an aesthetic primer and a basic textbook in structural engineering.[55] When the French crown created in the 1740s the first ever protoprofessional engineering school, the École des Ponts et Chaussées, Blondel's treatise became a core text in the curriculum.[56] Blondel reached the pinnacle of his career in the 1680s, when he was charged with the mathematical education of the dauphin. Combining mathematical instruction with education in war strategy—his pedagogy deployed hundreds of specially cast silver soldiers that allowed him to stage miniature war games for the young prince—Blondel became a visible and influential member of Louis XIV's court. His appointment to this lucrative and highly symbolic position also revealed the new synergy that was becoming operative in the 1680s between modern mathematics, strategic state and military science, and absolutist government. In the Academy, Blondel embodied the same orientation, working to push the culture of academic mathematics in this direction.

GIOVANNI DOMENICO (LATER JEAN-DOMINIQUE) CASSINI (1625–1712)

If Blondel was the first mathematician to directly link mathematics at the Royal Academy with the administrative governance of the French state, governmental calculations were also central to the new royal French astronomical complex that was integral to the new Royal Academy. Giovanni Domenico Cassini was in fact appointed to the Academy in the same year as Blondel (1669), and he shared much with him in his work as the de facto head of French royal astronomy.[57] Each became a mathematician through the traditional ties that bound mathematics to courtly states, with Blondel following the channel that tied mathematics to the military sciences and Cassini the one that linked mathematical astronomy with astrology. Cassini's astronomical career began in 1648, when the Marchese Cornelio Malvasia, a rich aristocrat and member of the Bolognese senate, invited him to work in his observatory at Panzano as a court astrologer. From this position, Cassini began to pursue rigorous astronomical research, and in 1650 the Bolognese senate offered him the chair of mathematics at the university, which had been left unfilled after the death of Bonaventura Cavalieri in 1647. Over the next decade Cassini developed a distinguished reputation as an astronomer, and a French royal memorandum from the middle of the seventeenth century listed him among the foreign mathematicians "who have [in their country] the greatest reputation."[58] This memorandum listed other savants worthy of French royal patronage, including Descartes's correspondent Princess Elizabeth of the Palatinate, and through these courtly channels Cassini's star began to rise within Louis XIV's monarchy.

Yet if courtly logics of patronage ultimately brought Cassini into the fold of French royal science, his reputation as a highly skilled instrumentalist and master of precision, quantitative mathematics was also important. His full persona, in fact, is revealed by his other duties performed before moving to France in 1669. As a mathematician in service to the Bolognese senate, he performed numerous astronomically related tasks for his patrons, but he also performed other mathematical work. Hydraulics was one example. In 1657, he served as a technical consultant in the settlement, directed by Pope Alexander VII, of a dispute between the cities of Bologna and Ferrara concerning the course of the Reno River. He carried out empirico-mathematical studies of water flows in the region, and wrote memoranda on the flooding of the Po as part of this work. Impressed with Cassini's efforts, Pope Alexander appointed him superintendent of fortifications for the Papal States, and then civil inspector for Perugia and superintendent of the waters for all the ecclesiastical territories. When he came to France in 1669, Cassini left behind a lucrative offer to join the papal court as a scientific savant responsible for all of these endeavors, and if his choice revealed his desire to focus exclusively on astronomy (and claim the exorbitant pension that Louis XIV offered him), his early career reveals the integration of this mathematical subfield within the wider pursuit of the mixed and mechanical mathematical sciences.[59] His lavish nine thousand livres pension also shows the high value placed on this kind of mixed mathematical expertise by the French crown.

PHILIPPE DE LA HIRE (1640–1718)

After 1669, Cassini set to work on what became a highly focused program of astronomical research. But the Observatory that he came to lead was involved in the full range of astronomically related mathematical endeavors, from geodesy and navigation science to instrument making, quantitative measurement and calculation, and spherical geometry. Such is the context in which the career of Philippe de la Hire needs to be situated, for he was a mathematician appointed to the Royal Academy in 1672 who went on to become a major figure at Cassini's Royal Observatory and within French academic mathematics as a whole.[60]

De la Hire was the child of painters and sculptors, another "mechanical trade" elevated to liberal professional status during the sixteenth and seventeenth centuries. His grandfather Étienne had established the family name and the success of their workshop, and he was succeeded by his son, Laurent, who joined with ten others to found the new Royal Academy of Painting and Sculpture in France in 1648. Philippe was raised to follow in his family's footsteps, and in 1660 he traveled to Venice to study painting and sculpture. Seventeenth-century artistic training was saturated with mathematics, however, including the geometrical science of perspective, the theory of proportions, and various other practices that bled into architecture and engineering, and as Philippe acquired mathematical training as part of his artistic education, he became drawn to it more than to the visual art that it was supposed to facilitate. Venetian *pittore* were also linked to the mathematical mechanics and engineers who inhabited the Venetian arsenal—Galileo's former haunt—and through contacts with this community, Philippe was launched into his adult career.

When he returned to Paris in 1664, he was for all intents and purposes a mathematician, even if no great leap across an imagined "Two Cultures" divide had occurred. De la Hire's family network was in fact populated with mathematician artists with strong interests in geometry, and this was the community he joined once returned to France. Central to his network was Abraham Bosse.[61] Bosse is best known as an engraver who among other works produced the famous frontispiece to Thomas Hobbes's *Leviathan*. He was also a member, with Laurent de la Hire, of the group that founded the Royal Academy of Painting and Sculpture. As an engraver, Bosse struggled with the reigning hierarchies that raised theorized visual representation with paint and canvas to a more elevated, liberal status than handiwork with burins upon metal, and his "mechanical" associations were also accentuated by his close relationship with another artist mathematician, Girard Desargues.[62]

Bosse learned the mathematics of artistic composition, and especially the geometry of pictorial perspective, at the classes in architecture that Desargues taught in Paris in the 1640s. In these lessons, Desargues taught an innovative form of geometry that completely eschewed Euclidean rigor in the pursuit of more effective ways to construct and manipulate geometric figures. His work has been judged by scholars to be astonishingly prescient, prefiguring such sciences as projective geometry, but situated in its seventeenth-century context it is best described as an aggressive form of mechanical mathematics, the kind favored by mathematical artisans and engineers. Bosse found Desargues's mathematics extremely powerful, and since Desargues was primarily a practitioner and teacher who published little, he became Desargues's advocate within the Republic of Letters, producing several books from 1643 to 1653 that disseminated his geometry.[63] Bosse also produced two manuals on engraving in the same years,[64] and this corpus marked him out as an aggressive advocate for mechanical ways of knowing, especially in mathematics.

The books also created a controversy between him and his fellow academic artists. Especially hostile to Bosse was Charles Le Brun, arguably the most distinguished painter in Louis XIV's entourage, and certainly the most powerful.[65] Le Brun was the overall director of the painting program at the royal palace at Versailles and the painter of many works in the courtly halls themselves. He was also the leading theorist in Louis XIV's Academy of Painting. Le Brun found two aspects of Bosse's artistic theory particularly disturbing. First was the engraver's claim that artistic works were man-made representations that mimicked nature through their illusionism but were otherwise thoroughly artificial. Le Brun adhered to a more direct mimetic understanding of the relationship between the artist and nature, and he argued that artists did not just mimic nature through their handiwork but actually achieved a fully natural representation, at least if their genius was sufficiently great. Le Brun's position allowed for a select group of exceptionally brilliant painters—Michelangelo was always the example—to leave behind their hands and bodies so as to actually re-create nature through the divine spirit of their artistry.[66] Bosse's more human and artisanal understanding challenged the liberal pretensions of Le Brun's theory by making art a man-made simulation of nature, one that approached the divine but never equaled it.

Equally problematic for Le Brun was Bosse's applied and practitioner-based conception of pictorial epistemology. Whereas Le Brun imagined liberal pictorial theory as a disembodying vehicle for joining the creative spirit of the artist with natural creation, Bosse saw visual representation as a skill to be learned and mastered like any other. He therefore advocated for an artistic training rooted in cultivating mechanical expertise. It was in this context that Desargues's geometry appealed to Bosse as the most powerfully effective instrument for applied artistic work, including pictorial illusionism. Le Brun saw Bosse's advocacy for Desargues as an affront to the liberal theoretical program that he hoped to institute at the French Royal Academy, and a bitter controversy between the two men ensued.[67] In the end, Le Brun prevailed, and Bosse was forced to resign his position in the Academy and to set himself up as an independent artist and teacher. This is where Hobbes found him in 1651, when he commissioned Bosse to do the frontispiece to *Leviathan*.

Although an exile from the Academy, Bosse's spirit lived on both inside and outside its walls. Laurent de la Hire was a Bosse supporter who remained active in the Academy despite Le Brun's victory. His son Philippe took the memories of these controversies with him to Venice, becoming a mathematician artist very much in the spirit of Bosse and Desargues. Bosse also lived until 1676, so when Philippe returned to Paris in 1664 he reconnected with his father's friend and colleague and continued his mathematical education with him.

De la Hire's subsequent mathematical work shows the deep imprint of Bosse and Desargues. In 1679 he published the third of a trilogy of books dealing with conic sections,[68] and René Taton notes the influence of Desargues via Bosse in all three, calling de la Hire's first book "a comprehensive study of conic sections by means of the projective approach." Of the second he says likewise that it "clearly displayed Desargues' influence." Taton also points to other influences worthy of note, especially de la Hire's use of Cartesian analytical geometry in these treatises. As he writes, de la Hire's was "not a work of great originality, [but] it summarizes the progress achieved in analytical geometry during half a century and contained some interesting ideas, among them the possible extension of space to more than three dimensions."[69]

From the perspective of the arguments here, the point to emphasize is de la Hire's assertive mechanical modernism as a mathematician. Open to whatever approach led to more effective solutions, de la Hire personified the new *mathématicien-méchanicien* who was as fluent in the classical tradition, and eager to develop its insights, as he was detached from ancient authority and the pristine purity that many found in it. De la Hire's admission into the Royal Academy in 1678 was predicated on precisely this combination of talents. In his *éloge* for de la Hire, Fontenelle claimed that it was his treatises on conic sections that captured Colbert's attention, along with its applicability to the minister's new plans for an Academy-led cartography project. Some historians find it hard to reconcile this claim with de la Hire's eventual appointment as an astronomer connected to Picard and the Royal Observatory,[70] but the connection in fact makes perfect sense when viewed in terms of seventeenth-century mixed mechanical mathematics. What de la Hire represented was not a geometer in the classical style of Roberval or Frénicle de Bessy, but a modern mathematician attuned to using mixed mathematics, including mechanical versions of it, in the solution of concrete problems. Such was also the character of the mathematics practiced at the Royal Observatory, and if de la Hire had no particular astronomical training to recommend him, he had exactly the right kind of mathematical orientation to find a home there.

And a home is exactly what he found.[71] His work as a royal mathematician in the 1680s illustrates the point. His first academic assignment was with the geodesic surveying teams who began, under Colbert's direction, to re-map the French kingdom. Together with Picard, he went to Brittany in 1679 and Guyenne in 1680 to perform surveying operations. Without Picard, he went to Calais and Dunkirk in 1681 and the coast of Provence in 1682 to pursue similar work. His actual labors involved everything from instrumental observation and spherical geometry to precise quantitative calculation, and in 1679 he and Picard established the first precision tidal measurements for the strategic royal port of Brest, recording data to within ten seconds of variation.[72] He also contributed quantitative data and technical expertise to the launch of *Les Connaissances des temps* in 1679, the French Observatory's pioneering almanac of astronomical, navigational, and meteorological data. A story recounted by Fontenelle characterizes well de la Hire's approach to such work.

"Scrupulously exact almost to the point of superstition," the secretary wrote, "M. de la Hire used to present to [the royal minister] Louvois lists of expenses drawn up day by day where even the fractions were not neglected. The minister habitually tore them up without looking at them, and then had the sums returned to him in whole numbers."[73] Though perhaps annoying to Louvois, such exacting quantitative precision was highly valued in Louis XIV's France, and de la Hire became a highly esteemed member of the French academic community as a result of such work.

He also became a powerful figure at the Royal Observatory, moving permanently with his wife and children to the new complex after it was officially opened in 1682, and ultimately producing a family dynasty of royal academicians who succeeded him in the eighteenth century.[74] He was also a close colleague of Huygens, and continued to correspond with him after the leader of the Royal Academy returned to Holland in 1684. He further obtained important professorial appointments, taking over the chair in mathematics at the Collège Royale left vacant by Roberval's death, and succeeding Blondel at the Royal Academy of Architecture in 1686. In these roles, he offered lectures on topics ranging from astronomy to mechanics, hydrostatics, dioptrics, and navigation while also contributing numerous papers to academic meetings. De la Hire was a prolific author as well, producing a practical treatise on the making of sundials and an introductory textbook on surveying along with more theoretical studies, such as his treatise on mechanics published in 1695.[75] He also edited and published the work of other mathematicians, including a treatise on leveling by Picard, a study of moving waters by Mariotte, and a compilation of ancient Greek mathematics that he selected, edited, and translated into Latin.[76] In all these ways, he embodied the modern-minded and application-oriented mathematics that was becoming ever more central in late seventeenth-century France.

MICHEL ROLLE (1652–1719)

Michel Rolle, who joined the Royal Academy in 1685, also personified the newest trends in seventeenth-century French mathematics even if his orientation was very different from de la Hire's. Rolle, the son of a provincial shopkeeper, acquired his mathematical foundations through his connections to bookkeeping and commercial calculation.[77] His primary mathematical focus throughout his career was therefore numbers and their relations, yet he did not cultivate an identity as a calculator or as a student of quantitative mathematics. Instead, he became a distinguished contributor to the new mathematical science of algebra, and to understand his career we must first understand something about this new and controversial form of seventeenth-century mathematics.

Determining the beginnings of algebra depends entirely on how you define the term. Algebra can be defined by distinguishing it from geometry as the science of number as opposed to spatial magnitude. Viewed this way, the origins of European algebra are found in ancient Greece since Euclid devoted several books of his *Elements* to the relations between numbers as opposed to spatial magnitudes. Diophantus also published a treatise called *Arithmetica* in the third century, which showed the solution to several standard algebraic problems. Save for the focus on discrete numbers as opposed to continuous spatial magnitudes, however, there is little to distinguish "Diophantine analysis," as it came to be called, or Euclid's books on number theory, from classical geometry. Each followed the classical paradigm of using postulates and definitions to define the mathematical content under consideration, and each used the standard Euclidean method of deductive demonstration from postulates and first principles to prove the general rules being shown.[78]

That these general rules look to us today as algebraic rules is an anachronism created by reading this ancient number-based mathematics through the lens of modern algebraic thinking. When we think of algebra today, we think of equations expressed with letters and other symbols such as +, −, and =, and of solutions expressed in arabic numerals. Greek mathematics did not have these symbols, and this difference in notation points us to a different origin of algebra. The word itself comes from Arabic, and many other features of algebra as we know it today also possess an Arab pedigree. If Diophantus is sometimes called the ancient Greek father of algebra, Muhammad ibn Mūsā al-Khwārizmī is often credited as the Arab father of the same. His seventh-century treatise *The Compendious Book on Calculation by Completion and Balancing* developed many of the core algebraic relations we recognize today. Al-Khwārizmī also deployed arabic numerals and symbolic algorithmic operations in his work—the words "algebra" and "algorithm" are thought to be bastardizations produced by the translation of his Arabic text into Latin in the twelfth century. He also distinguished his science from Greek geometry and arithmetic, sciences of which he was aware.[79]

The reception of Arab mathematics in Europe is a topic that is still too little researched, but the transplantation of the word "algebra" into all the European vernaculars after 1400 is a strong indicator of the important role that Arab sources played in creating this science in Europe. Yet the history of European algebra is no simple history of the reception and importation of Arab science. In medieval times, thinking about numbers was divided squarely across the liberal-mechanical boundary, with Greek arithmetic comfortably housed among the liberal mathematical arts, where it was treated as a partner of Euclidean geometry, while the use of arabic numerals to pursue arithmetical calculation was found outside the university in the work of the practitioners of the mechanical arts. Numerical manipulation and calculation were essential in mechanical trades ranging from bookkeeping and military engineering to architecture and artisanal industry, but since this brand of mathematics was neither concerned with nor disciplined by the standard of demonstrative rigor that defined mathematics as a liberal art, it existed completely outside the domain of legitimate mathematical science.

Derived from the Italian word for thing, *cosa*, which referred in this context to the "thing to be solved," the "cossist art," as it came to be called, emerged precisely out of the world of the European mechanical arts, and was another important source for modern algebra. By 1400, the cossist art, or the art of the abacus, as it was also called in reference to the calculating device that was a fixture of it, was a vibrant mathematical subdiscipline in Europe, one lavishly supported by the growing mercantile community and the new sovereign states. It was also a mathematical discipline increasingly located within its own institutional system, composed of "abacus schools," which became ubiquitous in commercially active European cities. Individual teachers, or "abbaco masters," also taught this mathematical art together with a host of new textbooks brought about by the advent of print.[80] Patronage for these quantitative mathematical practices was robust given their direct political and commercial relevance, and overall this nexus served as a crucial seedbed for the European development of a new algebraic science. Rolle's passage into the Royal Academy and the new science of algebra from the world of accounting and bookkeeping can be viewed as one illustration of the continued relevance of this particular historical connection as late as the seventeenth century.

The "cossist art" was already scientific in its search for general rules suitable for resolving discrete calculations in a universal, algorithmic manner. But it was not scientific according the standards of Euclidean demonstrative rigor, because quantitative accuracy, not deductive certainty, was the guiding epistemological standard of this mathematics. Mechanical practitioners rarely had any reason to worry about the Euclidean demonstrative rigor of what they were doing, but amid the wider shifts that began to blur the lines between mechanical and liberal mathematics, and the border crossing that the early modern spaces for scientific learning made available—print culture, the humanist Republic of Letters, princely courts, sociability in learned academies and societies, etc.—this situation changed.

Increasingly in Europe over the sixteenth century, algebra came to be recognized as a science unto itself, and to distance itself from arithmetic and the science of calculation. In this new, and more elevated, scientific mode, algebra started to become a general science of the relations between numbers and their arithmetic properties. This involved a step upward in abstraction that allowed algebra to move away from simple rules of calculation, and soon the "algebraist" became a mathematician focused on finding general principles applicable to the relations of numbers in general. This new algebra also grew more demonstrative, centered on proving the existence of certain extant relations, and developing proofs of general algebraic rules. The turn toward demonstrative algebra also allowed the new algebraists to fully occupy the liberal sphere of classical, Euclidean mathematics while working with objects—numbers, exotic new symbols, formulas and equations—that had no existence in the ancient Greek mathematical canon.

Fermat's famous "Last Theorem," left for posterity in the margins of his copy of Diophantus's *Arithmetica*, illustrates well what this new liberal algebraic science had become by the seventeenth century. "It is impossible to separate a cube into two cubes, or a fourth power into two fourth powers, or in general, any power higher than the second, into two like powers," Fermat declared, and "I have discovered a truly marvelous proof of this, which this margin is too narrow to contain."[81] Putting this discursive assertion into the algebraic notation that was still in its barest infancy when Fermat wrote these sentences in 1637, he was claiming that no three positive integers a, b, and c can satisfy the equation $a^n + b^n = c^n$ for any value of n greater than two. Leaving aside the details of this statement, not to mention the difficulty of trying to prove it, Fermat's assertion shows us how European algebra by the mid-seventeenth century had, for some elite practitioners at least, left behind its association with practical, solution-oriented calculation and become a fully demonstrative science focused on the production of proofs comparable in their intent and epistemological rigor to those found in classical geometry. Rolle's work was indicative of this new scientific brand of seventeenth-century algebra.

Yet Rolle's work was not friendly to another direction in which algebraic mathematics also moved in the seventeenth century, and since Rolle's later relationship to analytical mechanics was entangled with his relation to this other current of seventeenth-century mathematics, it needs to be briefly explored here. Historians often point to François Viète's *Introduction to the*

Analytical Art as the most important and seminal work in the emergence of this other strand of algebraic science in early modern Europe, the science that would soon come to be called "analysis." Viète's achievement was manifold, and while space does not permit a full discussion of his work here, its significance in relation to Rolle's requires a brief account of it.[82]

Most basically, but also most fundamentally, Viète clarified a new symbolism that consolidated the pell-mell of discursive and symbolic representational schemes available at the time into a standard algebraic nomenclature. Here Rolle followed almost all seventeenth-century algebraists in using the conventional notation that Viète had introduced. He also followed Viète's manner of distinguishing between arithmetic as a science of calculation and algebra as a general science of numbers and their relations. Viète was the first European to clearly mark out this distinction, using the classifications *logistica speciosa* and *logistica numerosa* to denote the difference. *Logistica numerosa* referred to arithmetic, and it involved simple work with numbers and their relations. *Logistica speciosa* referred to a more comprehensive domain involving "general species," or generic forms of numerical expression and their relations—things like the binomial theorem or the quadratic equation. *Logistica numerosa* was merely arithmetic and the rules of calculation, while *logistica speciosa* was the science of numerical relations in general, or algebra. With this distinction in place, Viète further went on to demonstrate numerous algebraic propositions, some ancient, others novel, that marked out his newly clarified algebra as a new and discrete science.

In all these ways, Rolle was an algebraist in the manner of Viète, and a staunch defender of the new discipline. What Viète's new symbolism, together with his epistemological clarification of the new discipline, made possible was a new algebraic science that could coexist alongside the ancient Euclidean science of geometric problem solving. Yet while this made possible a new place for algebra and algebraists in institutions like the Académie Royale des Sciences, it also subjected algebra to new controversies, many related to those that were already present in seventeenth-century geometry. The controversies with respect to algebra were often extra heated, because while Euclidean geometry possessed unimpeachable ancient authority, the new algebra had no such warrant. The very suggestion that an algebraic demonstration was comparable in epistemological rigor to a geometric demonstration was open to challenge, and given its lingering association with the lowly mechanical work of calculation, algebra also had to work doubly hard to gain credibility as a liberal science.

Fighting against these arguments, however, was the power of algebra itself to find solutions that other methods could not offer. Building from this position, Viète made the claim, soon to be echoed and reinforced by others, that algebraic mathematics was neither a lowly stepchild of classical geometry, nor even its peer, but a new and more universal science, one that revealed the underlying foundations of all mathematics. This claim turned the tables entirely on the traditionalist position by claiming algebra as both a wholly new and modern science, and one that advanced beyond the ancients in revealing the general truths present, if hidden, in all mathematics. Algebra from this perspective was not something that needed to be made more worthy by making it more Euclidean or geometric; it was the more universal mathematical science that should become the standard for evaluating the foundations of all the other mathematical disciplines including geometry.

We will see in the later chapters these universalizing arguments being made by French practitioners of the new "analysis," which algebraic mathematics in the mode of Viète soon came to be called. The "analytical" in analytical mechanics in fact derives precisely from its deployment of this kind of mathematics, and since Rolle was one the most outspoken and vehement opponents of analytical mechanics we should consider more fully these tensions as they developed within the new and growing field of seventeenth-century algebra and mathematics as a whole.

A further innovation central to the debate was the change that René Descartes made with Viète's "analytic art" in his *Géométrie* of 1637. This was the move that Rolle forever refused to make. Historians often make Cartesian analytical geometry the source for a seventeenth-century modernist mathematical revolution, and while the claim is overstated, the innovative significance of Descartes's arguments in his *Géométrie* should not be understated. Descartes's work cemented the foundations laid by Viète by securing a new conception of "analytic," or algebraic, mathematics as the cutting edge of modernist innovation. He also created a new science, analytic geometry, that made possible the infinitesimal calculus that Varignon used in the creation of his not accidently named "analytical" mechanics. Rolle stood vehemently against all these trends, even if he did so as a practicing algebraist. So we must briefly consider Descartes's *Géométrie* and its influence in order to understand Rolle's relationship to it.

The ambitious claim that opens Descartes's *Géométrie* marks the essential innovation of the work: "All the problems of geometry," he writes, "can easily be reduced to such terms that thereafter we need to know only the lengths of certain straight lines in order to construct them."[83] The first thing to note about this declaration is its focus on the domain of geometric problem solving, and its assumption, in step with the European geometry of the period, that problem solving implies geometric construction. Also present in this statement is the modernist desire to unleash new methods, including algebraic methods, for solving difficult geometrical problems with greater ease. These statements position Descartes among the modernizers in the seventeenth-century mathematical field. Following Viète, Descartes also pushes geometry in an even more novel direction. The key to solving "all the problems of geometry with ease," he states, is the reduction of those problems to "the length of certain straight lines." Stated more directly, this means solving geometric problems through recourse to discrete magnitudes, which is to say in terms of numbers. Following Viète, this means reducing the spatial magnitudes and ratios of geometry to numerical equations, and thus geometry to algebraic analysis. In a nutshell, this is precisely the pioneering innovation that Descartes's *Géométrie* unleashed.

We will return to the details of all these innovations later when we consider the development of the calculus. Suffice it to say now that Rolle found the whole complex distasteful, and he became an ardent critic of it. For our purposes, the point is simply to note here his hostility and to inquire into the possible reasons for it.

Rolle received no higher education, in mathematics or anything else, but having moved to Paris in 1675 to make a career for himself as a scribe and calculator, he read in the *Journal des savants* a challenge problem posed by Jacques Ozanam, a well-known public mathematician, that he was convinced he could solve. Ozanam asked contestants to find four numbers where the difference of any two is a perfect square and the sum of the first three is also a perfect square. Ozanam announced that the smallest of the four numbers would have at least fifty figures, but Rolle provided a solution in which the smallest number had only seven figures. It was a stunning result, especially from a mathematician that no one had ever heard of before. His solution was spectacular enough to attract the attention of Colbert, who is said to have given Rolle a royal pension as a prize. The minister died soon after, but Rolle's favor with the crown continued. Louvois appointed him as the family mathematical tutor for his son, and he also gave Rolle a position in the Ministry of War, which proved unsuccessful. In 1685, however, he extended further favor to his client by finding him a pensioned position in the Royal Academy.

Rolle entered the Academy as an *élève astronome*, a designation that fits with his social profile at this point in his career. Having made little mark in the mathematical world beyond his solution to Ozanam's challenge problem, and still only thirty-three years old, the rank of *élève* fit his status with respect to his academic peers. His four hundred livres pension (less than 10 percent of Cassini's annual salary) was also an appropriate marker of his social standing.[84] The designator *astronome* may have functioned, as it did for de la Hire, as a label locating him and his particular mathematical work in the zone of the Academy most appropriate for it, namely the Royal Observatory. Yet whatever the reasoning behind his precise appointment, Rolle did not follow de la Hire and migrate to the Observatory, nor did he gravitate toward those doing quantitative mathematical work elsewhere. In fact, though classified as an *élève*, Rolle does not seem to have been teamed up with anyone even though this remained the norm for academicians of this rank. He instead set to work after 1685 establishing himself as an algebraic theorist and author of liberal scientific treatises.

Rolle's most important books appeared in 1690 and 1691, and they offered new methods for finding solutions for algebraic equations.[85] A third book, published in 1699, added more to this legacy,[86] and for our purposes the details of his mathematical achievements are less important than what they reveal about his characteristic mathematical style. Michael Mahoney has isolated three aspects of what he calls the characteristic "algebraic mode of thought" of the seventeenth century, and Rolle embodies each perfectly.[87] First was the algebraist's use of an operative symbolism in his mathematical work. Rolle not only cultivated this symbolic approach, he innovated new symbols that became standards for the field. Second, says Mahoney, is a relational approach to mathematics as opposed to an object-centered orientation. This tendency is related to the first in that mathematicians who conceive of mathematics symbolically saw the relations between these terms as their object of study, and further developed a symbolism to facilitate this mode of thinking. Rolle's famous "method of cascades," which sits at the historical origins of what has come to be called "Rolle's Theorem," and is still taught to introductory calculus students, illustrates the point. It offers a precise rule for ordering and manipulating a symbolic equation so as to produce the desired solutions. The result is also dependent on its symbolization to generate new methods and solutions.[88] Third in Mahoney's classification is the absence among algebraists of any ontological commitment regarding the inherent meaning of mathematical objects conceived through symbols. Mathematical existence, Mahoney writes with respect to algebraic thinkers, "depends upon consistent definition within an axiomatic system." In short, what realities mathematical symbols may refer to does not matter to algebraists so long as the symbols themselves cohere into a rigorous axiomatic system.[89] All of these descriptions capture well Rolle's orientation as a mathematician.

Yet if Rolle was a seventeenth-century algebraist through and through, he also occupied a precise place within this new community of seventeenth-century European mathematicians. Part and parcel of the freedom from ontological commitment was a corresponding concern for appropriate rigor and method in mathematics. In Rolle's case, this general concern shared by all algebraists was also attached to an anxiety about his status as a fully liberal mathematician. This made him particularly susceptible to agitation along the shifting fault lines of seventeenth-century mathematics. A key source of worry for Rolle involved the claim, sometimes made by tradition-minded classical geometers, that algebra was a suspect mathematical discipline. Working via the manipulation of symbols that did not often have clear and natural referents (what was the quantity of a negative number, for example, or its square root?), algebraists often found themselves labeled as methodologically suspect or, worse, specious mathematical mechanics. Also worrisome to traditional geometers was the algebraist's comfort with finding new solutions while avoiding the demonstrative proofs that grounded such solutions synthetically. The seventeenth-century discourse that made algebra and analytical mathematics synonymous accepted an epistemological understanding that often made correct solutions to problems their own demonstrative justification. For classical geometers, this generated an epistemological foul, because demonstrative truth was something different from, and superior to, mere accuracy and correctness. Accordingly, classically oriented geometers could, and often did, dismiss the whole enterprise of symbolic, algebraic mathematics as nothing more than lowly analysis (or "mere problem solving"). These challenges created among algebraists, especially those aspiring to liberal status, a particularly strong urge to make evident their Euclidean epistemological bona fides.

For Rolle, who was an upwardly aspiring, and thus epistemologically anxious, algebraist of this sort, the solution to this tangle of worries was to chart a middle path between seventeenth-century mathematical innovation and traditionalism. Ancient Diophantine analysis, for example, was often his chosen field of mathematical practice, and while he innovated within this antique tradition he often did so in moderate ways that eschewed the most provocatively novel tendencies of the day. Euclid also was important to Rolle, and historians have described some of his work as a new and modern application of traditional Euclidean principles found in the *Elements*.[90] Rolle similarly steered clear of the fledgling discipline of Cartesian analytic (or algebraic) geometry. On the one hand, his algebra remained closely attached to the ancient tradition of geometric algebra and distant from that of Viète, Descartes, and other Moderns, and on the other hand, he also worked in the manner of Descartes to develop algebraic methods that could improve upon antique problem-solving techniques.

A case in point is a challenge problem that Rolle issued in 1693 via the *Journal des savants*. He offered sixty pistoles to any savant who could solve a particular problem in analytic geometry without using the methods developed in his own traditional algebraic books.[91] The point of the challenge was to assert the supremacy of Rolle's own algebraic methods when compared with those of Cartesian analysis. Johann Bernoulli believed that he had found a solution without recourse to Rolle's methods, but Rolle determined that no such solution existed. His dispute with Bernoulli, which involved two rounds of exchanges in the *Journal des savants*, ultimately resolved nothing. But the letters reveal much about Rolle's character as a mathematician.[92] Concerned to secure his own name and reputation as an esteemed liberal theorist, and eager to defend the hybrid traditional-modern approach to symbolic mathematics that he believed made his work distinctive and uniquely potent, Rolle was a particularly vigorous contestant in the still turbulent field of seventeenth-century mathematics. These same traits would be in evidence in his great dispute with Varignon about analytical mechanics, which we will consider in detail later in this book.

PIERRE VARIGNON (1654–1722)

Pierre Varignon's name has already appeared more than once in this book, and he will remain central to all that follows. Accordingly, the purpose of this final section is to bring this chapter on French academic mathematics to a close by introducing Varignon into the institutional nexus that was most deeply influential in all of his scientific work.[93]

Along with Rolle, Varignon was the other academic mathematician appointed to the Royal Academy by Louvois. Like the others already surveyed, his appointment also reflected the turn toward more modern mathematics that was characteristic of the second generation of French academic mathematicians.[94] Varignon was also an upstart, like Rolle, who achieved academic prestige despite starting from humble beginnings. He was born in Caen in Normandy, a region as we will soon see that proved particularly fecund in the production of royal academicians. Yet he was not the child of a learned family. His father was an *architecte-entrepreneur*, the Old Regime French equivalent of a building contractor, and his special expertise was masonry. Varignon said of him that he owed only his comfort with difficult technical matters to the legacy left him by his father. In the 1670s, before his twentieth birthday, Varignon left home, leaving his two brothers in Normandy to continue the family business. His first stop was the Jesuit college in Caen, and while he acquired his fundamental mathematical training there, he was also ordained as a priest in 1676, and after further study became the head of a parish near his home in 1682. The priesthood was a common path of social elevation in Old Regime France, and given the ubiquity of the clergy within all domains of French society, including the sciences, it was a career choice that opened rather than closed professional doors.

Possessed now of a steady income, Varignon began to lead the life of an Old Regime cleric, and it was during these years that his attentions turned toward mathematics. As Fontenelle told the story in his eulogy, transforming what was likely acquired in his Jesuit education into a compelling individual story, Varignon happened one day upon a copy of Euclid's *Elements*, and "he was immediately struck not only by the systematic order of the ideas, but also by the ease with which he entered into its presentation." He took the book home with him, and helped by his experience with "the eternal uncertainty, the sophistic confusion, and the useless and sometimes affected obscurity of the philosophy of the schools," he was able to "taste the clarity, the interconnectedness, and the surety of geometric truths."[95] This recognition launched him on a mathematical journey that passed through the analytical geometry of Descartes and then on into the most recent and innovative mathematical work of the period. Varignon soon resigned from his clerical positions and moved to Paris, and in 1687 he published his first mathematical book, a project for a new mechanics that engaged with the Italian Borelli's work, as was noted above. From this date forward he was a fully committed mathematician with no attachments whatsoever to the church or his priestly vocation.[96]

At the same moment, Varignon also installed himself permanently in Paris, taking advantage of the generosity of an affluent college friend who was also destined for fame as a Parisian savant. Charles-Irénée Castel, abbé de Saint-Pierre, was born at Château de Saint-Pierre-Église near Cherbourg to a family of well-established provincial aristocrats. He met Varignon at the Jesuit college in Caen, and "possessed of a shared taste for matters of reason, be they in *physique* or *métaphysique*, and an affection for argument," the two young students became close friends.[97] Saint-Pierre was born into Old Regime privilege, and upon graduation he headed to Paris, where he began to establish himself as a diplomat and a sociable man of letters. He was especially connected to the salon hosted by the Marquise de Lambert that was a gathering point for many of the leading lights of Parisian society, including the members of the royal academies. Saint-Pierre retained close attachments to Varignon as well, and at some point in the 1680s he agreed to provide the increasingly distracted priest with an annual pension of three hundred livres, encouraging him to leave the church and join him in the city. As Fontenelle, another Norman friend of both Varignon and Saint-Pierre, explained, "Paris was the only place for reasonable philosophers to live." Varignon accepted the offer and made the move, and the friends set themselves up in an apartment in the Faubourg Saint-Jacques in 1686. Fontenelle wrote wistfully of the intellectual vitality that reigned in their *petite maison* during his many visits to it.

Amid their recreations, Saint-Pierre and Varignon also began to lay the foundations for their individual careers. The first used his aristocratic connections to find posts within the French state, becoming an astute observer of public affairs and a pioneer of Enlightenment political science. Varignon focused his energies entirely on mathematics. "He passed entire days working without any distraction or recreation," Fontenelle reminisced. "I once heard him say that he was sometimes surprised when working after dinner, as was his custom, to hear the bells ringing two hours after midnight. Yet this made him happy since it meant that he would not have to wake himself from bed in four hours to start working again." "He smiled continually when speaking about geometry," Fontenelle added, "and for him it was as if he needed to be studying to be most entertained."[98]

This obsessive research led to Varignon's debut mathematical work. In 1687, as Newton's *Principia* was going to press in England, he published a small piece on mechanics in Pierre Bayle's *Nouvelles de la république des lettres*.[99] Later the same year his *Projet d'une nouvelle méchanique* was published in Paris.[100] The substance of this work, which was more directly empirical in orientation than his later analytical work, is less important for this discussion than the book's dedication. Varignon presented his work as a gift to the members of the Royal Academy. "There is not any point of science that you have not perfected or enriched with your work," Varignon wrote, "and who does not wait for more from you, animated as you are by a great sovereign who wants his reign to be as glorious in the arts and sciences as it is already prodigious in conquests and heroic actions? Under the protection of such a wise and vigilant minister, is there anything to which you should not aspire today?" Having started his dedication by expressing the modesty of his own scientific achievement and the titanic accomplishments of the Royal Academy, he ended by situating his own humble efforts squarely within theirs. "You are like the source of all the human sciences," he wrote, ". . . and I dare to offer you, and to the public, this that I have drawn from that source. In trying to follow you and to imitate you, I hope to profit from your enlightenment, and I assure you, my sirs, that with a perfect veneration I am your very humble and obedient servant."[101]

This was the servile rhetoric of Old Regime clientelism, yet no one before Varignon had used such rhetoric in print in a treatise on mechanics to address a patronage appeal to the members of the Royal Academy as a company. The ploy worked. In 1688 Louvois appointed Varignon to the Academy while also giving him a teaching post in mathematics at the Collège Mazarin. The appointment secured Varignon's position in Paris, and with it he set to work transforming his project for a new mechanics into his new science of motion. These developments will be the focus of the chapters to come.

PART II

Beyond the Continental Translation of "Newtonian Mechanics"

The Intellectual Roots of Analytical Mechanics

On the rare occasions when modern historians have analyzed Varignon's work in the development of analytical mechanics, the legacy of Newton's *Principia* has loomed large in the discussion. One pervasive (and still stubbornly persuasive) historiographical tradition in fact makes Newton the primary author of Varignon's new science of motion. In this understanding, Newton's *Principia* brings all the essential elements needed for his science into the world in 1687, and the French origination of analytical mechanics is reduced to Varignon's transcription and then translation of Newton's science as found in the *Principia* into idioms more familiar to his Continental mathematical colleagues. E. J. Aiton articulates this view when he writes that "Varignon was in effect a Newtonian. His real achievement was the interpretation of Newtonian planetary theory to Continental mathematicians more conversant with the language of the differential calculus than the geometrical style of the *Principia*."[1]

This book offers a very different account of this history. Varignon's analytical mechanics is misunderstood, I argue, if it is conceived as the rationally determined offshoot of Newton's prior work in the *Principia*. It is also misconstrued if it is not viewed as the contingent and locally produced historical outcome that it was. Analytical mechanics is especially misrepresented when it is described as a mere dissemination of the oracular scientific light said to have begun radiating out of Newton's *Principia* after 1687. Newtonian influences were certainly crucial in the germination of Varignon's science, and what is needed is what the historiography currently lacks: a rigorously historicist scientific genealogy of analytical mechanics that includes Newton as one, but only one, of its ancestors. This genealogy must also trace all of the influences, including the local French ones, which converged after 1700 in making this new form of mathematical physics possible and then successful. That is the project of this book, and the chapters in this second part develop this interpretation by examining three clusters of intellectual influence that were each crucial to the formation of Varignon's new science of motion.

Chapter 4 takes on the question of the Newtonian origins of analytical mechanics directly, looking at all the ways that the *Principia* did and did not shape Varignon's work. Chapter 5 turns to the role that Leibniz and his new infinitesimal calculus played in the making of Varignon's science. It argues that the calculus was perhaps the single most important intellectual influence in the making of Varignon's science, but also the source of many of the controversies that it would provoke. Chapter 6 completes part 2 by looking at the role that Nicolas Malebranche played in this history, both scientifically and culturally. It suggests that a uniquely French complex of influences centered upon Malebranche's thought and its widespread influence congealed around 1700 in a way that proved conducive to the formation of this particular science in this particular place at this particular moment in time. Out of this ferment, analytical mechanics was born, and having traced its institutional origins in part 1 and its intellectual origins in part 2, the final section of the book will look at the actual development and institutionalization of this science in the first decades of the eighteenth century.

CHAPTER 4

The Newtonian Sources of Analytical Mechanics

A revealing document, a kind of time capsule taking us back to the first moments when Newton's *Principia* entered the scientific world and began to change it, can lead us into the arguments of this chapter. The document is an anonymous book review published in the most widely read and highly regarded French learned periodical of the day, the *Journal des savants*. The review appeared little more than a year after the first printing of the *Principia* in London,[1] and the text was astonishingly brief—roughly three hundred words spread over one and a half folio pages. Yet it captured in a nutshell the understanding of this book that would define the French reception of it for the next quarter century. Short though it was, it also condensed into a few paragraphs the essential relationship between the *Principia* and the new mechanics that Varignon began developing a few years later. For these reasons, we will use this text to enter into this history.

The review began by praising the work of Newton as "the most perfect work of *Mécanique* that one can imagine." "It is not possible," the reviewer declared, "to make either more precise or more exact demonstrations about weight, lightness, elasticity, the resistance of fluid bodies, or the attractive and repulsive forces which are the principal foundation of *Physique* than the ones he gives in the first two books."[2] Immediately qualifying this praise, however, the reviewer continued that, "it must be avowed that one can only regard these as mechanical definitions since the author (as he recognizes himself at the end of page 4 and the beginning of page 5) has not considered their principles like a *physicien*, but only as a simple *géomètre*. He avows the same thing at the beginning of the third book, where he nevertheless tries to explain the system of the world. Here, however, he offers hypotheses that are for the most part arbitrary and which serve only to ground a treatise of pure *mécanique*." Citing Newton's explanation of the tides, "which he bases on the principle that *all the planets gravitate* [*pesent*] *reciprocally one toward the other*,"[3] the reviewer then isolated the key fallacy in Newton's reasoning. His argument, the reviewer wrote, "is unassailable according to his supposition. But since the supposition itself is arbitrary, not having been proven, the demonstration which depends upon it is only proven mechanically." "In order to make the most perfect work possible," the reviewer concluded, "Mr. Newton only needs to give us a work of *Physique* as exact as his work of *Mécanique*. He will achieve this when he substitutes true motions for those he has supposed."[4]

The author of this review is not known, but given the editorial practices of the journal at the time, it is almost certain that it was written by either a member of the Royal Academy or someone who moved in the same circles as the royal academic mathematicians. Varignon could very well have been the author, and while no evidence supporting this attribution exists, the review articulates a general judgment of the *Principia* that he and his fellow French academic colleagues would have shared. Most important is the framework that sees the *Principia* first and foremost as a treatise in mechanics. This is alien to modern understanding, which sees Newton's primary intention residing in physics and natural philosophy, with the mathematical mechanics of the book serving as a means to achieve these other natural philosophical ends. The review also offers an accurate understanding of what the *Principia* aspires to accomplish, for as the reviewer understands clearly, Newton does place an innovative mechanics at the base of a new set of theories about physics and cosmology (or natural philosophy as the early moderns would have called it). To understand these issues, and the particularly influential reading of the *Principia* that was present in it, let us briefly consider the innovations that Newton's treatise offered to its seventeenth-century readers.

Newton's fundamental argument is inscribed in the full title of his work: *Mathematical Principles of Natural Philosophy*. In the 1680s, such a project would have struck informed readers as both resonant with progressive mathematical currents and provocatively innovative at the same time. Aligned with extant modernizing trends was Newton's extension of advanced mathematics further into the expanding domain of physico-mathematical science. Galileo's geometrization of the science of motion had opened the door to this new and more systematically mathematicized conception of mechanics, and after Descartes's attempt to establish universal mathematical laws of motion at midcentury, and Huygens's work on pendulum motion, centrifugal force, and the science of motion in the 1650s and '60s, physico-mathematics had become a thriving field with a tendency toward ever more aggressive mathematicization present in it.[5] Newton's *Principia* catalyzed these emerging trends in exciting new ways, not least through his demonstration that celestial and terrestrial mechanics could be unified through one set of mathematical concepts and laws. The reviewer for the *Journal des savants*, who was likely linked to the vibrant community of mathematicians devoted to precisely these new scientific currents at the Académie Royale des Sciences, acknowledged this innovation when devoting high praise to the *méchanique* of the treatise.

Yet the *Principia* did not offer new mathematical principles of mechanics alone; it also claimed to offer mathematical principles *of natural philosophy*. This was to assert a much more provocative claim, because according to traditional epistemological canons, mathematical mechanics was not warranted to speak about the actual nature and motion of physical bodies. To claim that it could was therefore to commit an epistemological category error. To practice natural philosophy meant discerning the physical causes operative within the motions we see, and while mathematics could describe those motions, and even predict them with great precision, it was not warranted within the epistemological canons of the day to offer mathematical accounts of motion as substitutes for causal, physical explanations. In a letter to Marin Mersenne in 1638, Descartes marked out this distinction with respect to Galileo's mathematical mechanics, writing, "Without having considered the first causes of nature, he has merely looked for the explanations of a few particular effects, and he has thereby built without foundations."[6] Descartes's own mechanics and physics was conventional in this respect, since it was not an application to natural philosophy of principles derived from mathematical mechanics, but an account of the causes underlying natural change derived deductively from physical first principles. Huygens's *Horologium oscillatorium* of 1673, with its combination of empirical and experimental data and mechanical mathematical analysis, offers a better model for what Newton offered in the *Principia*. But Huygens made no pretension of doing natural philosophy with this kind of work while this was the central argument made by Newton in his *Principia*.[7]

These disciplinary distinctions were front and center in the mind of the reviewer for the *Journal des savants*. They were deployed clearly when the *méchanique* of the *Principia* was declared a brilliant success while the *physique* was called a failure. The reviewer also pointed to these disciplinary tensions when he asked Newton to remedy his failures by offering a full account (i.e., a causal explanation) of the motions that he had only demonstrated mechanically, which is to say mathematically and without recourse to evident physical principles. Later historians looking back on the *Principia* often locate its revolutionary innovation in precisely its transgression of this early modern disciplinary distinction separating causal, demonstrative natural philosophy from empirical and quantitative physico-mathematics.[8] Yet what appears clear to twenty-first-century observers should not guide our interpretation of what the first seventeenth-century readers of the text saw. For our historical purposes, what is most valuable about the *Journal des savants* review is not its failure to foresee what later commentators would take the *Principia* to have done, but rather its articulation of what a sophisticated reader in the late seventeenth century took it to be doing, both for good and for ill.

From this perspective, the description of the treatise as a work of mechanics stitched together problematically with a work of physics all in the service of a new set of claims about natural philosophy in fact describes fairly accurately the book that Newton actually produced. The *Principia* is not, for example, a single book with one sustained argument, but a collection of three books, each with its own particular agenda. There is also no overarching synthetic statement pulling all three books into a single demonstration. Rather, the three books stand alone, and while they can, and were, read as a linked set advancing a single argument, they can also be approached separately without violating any declared statement by the author counseling the reader not to do that. The French reviewer in 1688 was therefore offering readers an accurate account of Newton's work, even if it was a particular assessment as well. To understand its specificities, let us briefly compare it with another contemporary interpretation of the *Principia* that was no less accurate if also very different. This second interpretation of the text was not, at least at first, influential in France, so it can serve as a contrast clarifying the reading of the *Principia* that was pervasive in France.

The Influence That Wasn't: The *Principia* as a Work of Anti-Cartesian Physics

Edmund Halley articulated this other interpretation in a prepublication review of the *Principia* published in the *Philosophical Transactions of the Royal Society of London*.[9] Unlike the concise three-hundred-word assessment found in the *Journal des savants*, Halley's review is long on summary description and short on scientific judgment and analysis. His review also describes a very different book than the one found in the French review in the *Journal des savants*.

Central to the difference is how each review treats Book II of Newton's *Principia*. Stated simply, the three books of the *Principia* can be classified into two groups. Books I and II offer geometrical demonstrations of the mathematical behavior of bodies in motion, and Book III offers an array of empirical and experimental results that purport to prove empirically the principles demonstrated mathematically in the first two books. This was the dichotomy that the French reviewer had in mind when he spoke of separate works of mechanics and physics, and also the split that led him to claim a deep epistemological flaw in the bridge between them. In one of the few moments of self-conscious metareflection offered in his treatise, Newton also spoke about this distinction and the reasons for it in his book. "I composed an earlier version of Book III in a popular form," he explained, "so that it might be more widely read." But fearing misreading and "lengthy disputations" triggered by an inadequate grasp of the argument's underlying principles, he instead "translated the substance of the earlier versions into propositions in a mathematical style so that they may be read only by those who have first mastered the principles."[10] Thus was born, or so this introduction suggests, the first two purely mathematical books of the *Principia*.

Newton also pointed in the same introduction to his addition of some "philosophical scholiums" (i.e., scholiums dealing with natural philosophy) that were designed to soften the "seeming steril[ity]" of these initial geometrical books. These, Newton explained, also offered readers suggestions, however brief and disjointed, about how to interpret the two-part geometrical and then empirical architecture of his book.[11] Yet Newton did not stop there. Recognizing that the mathematical demonstrations in Books I and II contained "a great number of propositions . . . which might be too time-consuming even for readers who are proficient in mathematics," he also outlined a sort of "executive summary" of his text, one that would give any reader who mastered only these parts a "sufficient" understanding of the book's overall argument.[12] "I am unwilling to advise anyone to study every one of [the propositions in Book I and Book II]," he wrote. It would be enough, he conceded, to "read with care the Definitions, the Laws of Motion, and the first three sections of Book I" before turning to the exposition of the empirical conclusions about the system of the world offered in Book III. He left it to his readers to decide whether the other propositions of Book I and Book II were worth their time.

The analysis in the *Journal des savants* suggests that the reviewer might have followed Newton to the letter when reading his treatise, even if he came to a different conclusion about the adequacy of the overall argument. In his review, however, Halley took a different approach. He treated each book on its own terms, and this led to an assessment of Book II that has no counterpart in the French review. "The last Section of [Book II]," Halley's inventory notes, "is concerning the Circular Motion of Fluids, wherein the Nature of their Vortical Motions is considered, and from whence the Cartesian Doctrine of the Vortices of the Celestial Matter carrying with them the Planets about the Sun is proved to be altogether impossible." John Locke gave French readers access to a similar presentation a year later in a review published in Jean Le Clerc's pioneering periodical *Bibliothèque universelle et historique*.[13] This anonymous review asserted that, based on the arguments of *Principia* Book II, "The Author concludes . . . that the hypothesis of the vortices [*tourbillons*] does not serve at all to explain the movements of celestial bodies." The review also explained that Newton offered an alternative hypothesis for explaining these motions in Book III, although the review did not elaborate on what that was.[14] Halley and Locke were echoing Newton in this assessment, for in the final sentences of Book II, Newton affirmed this conclusion, writing that as is shown by the previous demonstrations, "The hypothesis of the vortices can in no way be reconciled with astronomical phenomena and serves less to clarify celestial motions than to obscure them. How these motions are performed in free spaces without vortices will be shown more fully in Book III on the system of the world."[15]

These assessments of Book II of the *Principia*, focused as they are on Newton's proposed critique of the Cartesian system of celestial mechanics, namely his "theory of the vortices," introduces a topic not found in the 1688 French review of the text. In doing so, these other reviews also introduce a different frame for interpreting the *Principia*, one that displaces the epistemological and disciplinary gymnastics of mechanics and physics in the treatise, and substitutes for them an agenda focused primarily on physics. Like the understanding articulated in the French review, this different interpretative frame has also exerted an important influence on the subsequent historiography. In this understanding, the primary agenda of the *Principia* is not the introduction of a new and innovative application of mathematical mechanics to the practice of natural philosophy, but the use of this innovative new method, the validity of which is taken for granted, to challenge the prevailing "vortical-physical" understanding of the cosmos, and to replace it with a new and more scientifically grounded alternative. From this perspective, Newton is also understood to have had a particular scientific target in mind when drafting his treatise: the system of celestial and terrestrial mechanics that René Descartes developed in the 1640s.

In this Cartesian understanding, the universe is conceived as a plenum consisting of swirling oceans of fluid matter (fig. 4). Planets are said to move in their orbits by swimming in the vortical currents produced by this fluid matter. The phenomenon of terrestrial weight is also said to be produced not by the universal attractive force of gravity acting between bodies, as Newton would suggest, but by the centrifugal force exerted by these fluid

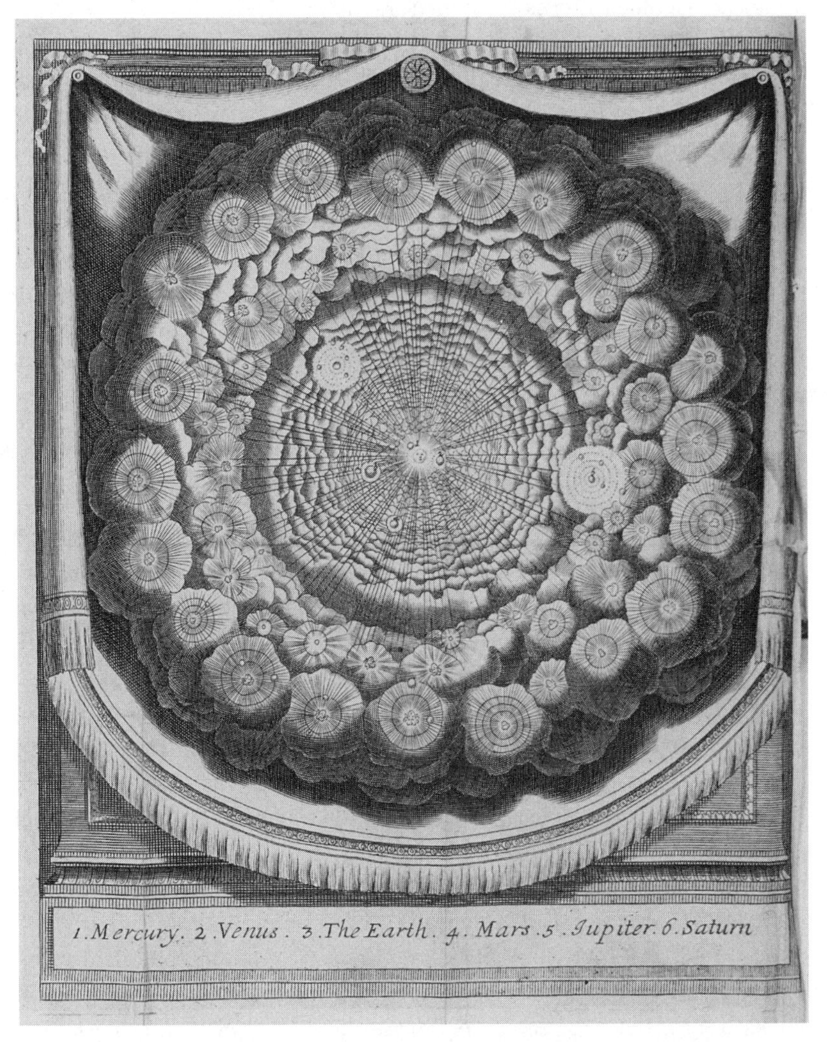

FIGURE 4. *Frontispiece, Bernard le Bovier de Fontenelle,* Entretiens sur la pluralité des mondes habités *(Amsterdam: Pierre Mortirer, 1701). Courtesy of O. Meredith Wilson Library Special Collections, University of Minnesota.*

vortices as they resist bodies moving up from the earth into them.[16] Newton makes clear that Book II is directed at undermining precisely this understanding of celestial mechanics, and the book is accordingly often referred to as the "anti-Cartesian" section of the treatise. Yet since it is, like Book I, a work of geometrical mechanics alone, its argument is also susceptible to the same epistemological critique leveled by the French reviewer against Newton's overall use of mathematics to claim revisions of physical theory. For a seventeenth-century scientific thinker such as the French reviewer, the reasons for accepting (or not) Cartesian celestial mechanics stemmed from the rigorous natural philosophical demonstrations that Descartes used to build his system. To challenge Cartesian celestial mechanics, a clearly demonstrated alternative would be required, and no matter how brilliant Newton's mathematical demonstrations might be as works of mechanics, his arguments would remain irrelevant to these physical discussions, at least according to seventeenth-century epistemological canons. This was because mathematics had no epistemological warrant to challenge rigorously demonstrated natural philosophical explanations. In short, Newton's critique of Cartesian physics is built on epistemological sand, or so one strand of seventeenth-century scientific thinking held.

Halley saw things differently, as did Newton obviously. But their conviction that mathematical description could and should serve as the basis for natural philosophical understanding should not lead us to conclude that they evaluated the *Principia* correctly while the French reviewer did not. What this difference of opinion reflects is the radical innovations found in Newton's treatise and the multifaceted reception that this complexity produced.

One thread of reception followed Halley in viewing the work as a challenge to both the prevailing cosmological understanding of the day, namely that of Descartes, and the methods for determining this understanding, in particular the epistemological relationship between mathematical description and causal, physical explanation. In this view, the *Principia* was nothing less than a monumental revision of the very foundations of natural philosophy. By the 1720s, Halley's view had become the general view, and the "Newton Wars" that erupted, which I have analyzed elsewhere, were nothing less than an expansive and vigorous debate about the fundamentals of natural philosophy of the kind that Halley expected the treatise should provoke. Since these later eighteenth-century debates also cast the mold out of which modern historical understandings of Newton's work and legacy would be fashioned, modern scholars have generally operated by seeing these later debates as indicative of the *Principia*'s overall reception, including its initial reception in France in the late seventeenth century.

Historical thinking in this vein sees the *Principia* as a revisionist work of physics, one that is centered on the argument that a universal force of attraction exists and is operative in all matter. In this view, bodies, situated in otherwise empty space, are moved by one another according to a universal law relating the force of their relative attraction to their relative mass and the square of the distance separating them. Especially illustrative of these universal principles are the motions of the planets, and by demonstrating the quantitative law of universal gravitation in the *Principia*, Newton is seen to have shown the errors of Descartes's radically different cosmology, especially its causal, contact-point mechanisms and plenist conception of cosmological space. Viewed from this historiographical perspective, the natural consequence of the reception of the *Principia* was therefore a battle between "Newtonians" and "Cartesians" over the proper foundations of natural philosophy and cosmological understanding.

Pierre Brunet's pioneering history of the introduction of Newtonian ideas into France adopted this physicalist understanding, and he set the pattern for later historians. Brunet argued that "the theories of Newton encountered over the course of the eighteenth century particularly violent resistance in France because they clashed there with Cartesian doctrines already solidly established."[17] Other historians have echoed this understanding.[18] Yet problems arise when one tries to understand the initial French reception of the *Principia* in the decades around 1700 in similar terms, because no evidence of this battle or its rallying cries are present. Our French reviewer in 1688 made no use of terms such as "attraction," "the void," "the vortices," or "Cartesianism" when discussing Newton's work, and in this respect he was typical of the initial discussions of the *Principia* in France because no one else used these terms either.

One in fact looks long and hard in the archive of French science to find evidence from the decades before and after 1700 indicating that a great battle pitting Newtonians against Cartesians had immediately erupted. The absence of such a discourse in France before 1710 has sometimes led historians to extend Brunet's thesis about a pervasive and blindingly slavish French Cartesianism to claim that Newton's work was initially ignored altogether. But our French reviewer shows the fallacy of this understanding. Varignon does as well, for as Aiton saw, Varignon was an avid reader and user of Newton's work even if he never proclaimed himself a Newtonian or a Cartesian. Aiton's claim that Varignon was "in effect a Newtonian" because he translated "Newtonian planetary theory" to "Continental mathematicians more conversant with the language of the differential calculus" illustrates perfectly the distortions that ensue when the categories that would become operative during the Newton Wars three decades later are used to interpret Newton's reception in turn-of-the-seventeenth-century France.[19] As we will soon see, the difference between Newton's geometrical approach in the *Principia* and Varignon's use of the differential calculus to pursue similar questions was indeed decisive. But to connect all this to an interpretation of Newton's physical claims about gravitational attraction and planetary theory is to miss the point entirely.

Returning to our French reader in 1688, he talked about the *Principia* in terms of the relationship between mechanics and physics, and while he mentioned planetary theory, namely Newton's claim that the planets gravitate (*pesent*) one toward the other, his point was conceptual and epistemological, not natural philosophical and physical. Varignon's relationship to the *Principia* was the same. Like the French reviewer (who he very well might have been), Varignon did not see in the *Principia* a new planetary theory or a work of anti-Cartesian physics; he saw a brilliant new treatise in mechanics, albeit one with misguided pretensions to being a new kind of natural philosophy. As a mathematician and *méchanicien* in the mold of the great tradition of French mathematical mechanics personified by Huygens, Varignon saw no reason to indulge these pretensions. But this did not lead him to dismiss the *Principia* altogether. Quite the contrary, there was much to think about and do with Books I and II of the *Principia*, the ones devoted to mathematical mechanics, and if Book III was oriented very differently, with its experimental demonstrations and use of empirical data to ground a new kind of mathematical natural philosophy, better to just ignore it and focus instead on the unquestionably brilliant mechanics found in the first two books.

This is in fact what Varignon did, and in pursuing a partial, if no less focused, approach to the *Principia*, one that emphasized the first two mathematical books, he followed the general pattern for French readers overall during the two decades from 1690 to 1710. The reviewer at the *Journal des savants* articulated the frame of this reception in his separation of Newton the brilliant *méchanicien* from Newton the specious physicist and natural philosopher. No one in France before the 1720s embraced the unity between these positions that Halley celebrated, and even those who were inclined to find more of interest in Book III—namely the astronomers and other empirical mathematicians at the Royal Observatory—approached the arguments in this book in isolation of those in the other two. Only later would all the cross-disciplinary provocations of the two-part mathematico-empirical architecture of the *Principia* be scrutinized fully, but this did not mean that before 1720 the treatise did not exert an important influence, even if it did so in a partial and seemingly indirect way.

Analytical mechanics was one outcome of this particular French reading of the *Principia*, for it was profoundly shaped by Newton's work even if it was neither determined by it, nor ever called "Newtonian mechanics" by any of its originators. The term "Newtonian," in fact, was another invention of the later, more polemical phase of the *Principia*'s French reception, and before 1715 one finds neither the term, nor any of the polemical heat later associated with it, in any of the scientific discourse of the period. What one does find are references, albeit infrequent ones, to Newton and his scientific work, along with mixtures of admiration and befuddlement about the nature and significance of his science. Since our focus is on the emergence of analytical mechanics in France, we will concentrate our attention on the strand of the *Principia*'s reception that leads here.

The Newtonian Sources of Analytical Mechanics

Central to the development of analytical mechanics in France was a particular reading of the *Principia* that treated it as a brilliant work of mechanics tout court. This involved more or less ignoring Book III and focusing on the mathematical conceptualization of bodies in motion found in the first two books. This was certainly a partial reading of the text, but once the fundamental excision had been made, it was actually easy to find support for this approach in the text itself. This was because Newton often talked in Books I and II as if his mathematical mechanics should be read as inferring no physical correlate whatsoever. To sustain the two-part mathematical, then empirical, architecture of the treatise as a whole, Newton insisted on a separation between mathematics and physics that, while ultimately intended to secure their reattachment as "mathematical principles of natural philosophy," isolated them in new and fruitful ways. Because of this isolation, knowledgeable students found in the first two books of the *Principia* a new kind of mathematical mechanics, one where the mathematics led to new theory irrespective of any physical assumptions.

As an example of these moments, consider Newton's statements at the end of his definitions when after distinguishing *vis insita*, or the innate force of matter, from the motive and centripetal force that bodies may acquire, and after speaking in very physicalist terms about the difference between the absolute, motive, and accelerative quantities of these forces, Newton clarified his meaning. He wrote:

> It is in the same sense that I call attractions and impulses accelerative and motive. Moreover, I use interchangeably and indiscriminately words signifying attraction, impulse, or any sort of propensity towards

> a center, considering these forces not from a physical but only from a mathematical point of view. Therefore let the reader beware of thinking that by words of this kind I am anywhere defining a species or mode of action, or a physical cause or reason, or that I am attributing forces in a true and physical sense, to centers (which are only mathematical points) if I happen to say that centers attract or centers have forces.[20]

Statements such as these evacuate the seemingly physical meaning that Newton appears to suggest in the first two mathematical books of the treatise, and when they are read by themselves, without reference to the arguments of Book III, they work to define a new kind of mathematical mechanics, one in which mathematical theory is free to operate without physical constraints.

If one were to isolate the single most important influence of Newton's *Principia* on the mechanics that Varignon ultimately developed, it would be the model it offered of a mechanics capable of theorizing motion mathematically without recourse to constraining physical assumptions: in short, the model of what would much later come to be called theoretical mathematical physics. Yet the leap to this very modern conception of the relationship of abstract mathematical theory to physical conceptualization and explanation did not happen all at once, and it was certainly not present fully formed in Newton's *Principia*. Newton in fact had a very different agenda, and Varignon's conception of mathematical mechanics was not derived directly from Newton but assembled ironically and contingently through his very particular reading of the *Principia* within the French context where he worked. It also stemmed from the way that he took what he found in Newton's treatise and then transformed it into something Newton had neither done nor would have sanctioned.

Most problematic from Varignon's point of view was the peculiar mathematics Newton had deployed in producing his newly mathematicized approach to mechanics. Thanks to the magisterial scholarship of Niccolò Guicciardini, Newton's highly personal, abstruse, and historically idiosyncratic mathematical thinking has now been meticulously documented. Guicciardini's scholarship allows us to sum up the important aspects here as they pertain to Varignon's relationship to it. To start, one should stress Newton's traditionalism with respect to the newer mathematical currents of the day. These modernizing trends will be explored in more detail in the next chapter, but suffice it to say that Newton was by and large a vigorous "Ancient" in the Ancients-versus-Moderns battle that occupied seventeenth-century mathematicians no less than other savants. His reverence for ancient standards and traditions especially led him to believe that geometry was the highest form of mathematics because of its impeccable demonstrative certainty and synthetic rigor. It was also the only mathematics suitable for "public," which is to say published, mathematical work. For this reason, he insisted on a geometric presentation of his ideas in the *Principia*, and it was here that he parted ways with many in the wider mathematical community.

As was discussed in the previous chapter with respect to Michel Rolle, geometry was undergoing radical change in the seventeenth century thanks to the emergence of the new algebra, or analysis, as it was often called after Viète. Especially influential was Descartes's use of algebraic methods to define a new kind of analytical geometry in which problems once pursued geometrically using Euclidean terms and principles could be solved using numbers and algebraic equations. Newton was by and large an enemy of these new mathematical methods, or at least an opponent of their elevation to epistemological equity with traditional geometry. Also bound up in the term "mathematical analysis" as it was originally developed after 1600 was an epistemological meaning that also shaped Newton's relationship to the new "analytical mathematics."

Mathematically, "analysis" has two related meanings. One refers to an algebraic as opposed to a geometric approach to mathematical problem solving, and to the mathematical domain of numbers and equations as opposed to spatial magnitudes and figural constructions. A second epistemological meaning of analysis is also connected to this in that geometry is also associated with its "synthetic method," or its demonstrative manner of deducing conclusions from indubitable first principles. Analysis is the epistemological opposite of geometrical synthesis in this understanding, for it refers to a rival method of reasoning that starts by positing the conclusion that one wants to prove, and then works inductively to locate the general principles that ratify this conclusion. Since the method of analysis works backward from a conclusion to its underlying principles, it was often called the method of discovery as opposed to the synthetic method of proof. Furthermore, since the two methods were seen to be reciprocal, it had become common in seventeenth-century epistemological discussions to treat the two methods as a pair and to celebrate the value of each.

Newton was no enemy of analysis in principle, and as a method of discovery that often led to the illumination of new and fruitful results, he used analytical mathematics and its methods routinely in his work. His differential fluxional calculus, which he developed more or less simultaneously with, if separately from, Leibniz's infinitesimal calculus, was in fact a result of

Newton's own exploitation of the latest analytical methods in his mathematical work. Yet Newton's decision not to use his fluxions in his *Principia*, and his choice to develop instead a radically personal form of synthetic geometry to serve in their place, also illustrates well his relationship to the new analytical mathematics of the day. For him, analysis was simply a tool of discovery and a helpful resource when pursuing private mathematical research. But since it did not produce certain results in the manner of synthetic geometry, it was appalling in his view to offer analytical solutions as formal mathematical presentations. To use the terms that the French reviewer of his *Principia* used to describe his reasoning as a physicist, for Newton a mathematician who offered an analytical mathematical treatment of a problem in the place of a rigorous synthetic demonstration would be offering a mere mechanical account as opposed to the scientific one provided by a synthetic deductive demonstration. Such a substitution was intolerable in Newton's estimation, and he therefore insisted that his mathematical mechanics be publicly presented using the high epistemological standard of synthetic geometry while avoiding any merely mechanical and analytical (which is to say algebraic) presentation.

The problem with Newton's stance was that by adopting a strict traditionalism with respect to mathematical method, Newton was closing the door to some of the most innovative and powerful mathematics being developed in his day. What mathematicians gave up in rigor when they used analysis was often compensated by a new and potent capacity to simplify complex problems and economize mathematical reasoning. Moreover, since the reciprocal relationship between analysis and synthesis very often allowed for rigorous demonstration of analytical results after the fact, it started to become conventional to simply trust the results of analysis by themselves, and to treat analytical (aka algebraic) approaches as an equivalent method of reasoning despite the absence of any direct synthetic warrant for them. This was precisely the trend that Newton found troublesome, and he insistently deployed his more complicated, cumbersome, and abstruse synthetic method of first and last ratios in the *Principia* to secure the epistemological foundations of his work. This even though he possessed a more transparent and economical analytical approach that he could have used instead. He made this choice because in his mind to use the more modern mechanical (i.e., nongeometrical and analytical) methods would have been to transgress canons of scientific rigor that he held dear.

William Whewell offered a vivid image of the kind of mathematics that Newton's insistence on classical epistemological rigor produced. He likened it to the brutish cudgels and axes favored by warriors of yore, weapons that "make us wonder what manner of man he was who could wield as a weapon what we can hardly lift as a burden."[21] For readers of the *Principia*, not only on the Continent but also in England and Scotland, this made the *Principia* a very luminous book, but also a somewhat brutish one. Newton's idiosyncratic geometry first of all made a very complex and difficult corpus of arguments in any nomenclature even more difficult to read and understand. Many, even those with highly advanced mathematical skills, accordingly found the ultimate result opaque. These readers included Huygens, who was able to read and comment upon the first edition of the *Principia* before he died in 1695. In 1690, he wrote to the astronomer Ole Roemer likening the book to the darkness of a great ocean.

In his correspondence with English mathematicians such as David Gregory and Nicolas Fatio de Duillier, and Continental figures such as l'Hôpital and Leibniz, Huygens also repeated his desire to see a work from Newton that was "more accessible" and "less obscure."[22] Huygens was also at the center of talk, widespread in the 1690s, which expressed hope for a new edition of the *Principia*, one that would not only fix the errors but also develop its arguments in a more limpid mathematical style. L'Hôpital wrote to Huygens in this vein in 1692 expressing his hope for a new edition "in a style more accessible to everyone" (*plus à la portée de tout le monde*).[23] Gregory expressed a similar desire in 1694, penning a précis for a proposed commentary that would both correct Newton's errors and make his arguments more mathematically clear. In it, he proposed redoing some of Newton's work using Leibniz's differential calculus.[24] Newton ultimately eschewed any such revisions, and when a second edition of the *Principia* was published in 1713, it contained many corrections, but no break with the cumbersome and by this date even more archaic geometric idiom of the first edition.

The idiosyncratic geometry that Newton deployed was a general source of difficulty for readers, but even more perplexing was the peculiar way that Newton deployed this mathematics in the solution of complex problems. Most relevant to the later development of analytical mechanics was his method for resolving the incommensurable relationship between discrete and continuous magnitudes. Accomplishing this resolution was crucial for the advanced mathematical mechanics that Newton developed in the *Principia*, and by the 1680s a large body of mathematics had been developed that was available to Newton. It has become conventional to lump all these mathematical developments together under the rubric of "the history of the calculus" and then to treat each of them as preparatory steps—large, small, or in the wrong direction—leading toward the singular advance that is said to complete this development in the 1680s, namely the discovery of the algebraic algorithms of the differential calculus.[25] Yet the rules for manipulating algebraic equations that we call "the calculus" were neither the only, nor the preordained, result of these developments within seventeenth-century European mathematics. The complexities of this history are crucial to the history of analytical mechanics, and they will accordingly be explored in detail in the next chapter in relation to Leibniz's work in this area. For now, the point to emphasize is that Newton did not use his analytical calculus in the *Principia*, even if the geometry he did employ nevertheless included methods for reconciling discrete and continuous magnitudes that were akin to those at the center of the calculus even if they were not those of his "fluxional calculus" per se.

To state that Newton did not use his calculus in the *Principia* is, therefore, both to speak a truth and to mislead at the same time. A precise answer to this proposition depends entirely on what we mean when we say "the calculus." If by that term we mean algorithmic rules for deriving discrete solutions from algebraic equations that represent continuous curves in Cartesian analytical geometry, then Newton emphatically did not use the calculus in his *Principia* since he avoided on principle the algebraic formulations of the new analysis and used only traditional synthetic geometry in it. Nevertheless, if by "Newton's calculus" we mean his methods for reconciling discrete and continuous magnitudes within geometric problem solving, then the answer is an emphatic yes since his method of first and last ratios developed explicitly for the *Principia* was a kind of calculus, or at least a mathematical tool that did precisely the same work while eschewing algebra altogether in favor of traditional geometry.

Moreover, Newton's adoption of this seemingly idiosyncratic, and to some eyes retrograde, approach in the *Principia* should not be surprising once his work in the *Principia* is situated within the history of the calculus as a discrete mode of algebraic algorithmic mathematics. Leibniz's 1684 paper, which gives him credit as the first published author to articulate the fundamental rules of the calculus, was published in the *Acta Eruditorum*,[26] and while manuscript evidence makes clear that Newton was already in possession of his own calculus before this date, it would involve the worst sort of Whiggishness to ask, "Why didn't Newton use his calculus in his 1687 treatise?" Even if Newton had been fond of the algorithmic crystallization of the method that Leibniz offered (he in fact abhorred it), it is still grossly anachronistic with respect to the actual mathematical climate of the late seventeenth century to have expected him to have deployed his own method of fluxions within the formally demonstrated geometrical arguments of the *Principia*. To have done so would have been akin to publishing his scratch sheets, not the formal demonstrations worked out from them. No one in 1687 thought of the calculus as anything more than an innovative new tool for mathematical problem solving, and the failure to find this mathematics in the *Principia* reveals little more than the impossibility of mathematical science being transformed all at once through singular lightning bolts of progressive innovation.

These expectations are also misleading because they imply that the calculus can and should be reduced entirely to either the algebraic algorithms of Leibniz's 1684 paper, or Newton's comparable, though different, method of fluxions. Such thinking also assumes, wrongly, that mathematicians should be measured by their decision to use, or not, one particular tool as opposed to another. The calculus certainly looks to us today like a set of instrumental algebraic algorithms, but understood historically, these instrumental understandings were but one possibility within a large array of new mathematical ideas and tools developed in the seventeenth century.

Viewed schematically, what we call the calculus stems from the convergence of at least three historical transformations, and they might fruitfully be isolated here in order to see the presence of each in the formal mathematics of the *Principia*. First, the calculus for us assumes the use of algebraic equations to represent and then solve geometric problems. Second, the calculus assumes that solutions to the problem of reconciling discrete and continuous magnitudes such as in quadrature (i.e., finding the equality between rectilinear and curvilinear figures) can be taken for granted such that geometric techniques can be contained within algebraic rules of reasoning. Finally, to talk of the calculus is to assume a comfort with obtaining algebraic solutions to geometric problems such that a solution to an equation can legitimately stand in for a geometric construction and demonstration of a figural relation (think of the Pythagorean Theorem, or Proposition 47 from Book I of Euclid's *Elements*, being represented by the equation $a^2 + b^2 = c^2$). Accept all three assumptions, and you get Leibniz's and Newton's algorithmic rules for problem solving and the substitution of algebraic equations for the geometric techniques of traditional mathematics, which undergird what we call the calculus.

Using this template, we see none of the first and third aspects in the *Principia*, but much of the second. Newton's traditionalism about the geometric character of formal, public mathematics led him to avoid any use of algebraic symbols or equations in the *Principia*. In Book III he used numbers as discrete markers of quantitative measurement, developing sophisticated quantitative arguments as a result. But in the demonstrative mathematical books that preceded Book III, he avoided numbers altogether, developing his ideas geometrically according to his method of first and last ratios. His need to innovate within the traditional methods of quantitative geometry reveals the difficulty he experienced adapting ancient geometry to the new quantitative physics and mechanics he was developing. Yet his convictions about the legitimacy of this time-honored approach were steadfast. No algebra at all appears in the *Principia*, and all of the book's conclusions are demonstrated geometrically using traditional canons of synthetic rigor.

Within this synthetic, geometric frame, however, the analytic and algebraic calculus was nevertheless present in the *Principia* in all sorts of ways. The early historian of mathematics Jean-François Montucla captured well the nature of this presence. "Although Newton's *Principia* offers us in many places examples of the ancient procedure, in general the calculus pierces through the disguises with which Newton has covered it. This is a defect, which is common in those books presented as written according to the ancient method, but which are really only algebra in disguise."[27] What Montucla was pointing to were all the ways that Newton's actual work as an analytical mathematician, work that was completely legitimate in his eyes as long as it never left his study, in fact appeared in the formal arguments of the *Principia* as background assumptions and arguments that were not evident on the geometric surface, even if they were discernible to skilled mathematicians capable of reading through this veneer.

Guicciardini, as always, documents the phenomenon with meticulous care,[28] and since this aspect of Newton's work was central to its French reception, two tendencies of it need to be noted. One involved dismissing without demonstration the derivation of solutions that would otherwise have required tedious geometry to be fully demonstrated synthetically. Rather than stay true to his rigor and clutter the propositions with lengthy proofs, Newton simply announced that the solution could be obtained, and then he moved on. His conviction in operating this way was the same as the one that led analytical mathematicians to defend algebraic solutions as legitimate solutions despite the absence of the synthetic proof supporting them. Newton also inscribed analytical solutions into his geometrical reasoning in the *Principia*, especially letting algebraic solutions to limit problems and questions of quadrature stand in for the rigorous pursuit of geometric solutions. Economy was again the justification for this slip from strict rigor, and in this way Newton deployed his calculus in the *Principia*, which is to say his full toolbox of analytical mathematical techniques, even as he explicitly avoided its explicit use in the text and seemingly celebrated his avoidance of it through traditional geometry.

The upshot of Newton's hybrid and idiosyncratic mathematics in the *Principia* was a work that simultaneously presented itself as mathematically innovative and conservative at the same time. To achieve the new mathematicization of physics that was his central goal and achievement, Newton had to deploy all of his mathematical powers and tools, including those that were most innovative. But because Newton was a traditionalist with respect to public mathematical presentation, he also framed his arguments as much as possible within the terms of classical geometry.

Varignon was one of many who found in this result a mixed bag of brilliant innovation and frustrating confusion and caution. To call his analytical mechanics by consequence the Continental translation of Newton's mechanics in the *Principia* is to erase all the complexity of Newton's actual work, and then all the complexity of Varignon's own multifaceted encounter with it. The *Principia* certainly pushed Varignon down the path toward his new science of motion, but he charted his course as much by leaving Newton behind as by following in his footsteps. Other influences also entered that were crucial as well, and having outlined the particular, if oblique, influences that the *Principia* provided in shaping Varignon's work, we need to turn now to the other intellectual encounters that were just as important, and some that were even more decisive.

CHAPTER 5 *The New Infinitesimal Calculus and the Leibnizian Origins of Analytical Mechanics*

A central feature of Varignon's new science of motion was his aggressive deployment of the differential calculus, first articulated in Leibniz's 1684 *Acta Eruditorum* paper, to capture mathematically the motion of moving bodies. By deploying the Leibnizian calculus, Varignon was breaking fundamentally with Newton and the *Principia*, yet in the traditional scholarship this break is erased through the idea of his work as a "Continental Translation of Newtonian Mechanics." Within this frame, Varignon is said to have found the substance of his new science of motion in the *Principia*, albeit clothed in a strange geometric dress. He is then said to have refashioned it into the beginnings of "classical Newtonian mechanics" by translating Newton's science into the Continental idiom of the Leibnizian calculus. We have just seen all the ways that this account defies the actual history of what the *Principia* contained and presented to its initial readers, and the relationship of Varignon to Leibniz and his calculus was equally complicated.

Overall, Leibniz was probably more influential than Newton in shaping the actual content of Varignon's science, but Newton's *Principia* offered him a model for a new mathematical physics that was also crucial to his conceptualization. Neither influence was singularly decisive, however. Analytical mechanics was not planted in Varignon's head like a seed ready to germinate; rather it was developed through his own work, including his particular study of Newton and Leibniz, along with other influences which we will explore, in the context of his life and work and the events that shaped each in the decades around 1700.

Intellectually speaking, Michel Blay has already laid the groundwork for the arguments I will make. Building his understanding of Varignon's science upon the seminal work of Henk Bos, who has studied in great detail the scientific innovations initiated by Leibniz's new infinitesimal calculus, Blay shows that Varignon's real breakthroughs were mathematical, and that they stemmed from his capacity to derive a new physics by extending to the study of bodies in motion the innovative mathematical conceptualizations of Leibniz's differential calculus. "By allowing the science of motion to benefit from the recent progress of analysis," Blay writes, "Varignon . . . truly paved the way for the immense development of mathematical physics in the eighteenth and nineteenth centuries."[1] In particular, he showed: "In an exemplary, and, at long last, inaugural way, that scientific work must aim above all at obtaining, and rigorously manipulating, rules and formulas. The field of mathematical physics was now entered into once and for all, and that of the old science of motion, with its ontological and geometric ambitions, was left behind."[2]

I have no argument with Blay's analysis, which I will follow closely in all that follows. But I depart from Blay in my conviction that the sources for these innovations were not as singularly located in Leibniz's mathematics and Varignon's use of it as his internalist analysis would lead us to believe. Newton's *Principia*, for example, played an important role in the development of analytical mechanics, and even if Blay is right to dissolve the old story that makes the *Principia* the singular intellectual source for it, the treatise and its reception cannot be removed altogether as an influence. Blay's thesis is in fact reinforced and complicated in ways sympathetic to my argument by the immediate inheritors of Varignon's legacy: the eighteenth-century French mathematicians who first recounted the history of the mathematical physics that they had come to practice as distinguished members of the Académie Royale des Sciences after 1750.

Jean-Sylvain Bailly's *Histoire de l'astronomie modern*, published in 1779, is a case in point. Writing about the seventeenth-century "revolution" (his term) that had produced the modern astronomy that he practiced, Bailly credited Newton's "most powerful genius" with playing the key role in effecting this change.[3] Newton's physical theories were not what he stressed, however. He never mentioned Descartes or the battle between gravitational attraction and vortical mechanics in his account. Instead, Newton's innovative mathematical approach to astronomy was the key theme, along with the innovative mathematics, developed in the *Principia* he claimed, that made this mathematical astronomy possible. "Geometry, of which he was a master and possessed in all its detail, received from him a new form," Bailly explained. "It became in his hands a more subtle instrument, one more suited to profound research."[4] The key innovation was Newton's infinitesimal calculus, or "*calcul de fluxions*" as Bailly called it. "This invention caused a revolution in the exact sciences. Like the application of lenses to instruments and the invention of micrometers in the practical sciences, instruments that gave man organs that permitted him to penetrate into the knowledge of causes, the calculus of fluxions, or differential calculus, serves as a sort of micrometer that the mind [*esprit*] uses to see in a more intimate and certain manner the relationships [*rapports*] between things."[5]

Bailly's academic colleague Joseph-Louis Lagrange made the same point in a way that connects it even more directly to Varignon when he singled out Newton's *Principia* as the revolutionary founder of the science of *mécha-nique analytique*, the science which he set out to formalize in his so-named treatise of 1788.[6] "Mechanics became a new science in Newton's hands," Lagrange asserted, "and his *Principia* which appeared for the first time in 1687 was the agent of this revolution. Ultimately, the invention of the infinitesimal calculus put mathematicians in a position to reduce the laws of moving bodies to analytical equations, and since that time the study of the resulting moving forces has become the principal object of their work. I have proposed here a new means to facilitate this same research."[7]

What is significant about these two eighteenth-century accounts is the way that they echo Blay in making the infinitesimal calculus the central source for the new mathematical physics developed in the 1690s while also parting with him in making Newton's *Principia*, which technically speaking did not contain this calculus at all, the source for these revolutionary changes. Lagrange was no doubt aware of this historical anomaly, and his subtle use of the word "ultimately" (*enfin*) to insinuate a gap between the *Principia* itself and the revolutionary new mathematical science that was built from it offers us a telling point of entry into the more complicated history that this book is exploring. In his later expansion upon Bailly's history of modern astronomy, the royal astronomer and academician Jean-Baptiste Joseph Delambre noted this history explicitly, writing that, "mathematical analysis was not sufficiently advanced in the time of Newton to be fully deployed . . . [and] was often limited to offering ingenious flashes of insight [*aperçus*]. It was left to his successors Euler, d'Alembert, Clairaut, Lagrange, and Laplace to resolve the great questions of the system of the world, and to give astronomers the [mathematical] theories needed to construct planetary tables."[8] One of the figures listed here, Alexis-Claude Clairaut, also pointed out the same truth in his commentary to the first (and only) French translation of Newton's *Principia*, which Clairaut coauthored with the Marquise du Châtelet in 1754. Writing about Book I in the first section of their commentary, Châtelet and Clairaut wrote that Newton's *Principia* "explains in eleven lemmas [Newton's] method of first and last ratios. This method is the foundation of the geometry of the infinite, and with its help this geometry can be given classical certainty [*la certitude de l'ancienne*]."[9] Joining these assessments together, the point is that while the *Principia* does not contain the calculus per se, the fundamentals of it that are present, albeit in a different and more traditional geometrical form, and that the treatise was influential in generating the analytical turn essential to making eighteenth-century mathematical physics.

In another text, a formal eulogy for Leibniz that won the Berlin Academy of Sciences prize in 1768, Bailly also indicated his appreciation for these same historical complexities.[10] Echoing verbatim his remarks in his history of astronomy about the revolutionary significance of the differential calculus, particularly his description of it as a new "tool for expanding the human intellect" akin to the microscope and the telescope, Bailly acknowledged Leibniz's role as the parallel inventor of this same mathematics.[11] "Leibniz published his principles of differential calculus in 1684 in the journal from Leipzig" while "Newton deposited the fundamentals of the calculus of fluxions in 1687 in his immortal work on the mathematical principles of natural philosophy." "In effect," Bailly continues, "the glory of the invention resides with each of them," and yet it was Leibniz's calculus that was disseminated and adopted, and his name that "soared across Europe." Bailly also joined with Clairaut, Châtelet, Lagrange, and others in describing the "almost calculus" of the *Principia* as the fundamental influence in the creation of the Leibnizian-calculus-based analytical mechanics that all agreed was the major consequence of the *Principia*'s European reception.[12]

This is an odd distillation of a much more complicated history, because only in retrospect does the science that came to be called analytical mechanics, and, eventually, Newtonian mechanics appear to derive directly and seamlessly from Newton's *Principia*. For one, the mathematics that all of these commentators agreed was the decisive element in the making of this revolutionary transformation (their description) was not in fact to be found in the treatise at all. The suggestion of it could be found buried within a very different kind of mathematical clothing, yet if the differential calculus was the decisive innovation in the making of analytical mechanics (and I follow Blay and the eighteenth-century historians above in thinking that it was), then the fact that it was not found in the *Principia*, and could not possibly have been found there given the scientific attitudes and epistemological convictions of the book's author, suggests clearly that the *Principia* cannot be the singular determining source of this science. Yet at the same time, if these sophisticated eighteenth-century mathematicians, in addition to others from the 1690s forward, found these innovations in Newton's treatise nevertheless—Fontenelle described the *Principia* in 1696 as "all about the new calculus"—then we also need to consider the reception of this text together with the reception of the Leibnizian differential calculus in France when conceiving this history.[13]

In the end it was neither Newton, nor Leibniz, nor Varignon by himself who brought analytical mechanics into the world, but the historical convergence of all of them in France in the decades around 1700. To understand this convergence, let us now consider the arrival of the Leibnizian strand within it.

The Reception of the Leibnizian Calculus in France

Varignon was at the very center of this history for he was one of a very small number of French mathematicians (and only a slightly larger group of Europeans overall) who learned of Leibniz's new mathematics soon after it was introduced. He was also among the very first to begin to use the new calculus to solve mathematical problems. The Swiss mathematician Johann Bernoulli linked Varignon to the new calculus in 1691, when he made what in retrospect was a seminal visit to France during the same months that the newly ascendant Pontchartrain ministry was beginning to alter the institutional environment of the Académie Royale des Sciences. As he recalled almost three decades later, he had the opportunity during his visit to "meet Father Malebranche in the company of a large number of savants and other people of distinction."[14] The company included the Marquis de l'Hôpital, Varignon's future colleague and ally in the Royal Academy, and it was this blue-blooded aristocrat turned full-time mathematician who served as the key conduit for the essential mathematical exchange.

That it was l'Hôpital of all people who played the key role in this commerce reminds us of how the hierarchical social structures of Old Regime French society still shaped the mathematics that was produced there. L'Hôpital's social position is best captured by listing his full name: Guillaume-François-Antoine de l'Hôpital, Marquis de Saint-Mesme et du Montellier, Comte d'Etremont, Lord of Ouques-la-Chaise, le Bréau, and other cities, and governor of the cities and castles (*châteaux*) of Dourdan. In these capacities, he filled an inherited dynastic role that could be traced back to the thirteenth century, one that tied him directly to the house of Orléans, a lineage second only to that of the royal Bourbon line itself.[15] L'Hôpital's development as a mathematician also illustrated the place of this science within these still foundational Old Regime aristocratic social structures

L'Hôpital acquired his fundamental mathematical education from the private tutor who formed him—college and university education was beneath his station. And while he demonstrated precocious talents in this area, earning acclaim as a fifteen-year-old when he easily solved a difficult problem before a group of able mathematicians assembled at the home of the Duc de Roannés (Antoine Arnauld was present), a career as a mathematician was completely unthinkable for him given his rank and titles. Choosing the path that his social position charted, l'Hôpital instead followed his ancestors into the king's armies, obtaining a post as a cavalry captain. Yet according to Fontenelle, who narrated these and other stories in the academic eulogy he delivered for l'Hôpital in 1704, the young aristocrat did not give up his passion for mathematics while in the army. "He would study mathematics in his tent," the secretary reminisced, and grabbing an opportunity to poke at the social mores of the day, he added, "it was not just for study that he retreated there. For it must be admitted that the French nation, although as well-mannered as any, is still prone to that sort of barbarism that wonders whether the sciences, taken to a certain point, are incompatible with nobility, and whether it is more noble to know nothing." "I have personally seen some who served at the same time," Fontenelle added, "who were greatly astonished that a man who lived like them," that is, as a blue-blooded aristocrat, "was one of the leading mathematicians of Europe."

The image of a gloriously titled aristocrat choosing to occupy himself with mathematical science was indeed a strange one in Louis XIV's France, and l'Hôpital's case reminds us of the work that still needed to be done to erase the mechanical aura from the discipline and make it seem worthy of liberal and gentlemanly regard. Fontenelle also did not mention, but could have were it not contrary to his intellectual agenda, that the aristocratic prejudice he exposed was also reinforced by the presence of mathematicians among the engineers and other military mechanics in the king's armies. These men were skilled savants and soldiers, but from the perspective of the officers for whom military service was an ancient chivalric bond tying the sovereign to his sword-wielding knights, no commonality across this social divide was possible. No heroic crossing of social divides occurred in l'Hôpital's transition from a military officer to a full-time mathematician either. Plagued with acute nearsightedness that made it impossible for him to see anything farther than ten yards away, l'Hôpital's career as a mathematician began when he retired from the army in his late twenties and returned to his estates to live a life of leisure. He devoted himself fully to mathematics in this setting, acquiring his reputation through the brilliant solutions he offered to the challenge problems that were starting to appear frequently in learned journals such as the *Acta Eruditorum* and the *Journal des savants*.

Fontenelle's account of his work emphasizes his brilliant acumen as a mathematician, and the ease with which he penetrated the most difficult problems, discerning solutions that no one else could perceive. No whiff of the mechanical was anywhere present in his oeuvre, or in the reputation that l'Hôpital acquired from it, and as Fontenelle stressed, his acclaim as a mathematician, which made him the equal of Huygens, Newton, Leibniz, and Bernoulli, was achieved while remaining fully attentive to the demands of his rank and title. "His birth demanded engagement in many affairs," Fontenelle summed up, and this gave him little time to pursue his mathematical work. Yet "his great mathematical genius seemed to require no payment [of time and labor] in return." He also exhibited in his ordinary manners all the natural markers of nobility. "Judiciousness, solidity, and, in a word, a geometry of spirit" characterized him, Fontenelle recalled. He was also "quick to declare his ignorance and to receive instruction, even on mathematical matters," and "never conveyed jealousy, not because of his own sense of superiority, but because of his natural sense of equity." "Easy to talk to [*d'un commerce facile*] and possessed of a perfect, open, and sincere probity along with a veritable modesty," l'Hôpital's manners were nothing less than the personification of what he was: "*un grand homme*."

To understand the resonance of this description in Louis XIV's France, one needs to remember that while men could make themselves great in Old Regime France through their own effort, the greatest were still those that were born to the role and who then used their life to achieve what their pedigree suggested they deserved. L'Hôpital's mathematical career exemplified this entanglement of birth and merit in the evaluation of quality, for while he was without question a brilliant mathematician, his acclaim was inseparably tied to the social position that anchored it. The relationship with Bernoulli that he formed in the summer of 1691 further illustrates the point. It began because of l'Hôpital's prior connection to Father Nicolas Malebranche. Pinning down when exactly their relationship began is not possible given the sources, but Fontenelle relates that l'Hôpital made contact with Malebranche soon after his retirement from the army after judging from his book *De la recherche de la verité*, which first appeared in 1674, that "the author would be an excellent guide in the sciences."[16] They remained closely tied until l'Hôpital's death in 1704, and however old the relationship was in 1691,[17] it was through Malebranche that Bernoulli was introduced to the recently liberated warrior-aristocrat turned advanced mathematician.

Bernoulli recalled that at their first meeting l'Hôpital expressed his passion for mathematics, and demonstrated himself to be "a good mathematician at least when it came to ordinary geometry." Bernoulli added, however, "He knew virtually nothing at all about the differential calculus, save its name, and even less about the integral calculus, which had only recently been developed." What little of this calculus that was available was found in the *Actes de Leipsic* (i.e., the *Acta Eruditorum*), yet Bernoulli reported that l'Hôpital had not seen these volumes because of the war, during which the circulation of German books in France was prohibited. The Swiss visitor obliged his new acquaintance with a copy of one of his recent mathematical papers, and the exchange facilitated a new relationship. Bernoulli recalled that l'Hôpital "immediately recognized that the solution to the problem in question required a completely different kind of mathematics than that provided by Cartesian analysis, and he was amazed to see such a young man (for I was twenty-four years old at the time) possessing such sublime mathematical knowledge." L'Hôpital began visiting Bernoulli four times a week to discuss the new mathematics, and when summer arrived the marquis invited his new teacher and colleague to reside at his country estate near Blois. Bernoulli returned to Basel at summer's end, just as Bignon was beginning his new management of the Académie Royale des Sciences, but a correspondence between the two men began and continued without interruption until l'Hôpital's death in 1704.[18]

The influence of the Bernoulli-l'Hôpital meeting was enormous for the subsequent development of French mathematics. It was also especially influential for Varignon because it was through this partnership that he was initiated into the new mathematics as well. Varignon started his own correspondence with Bernoulli in 1692, and their early letters indicate that his collaborations with l'Hôpital had already begun by this date.[19] In June 1693, l'Hôpital wrote to their Swiss colleague that "Mr. l'abbé Bignon has extended to me an invitation to attend the meetings of the Academy of Sciences," and from this date forward l'Hôpital and Varignon would practice their partnership as royal academicians.[20] No evidence describing Bignon's deliberations in making l'Hôpital an academician exists, and it is worth noting that almost no published mathematical work supported the nomination even if a massive reserve of aristocratic privilege and honor did. L'Hôpital's budding reputation as a mathematician was certainly crucial in the decision making, but David Sturdy is also right when he says with respect to the political calculation that "Bignon envisaged the entry of l'Hôpital as a proclamation that membership in the Académie was fully commensurate with aristocracy."[21]

Whatever the path of entry, all three correspondents considered Malebranche an intimate, and his role in the developments that ensued will be considered in detail in the next chapter. Leibniz, who became l'Hôpital's correspondent in 1692 and Varignon's in 1702, was also a close collaborator.[22] Huygens, the doyen of French academic mechanics and physico-mathematics, was also the correspondent of l'Hôpital and Leibniz until his death in Holland in 1695.[23] Newton was also present within this community, and all these individuals knew about the *Principia*. His aversion to the kind of sociable commerce that was the lifeblood of this Continental mathematical circle, however, kept him distant from their regular discussions, and even if no member of this group doubted the brilliance of the *Principia*, they generally found more to puzzle over in the idiosyncratic mathematical science of this obscure and rather reclusive Englishman.[24]

Bernoulli further solidified his French ties by negotiating a contract with l'Hôpital in 1694 that appears odd by modern standards, but was in fact a common arrangement in the Old Regime given the patron-client relations that still governed all social life. The contract is also unsurprising given the excess of l'Hôpital's *crédit* and Bernoulli's youth and massive ambition to acquire some *crédit* of his own. In brief, Bernoulli agreed to become l'Hôpital's client, providing him with mathematical material in exchange for a payment of three hundred livres annually. Aristocrats of l'Hôpital's stature routinely arranged contracts like this with talented men, and for a less socially elevated savant such as Bernoulli, the privilege and financial rewards that such an arrangement brought were often highly lucrative. In this case, the contract ultimately rendered the ownership of the mathematics that they were jointly developing opaque, and in later years Bernoulli would contest l'Hôpital's authorship of works such as *Analyse des infiniment petits*, claiming that he was the rightful creator. Whatever the actual situation, the arrangement also reveals much about the Old Regime nature of the "mathematical profession" around 1700, along with the wider social fabric that the Pontchartrains were working with in their effort to create a new kind of intellectual community within the Académie Royale des Sciences.[25]

Out of this Old Regime nexus, French analytical mechanics was assembled, and the new differential calculus introduced by Leibniz in his *Acta Eruditorum* article of 1684, and then disseminated in France by way of Malebranche, l'Hôpital, Varignon, and the networks that linked them, served as the initial trigger.[26] In short, the innovation at the center of Varignon's new science of motion derived from treating the mechanics of moving bodies, especially planetary bodies, in terms of the differential equations of Leibniz's calculus. As has been noted already, to think about mechanics in terms of infinitesimal calculus was anything but an obvious thing to do in 1692. To understand Varignon's choice and its consequences, we therefore need to first understand the nature and the cultural position of the new mathematics that Varignon put to use.

The Conceptual Innovations and Provocations of the Leibnizian Calculus

Leibniz's achievement in his 1684 paper was prepared by a century of mathematical transformations, and while his innovations were provocative, they also reactivated old tensions that had become conventional within the rapidly changing field of seventeenth-century mathematics. To appreciate Leibniz's innovations, and the controversies that they provoked, let us situate his influential 1684 paper in the context of the seventeenth-century mathematical developments that informed it.

Stated simply, Leibniz's "Nova methodus pro maximis et minimis" offered a set of algorithmic rules for manipulating algebraic equations representing geometric curves so as to allow for the quick determination of their maximum and minimum points. Cartesian analytical geometry had created the mathematical structures that allowed geometry to be pursued in this algebraic way, and Leibniz simply took the Cartesian analytical approach for granted. This was one source of controversy, since many still found algebraic approaches to geometry suspect, and some called the new calculus a mere tool of analysis, describing it as a clever contrivance but not a substantial mathematical discovery. In this way, the calculus reanimated the debates that had been going on for over a century regarding the validity of algebraic/analytic mathematics with respect to traditional Euclidean geometry.

Also provoked anew was the epistemological relationship between analytical mathematics, which is to say mathematics that deployed numbers and symbols, and the tried and true methods of synthetic Euclidean geometry. Were a set of analytical algebraic rules worthy of being considered a serious mathematical discovery? Or was Leibniz merely publishing clever problem-solving tricks and passing them off as formal mathematics? Here Newton's decision not to use his own fluxional calculus in his formally developed *Principia*, even as he made use of it when working out the solutions that he developed geometrically in the text, is illustrative of one widespread judgment made about the validity and value of Leibniz's new mathematics.[27]

Difficult methodological and epistemological questions were further sparked by the precise work that Leibniz's calculus purported to accomplish. His algorithmic rules not only streamlined the manipulation of algebraic equations they also allowed one to resolve quickly and precisely, or so it seemed, a knotty and ancient problem: the resolution of discrete and continuous magnitudes. This fundamental operation, which sits at the very heart of classical geometry, was manifest most visibly in the "problem of quadrature," or the problem of producing a rectilinear figure equal to a curvilinear one. Many problems of quadrature exist in classical geometry, but the *locus classicus* is the problem of "squaring the circle." Constructing a second rectilinear figure equal in area to a given square is easily done using the compass, straight edge, and other tools permitted in Euclidean geometry. Using the same tools to construct a square equal in area to a given a circle, however, is impossible. This is because the magnitude of a curvilinear line or figure is not exactly commensurable with the magnitude of a rectilinear one. This incommensurability is captured in the famously irrational magnitude π. Defined as the ratio of the discrete diameter of a circle to its curvilinear circumference, π is irrational because these two magnitudes are incommensurable.

Demonstrative proof of this irrationality, and the corresponding impossibility of squaring the circle, was only achieved in the nineteenth century, however, and accordingly efforts to find the elusive square that equaled the circle were made throughout the seventeenth century even as mathematicians such as John Wallis, Descartes, and Huygens rightly declared that a rigorous geometric solution would never be attained. A key phrase here, however, is "rigorous geometric solution," because there were many solutions to this problem available so long as one accepted solutions that challenged in one way or another strict Euclidean rigor.

An ancient solution was the most acceptable by the epistemological standards of Euclid. It is called "the method of exhaustion" and is associated with Archimedes. He showed that the area of a circle is equal to the area of a right triangle if one of the right angle sides of the triangle is equal to the radius of the circle and the other is equal to its circumference. To demonstrate the equality, Archimedes inscribed a square inside and outside a circle and then showed that the area of the circle was greater than the area of the inner square and less than the area of the outer square. He then shrunk the space separating the circle from these rectilinear figures by doubling the number of sides of each. He then showed that the area of the circle still fell somewhere between the areas of these two figures, while also showing that the magnitude of the difference had been diminished by more than half. He then continued to double the number of sides in the inscribed and circumscribed figures, reducing each time by more than half the magnitude of the space separating each. To complete the proof, Archimedes then deployed a series of *reductio ad absurdum* arguments to show that at the limit of this process of spatial compression (i.e., at the moment of "exhaustion" of the space separating the curve from its rectilinear approximations), the area of the circle can neither be greater than nor less than the area of the right triangle in question. Since it cannot be bigger or smaller than the area of the right triangle without generating an absurdity, Archimedes concluded that the areas must be equal, proceeding from that equality to construct a square, equal in area to this triangle, which was then, by demonstration, equal in area to the circle.[28]

In terms of rigor, there is nothing wrong with Archimedes's proof, and for that reason it remained a foundational demonstration invoked whenever quadrature of the circle was called for. From the perspective of early modern understandings of Euclidean geometrical purity, however, there was still a problem with this method. Archimedes did not, in fact, construct a square that was precisely equal in area to a given circle; he instead demonstrated that the area of a given square could not be anything but equal to it, because the contrary claim was demonstrably absurd. This might seem to us like a subtle and unimportant distinction, but it in fact annoyed zealous geometers who wanted to find a means of directly constructing a square equal in area to a given circle using only a compass, a straight edge, and the other acceptable tools of Euclidean geometry.

It is in fact impossible to do this, and for that reason the project of squaring the circle has come to stand as an emblem of human folly. If, however, one lets go of the strict Euclidean constraints, then solutions of a variety of sorts are possible. Many ancient mathematicians in fact produced solutions of this sort, and one illustrative example is the quadrature accomplished by an exotic ancient curve called the "quadratrix of Hippias." No original works of Hippias survive, but in the *Mathematical Collection* of Pappus, one of the seminal ancient mathematical books translated and published by Commandino, an account of this curve is offered. The details can be found elsewhere.[29] For our purposes here, it is simply useful to note that the curve is generated by the intersection of a uniformly moving circular curve sweeping along a planar surface and a rectilinear line moving uniformly across the same plane and cutting the circle at various points. Hippias called the locus of these intersecting points the quadratrix, and it is one of the class of curves, such as the spiral, the cycloid, the conchoid, and the cissoid, that Pappus described as having a "varied and forced origin" (i.e., it was not "naturally" produced, as all Euclidean curves were, from compass and straight edge alone), but one with "wonderful properties" as a result. One of its wonderful properties is that it can be used to construct squares equal in area to a given circle, which made it a tool for producing direct solutions to the problem of quadrature.[30]

The problem of quadrature was therefore caught at the crossroads between ancient and modern geometry in the seventeenth century. Without a proof showing that a direct solution from within the terms of Euclidean geometry was impossible, claims to have found a solution continued to appear throughout the period. These attempted solutions either masked a paralogism, or, more commonly, worked by breaking down the fence holding "mechanical curves"—that is, those that do not adhere to the restrictive Euclidean demands for curve construction—clearly out of bounds of properly rigorous geometric practice. Here one sees again how the complex term "mechanical" did various and important epistemological work in the seventeenth century.

To illustrate, consider the quadrature offered by the Dutch mathematician Willebord Snell in 1621. Wanting to give a visible illustration supporting his confidence that a discrete line existed that was equal in magnitude to a circle's circumference, he asked that we imagine a point on a circle tracing a line on a plane as it completes one full revolution along the surface. This line, said Snell, "argues, and as it were sets before the eyes, that a right line may really be exhibited equal to the periphery."[31] The line he described was a well-known mechanical curve, the cycloid, and once defined it could be explored and deployed using the methods of Euclidean geometry even though it was not, strictly speaking, a legitimate (or "natural") Euclidean curve. Seventeenth-century mathematicians ranging from Galileo and Torricelli to Mersenne and Roberval did just that, and these and other mathematicians often echoed Snell's appeal to the naturalness of their mechanically produced lines, as evidence to support their use of this curve in geometrical practice. Yet traditionalists could still deride such work as "merely mechanical" and mean by that two different but related things. One attack was directed at the nature of the mathematical object in question (a curve that was produced by other than strict Euclidean means), and another impugned the quality of the mathematical reasoning involved (its laxity with respect to the highest standards of Euclidean rigor). In each case, a contested border was drawn separating suspect innovation from time-honored standards, and it was across this frontier that many mathematical debates in the seventeenth century were contested.

Given the latent social understanding of the lowly mechanic haunting all of this mathematical discourse, the debates often became heated and complicated, and yet the mathematical changes that were in many cases most important in the seventeenth century often involved direct transgressions over this very line through the assertion of "mechanical mathematics" as something worthy of esteem. The dilemma confronting these would-be mechanical mathematicians was how to legitimate their use of mechanical curves and reasoning while also maintaining their claim to be practitioners of rigorous mathematics.

Henk Bos, who has studied this question most fully, points to a number of solutions deployed by geometers in the seventeenth century to achieve this double goal.[32] One involved hiding, or "black-boxing," the origin of the mechanical curve in a demonstration and then using it, once introduced, in strictly Euclidean terms. In this way, the demonstration presented itself as a rigorously Euclidean construction even if the curve that made it possible was technically not legitimate. Another strategy involved efforts to expand the class of legitimate geometric curves so as to include those that were generated in seemingly natural ways. The cycloid, which was generated through the exceedingly natural motion of a circle, a legitimate Euclidean curve, rolling on a line, is one case of a mechanical curve with strong claims to intrinsic natural, geometric status. A third compelling argument available to aspiring mechanical mathematicians was the empirical fit between a mathematical description and a testable empirical phenomenon. Huygens's achievement of superior regularity in time keeping thanks to a cycloid shaped pendulum is a good illustration since this discovery not only improved accurate clockmaking, it also offered a new argument for the naturalness of the cycloid as a non-Euclidean mathematical object.

Also available were claims to naturalness grounded in the production or character of a mathematical object or procedure. A curve produced through overly forced and artificial means was clearly mechanical, but curves produced through seemingly natural mechanisms that displaced willful human manipulation seemed more natural, or so some mathematicians began to argue. Descartes deployed arguments such as these to defend a machine for constructing curves that he invented called his "mesolabum compass." He claimed that this instrument was so exceptionally natural in its motions that it made curves worthy of automatic geometric status despite the instrumentality of their origins. It is worth noting that in this case the mechanical that was bad was that which manifested an excess of laborious activity and visible human handiwork, while the one that was good was justified by the freedom and ease with which it was seemingly realized and did its work. These aesthetic criteria were also used to distinguish the free and easy manners of the liberal gentleman from the willful bodily exertions of the mechanic, and here again we see how terms such as "economy," "elegance," and "amenability" contained social as well as epistemological meaning when they were used to judge early modern mathematics.

General criteria for distinguishing natural mechanical mathematics from its overly forced and artificial counterpart were, of course, never established, and given the strong desire for mathematical tools that went beyond those provided by Euclid, and the many pressures, not least those emanating from sovereign courts and states, to use mathematics in instrumental and empirically practical ways in the solution of concrete problems, a space was opened for mechanical mathematics to challenge the supremacy of Euclidean geometry as the epistemological overlord of the seventeenth-century mathematical field.

Leibniz's new differential calculus was a perfect example of the kind of mechanical mathematical innovations that these trends produced. If his was even more controversial than most, it was because he also incorporated into his new method arguably the most provocative seventeenth-century mechanical mathematical innovation of all: infinitesimals.[33] Leibniz's innovation was doubly provocative since he created an algebraic conception of infinitesimals suitable for use in Cartesian analytical geometry at a time when the use of infinitesimals of any sort in the solution of quadrature problems was a major source of Ancient versus Modern, and mechanical innovator versus traditional Euclidean rigorist, contestation overall. The method of indivisibles developed by Bonaventura Cavalieri, which in many respects stands as the direct geometric precursor to Leibniz's analytical calculus, can serve to illustrate the point.[34]

Summarized simply, Cavalieri's method of indivisibles offers a convenient way of finding the areas of geometric figures, and as such a powerful solution to quadrature problems. Imagine a bounded space to be measured, and then draw a line across this surface such that it cuts across the entirety of the figure. Then imagine a multitude of other lines drawn across the figure parallel to the first one such that a flow of parallel lines, or a "fluent" as Cavalieri called it, is produced that ultimately fills the entire shape. Cavalieri called this aggregate of parallel lines "all of the lines of a given figure," and his innovation rested in the suggestion that one could take "all of the lines of a given figure" to be equal to the area of that figure. With that equality assumed, ratios could then be created between one figure and another, and these ratios could then be used for comparisons of the areas of each. To cite one use of Cavalieri's method in the seventeenth century, both Roberval and Torricelli demonstrated that the area under a cycloid was precisely three times larger than the area under the circle that generated this mechanical curve by using "all the lines" of each figure to demonstrate the relation.

What is important to notice about Cavalieri's technique, however, is the complex interplay of classical and mechanical geometric techniques contained in the method. Generating lines across figures in the sequential, or "moving," way that Cavalieri demanded is perfectly acceptable within the postulates of Euclidean geometry. It was also common within rigorous Euclidean discourse to speak of rotating figures around axes, and lines around points, for so long as the motions in question conformed to those permitted by the Euclidean postulates, there was no problem. Cavalieri entered into heterodox territory, however, with his claim about the results generated by these moving lines. He simply took it for granted that the area of a given figure could be induced from the motion of "all the lines" he described across it, yet such an operation was not warranted within the Euclidean postulates. It also generated deep epistemological problems that Cavalieri's critics were quick to point out.

Among his fiercest critics were members of the Society of Jesus, a fact that is worth remembering when we encounter assertive Jesuit critics of modern, infinitesimalist mathematics in France later in this book. One especially vigorous antagonist, Father Paul Guldin, S.J., developed a powerful critique of Cavalieri's method of indivisibles that isolated clearly the modern mechanical provocations it contained. One issue involved the assumption, which Guldin found imbedded in Cavalieri's method, that a bounded surface could be imagined as a composite of infinitely many discrete lines. Cavalieri had carefully avoided any articulation of this idea, which did indeed fly in the face of the classical incommensurability of discrete and continuous magnitudes. Yet in speaking of lines as indivisibles, and of "all the lines" of a figure as a sum produced by the production (or "motion") of indivisible lines across a plane, Cavalieri appeared to be mathematically generating the area of his figures through a summing of the indivisible lines that composed them. Since he also offered no demonstration securing the move from indivisible lines to the total of "all of the lines of a given figure," Cavalieri further provoked suspicion.[35]

Seeing in these moves an unwarranted avoidance of inviolable standards of rigor, Guldin called Cavalieri's method fallacious and challenged the validity of the results. Cavalieri was not ignorant of these problems, for as Paolo Mancosu shows, much of Cavalieri's actual mathematical work centered on deriving ingenious ways to demonstrate from within classical Euclidean terms what in fact was a fundamentally non-Euclidean, or mechanical, set of mathematical innovations. Guldin, a rigorous classicist, presented a powerful challenge precisely because Cavalieri's departure from Euclidean standards was real and substantial.[36] Yet Cavalieri's refusal to surrender in the face of these classical challenges, and the adoption of the method of indivisibles by others, such as Roberval and Torricelli, despite Guldin's objections, reveals the power of the Modern position by the middle of the seventeenth century.

The Moderns were anything but secure in their position, however, and openly challenging Euclidean principles would remain a risky proposition throughout the seventeenth century even as much of the most innovative mathematics produced in this period proceeded by doing exactly that. A telling illustration of the tensions, and the peculiar outcomes that they produced, is the fact that while Torricelli developed two determinations of the area of the cycloid using Cavalieri's method of indivisibles, he also offered a third that used the "methods of the ancients" by proving the result indirectly through a double *reductio ad absurdum*.[37] This placement of classical demonstrations alongside innovative mechanical solutions so as to secure the validity of the modern approach would become a common ploy within seventeenth-century battles between classical and modern mathematicians. The deployment here of classical methods by an otherwise assertive modernist also shows how even among the innovators the epistemological call of the ancients remained strong.

Cavalieri, in fact, spent his entire life struggling to find secure foundations for his work, and one colleague, whose support he wanted dearly, namely Galileo, went to his grave skeptical about the rigor of the method.[38] Many others had qualms about it, and what this case reveals is the turbulent and disputatious nature of seventeenth-century mathematics and the heat of its contests. Geometry versus algebraic analysis; demonstration versus mechanical resolution; synthetic rigor versus analytical economy; ancient tradition versus modern innovation—each of these positions presented both opportunities and dangers for ambitious seventeenth-century mathematicians. Newton illustrates the point well, for as we saw in the previous chapter his mathematics exhibited an emblematic seventeenth-century fusion of mathematical classicism with mechanical innovation, a hybrid, especially in the version found in his *Principia*, that admits of no simple classification into either the Ancient or Modern camps. As he wrote in 1676 with respect to the quadrature of curves, "I consider mathematical quantities in this place not as consisting of very small parts, but as described by a continuous motion. Lines are described, and thereby generated, not by the apposition of parts but by the continued motion of points. . . . These geneses really take place in the nature of things, and are daily seen in the motion of bodies."[39] This is a classic articulation of the new seventeenth-century empirico-mechanical conception of geometry, and another is found in the preface to Newton's *Principia*. Here he writes:

> Mechanics is so distinguished from geometry, that what is perfectly accurate is called geometrical; what is less so is called mechanical. But the errors are not in the art, but in the artificers. He that works with

> less accuracy is an imperfect mechanic; and if any could work with perfect accuracy, he would be the most perfect mechanic of all; for the description of right lines and circles, upon which geometry is founded, belongs to mechanics. Geometry does not teach us to draw these lines, but requires them to be drawn; for it requires that the learner should first be taught to describe these accurately, before he enters upon geometry; . . . Therefore geometry is founded in mechanical practice, and is nothing but that part of universal mechanics which accurately proposes and demonstrates the art of measuring.[40]

These statements illustrate perfectly how mechanical mathematical thinking, rooted in the empirical study of the natural world, was starting to transform the practice of traditional geometry. Nevertheless, Newton asserted these views while also remaining a vigorous opponent of Cartesian analysis and a staunch defender of the epistemological superiority of classical geometric methods.

By the standard of Newton, Leibniz was an aggressive and unabashed modern innovator, and his calculus, although indebted to Cavalieri's method of indivisibles and a host of other seventeenth-century developments in infinitesimal mathematics, was a major step beyond these other approaches. For one, it broke emphatically with geometry by providing a purely algebraic algorithm (i.e., rules for manipulating symbolic equations) that allowed for the quadrature of continuous curves to be obtained within algebraic terms alone. It also offered these rules not as an approximation, but as a precise tool for producing quadratures of whatever curves one possessed. Previous methods of quadrature using geometrical notions of infinitesimals had expanded geometry, but Leibniz liberated algebra from geometry altogether, and showed how geometric solutions could be found through the manipulation of algebraic equations alone.[41]

To understand Leibniz's innovation, consider the so-called differential triangle, the classic case of the differential calculus (fig. 5). Assume a curve intersected at two points P and Q and ask what is the length of the curve between these two points? Speaking geometrically, no precise, rigorous answer is possible since discrete, geometric representations of continuous magnitudes do not exist. One can approximate an answer by using the ancient method of exhaustion, but since discrete and continuous magnitudes are incommensurable, no rigorous synthetic determination of the length of the curve is obtainable. The calculus, however, offers a solution to this dilemma. Imagine that the length of the curve in question is approximated by the length of the rectilinear

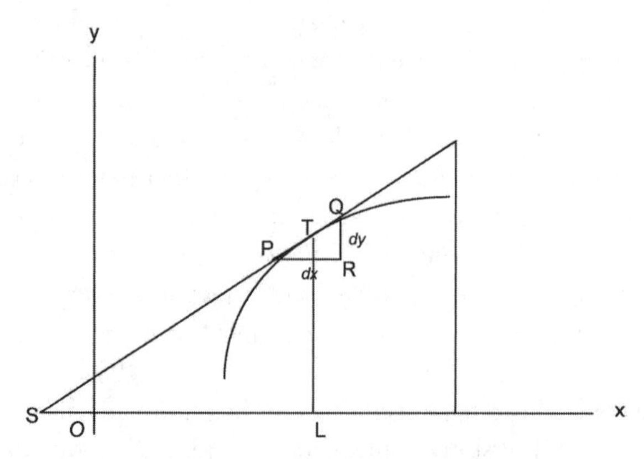

FIGURE 5. *The differential triangle of the Leibnizian differential calculus.*

line PQ on the straight-line tangent to the curve SPQ. Cartesian analytical geometry is built upon the use of discrete numbers to represent continuous geometric lines, and once approached this way the rectilinear length PQ can be calculated through the ordinary trigonometry of the triangle PQR. Also assume that the difference between the length of the rectilinear line PQ and the actual length of the curve between these two points is fleetingly small. Further assume that by conceiving of the curved line as a polygon composed of infinitely many rectilinear lines like PQ, each possessing an infinitesimally small trigonometric triangle consisting of infinitesimally small sides *dx* and *dy*, the disparity between the actual length and its rectilinear approximation can be reduced to nothing. Using this framework, one can develop an algebraic method that uses the infinitely small magnitudes *dx* and *dy* to mathematically calculate the discrete magnitude of the curve despite its continuity.

At the core of this solution, however, is an important and problematic assumption. The classical method of exhaustion allows one to approximate the length of the curve by showing how its length approaches, if never actually obtains, the length of a determinate polygon. Cavalieri's method of indivisibles, by contrast, finds the length of the curve by positing the existence of infinitely small yet indivisible parts (points in this case, lines in the case of spatial figures), and then by treating the whole as the sum of these parts. In the calculus, the difference between approximate and actual length, and between indivisibility and the infinite, is collapsed into the mysterious concept of the infinitesimal, or the *dx* and *dy* in the diagram. In his 1696 treatise on the differential calculus, l'Hôpital stated the key *Demande ou Supposition* of the new analysis this way: "It is asked that one treat indifferently, taking either one to represent the other, any two quantities that differ only by an infinitely small quantity; or, to say the same thing another way, that a quantity that is increased or diminished by another quantity, which is infinitely small, be considered identical to the original quantity."[42] By making this assumption, the calculus can translate the magnitudes of continuous curves into discrete calculable numbers.

But this success is bought at a steep epistemological price. Upon what logical ground does the infinitesimal postulate reside? For the calculus to work, the infinitely small magnitudes dx and dy must be considered at once as real quantities possessing positive magnitude and as nil quantities possessing no magnitude. But how can both be maintained simultaneously without provoking a contradiction? Voltaire, with characteristic wit, captured the dilemma perfectly when he described the calculus as the "art of exactly numbering and measuring that of which we cannot even conceive the existence."[43] The problems, however, do not stop there. If the foundational postulate of the calculus does not adhere to even the most basic laws of logic, then upon what epistemological ground does it lie? And how can mathematics, of all things, be rooted in any epistemology other than pure apodictic reasoning? These were difficult questions and later thinkers struggled with them while offering a range of different answers.

Intensifying the dilemmas was the shift from geometry to algebra that was essential to Leibniz's work. Cavalieri's method of indivisibles in many ways posed similar problems, and its critics, especially the Jesuit Paul Guldin, indicted the method for sustaining the untenable position that continuous lines (or figures) contained infinitely small parts that could be discretely summed. Leibniz's method claimed the same, but also went a step further. Not only did the calculus, like Cavalieri's method of indivisibles, eschew the ordinary canons of synthetic geometrical rigor, it did so in the name of a new algorithmic use of algebra that possessed no alternative rigor of its own. Cavalieri ultimately made his indivisibles more palatable by burying them within the Euclidean geometric veneer of his work overall, but Leibniz's analysis had no such reassuring geometric covering, Yet his calculus also obtained recognizably valid results in many cases, and even though the reasons behind its validity were not always clear, it rarely produced outright errors. It also offered a new algorithmic clarity and economy combined with a high degree of problem-solving potency. But was this new economy a sufficient warrant for calling the differential calculus, as Leibniz did, a new universal science?[44]

Some, such as Bernoulli, l'Hôpital, and Varignon answered yes to this question, but their position was by no means universally held. Corresponding with Huygens in the early 1690s, Leibniz learned that his former teacher and mathematical mentor found the new calculus opaque and less innovative than its inventor claimed. Huygens was not averse to using infinitesimal methods, for he used them often in his own mathematical work. What puzzled him, however, was Leibniz's claim that he had developed a new universal science capable of solving all manner of quadratures at once. Not only was Huygens skeptical of such universality, he did not see how Leibniz could sustain such a bold claim upon such limited foundations. L'Hôpital changed Huygens's mind somewhat when he showed him how the calculus could produce innovative solutions that were not at all apparent using ordinary methods. In the wake of this demonstration, Huygens praised his former pupil for discovering a new mathematical art with tremendous power. But he still failed to see in it a new mathematical science. He also worried about the limits and the foundation of these new infinitesimal rules. To l'Hôpital, he further expressed a private skepticism about the grandiose claims that Leibniz was making for his work.[45]

Newton also presents an interesting case in this regard, for while he also developed an equivalent set of algebraic rules for using "evanescent magnitudes" (his term for infinitesimals) in the solution of problems of quadrature, he was more reluctant than Leibniz to celebrate the innovation as a new mathematical science. He also kept his new calculus buried in his private papers, and avoided its obvious public use in his *Principia* as we have already discussed.[46] Newton's relationship to the new calculus was complex, but in the context of the 1690s his ideas put him in the mainstream of mathematicians who saw value in the new infinitesimal techniques while also worrying about their validity and epistemological rigor. From this perspective, Leibniz was extreme in his confidence about the new mathematics, as were those, like Bernoulli in Switzerland and l'Hôpital and Varignon in France, who each began to vigorously apply, develop, and publicly assert the value of the new calculus in the pursuit of innovative mathematical work.

On the side of the "infinitesimalists," as they soon came to be called (mostly by their critics), was the significant power that the new mathematics clearly offered. One reward was the ease with which it resolved previously difficult and tedious problems. Here support for the new method was provided by the agreement between its results and those achieved by traditional geometrical methods. The astonishing ease with which the calculus produced these solutions was also praised. Empowered by this economy, supporters of the new mathematics also began to apply it to mathematical problems previously deemed insoluble using existing methods. New and exotic curves, many of which were excluded from geometry classically conceived due to their mechanical origins, were brought under mathematical control through infinitesimal methods. Solutions to many previously intractable problems were also found. When the solutions so obtained permitted empirical or practical problems to be solved, these utilitarian rewards also provided an extra epistemic warrant for the new mathematics that had produced them. These outcomes mirrored the way that the link between the mathematics of the cycloid and Huygens's pendulum clock had mutually secured a new mathematics and a new mathematical physics simultaneously. Existing mathematical techniques were also simplified and made more general by applying the new infinitesimal method to them. In this way, the calculus attracted a small coterie of practitioners who believed that Leibniz's new universal science constituted a powerful new vanguard for future mathematical advancement.

Bernoulli was an originator of this program, and l'Hôpital and Varignon joined the team earlier than most. Their collaboration after 1692 amounted largely to extending and refining the new differential calculus by applying it to a host of new and old mathematical problems while also struggling to develop the new science's more difficult twin, the integral calculus. Leibniz likewise participated in these efforts, as did Newton even if he appears as a marginal figure when viewed from the perspective of the Continental center of these activities. From this work grew analytical mechanics. The influence of Newton and Leibniz was certainly instrumental in its development, but a French source of direct influence was also present in the mix, and he may have been the most decisive influence of all in the original creation of analytical mechanics. This was Father Nicolas Malebranche, the impresario of the circle that facilitated the introduction of Leibnizian analysis into France, and a thinker who made the new mathematics central to his scientific thought. All the key figures in the development of analytical mechanics looked to Malebranche as an intellectual guide, and his influence within French culture overall in the decades around 1700 was immense. This local Malebranchian influence played a crucial role in fusing the various intellectual sources flowing through France in the 1690s and making of them a new and coherent intellectual whole. Accordingly, the next chapter completes part 2 by providing a full examination of Malebranche's influence in the making of analytical mechanics.

CHAPTER 6 *The Malebranchian Moment in France and the Cultural Origins of Analytical Mechanics*

The previous chapters set the stage for showing how an encounter with the complex work of Newton's *Principia*, together with the new Leibnizian calculus, provided the scientific context for the initiation of Varignon's new science of motion in the 1690s. The changing understanding of physico-mathematics was indeed crucial in pushing Varignon toward his innovative new science, but his work was not conceived in a mathematical isolation chamber hermetically shut off from the wider currents of society. Mathematicians in seventeenth-century Europe did not possess any of the professional autonomy characteristic of today's specialized, disciplinary mathematical scientists, and they accordingly pursued their work while swimming fully within the mainstream culture and society of their times. Varignon was no different, and analytical mechanics was not only a product of French society circa 1700, it was, I argue, unusually and decisively shaped by the wider cultural dynamics of French society during this period. The goal of this chapter is to trace these broader cultural influences generative of Varignon's innovations.

Stated simply, the dynamic that mattered most was the realignment around 1700 of French political structures with new habits of life and sociability. Out of this social ferment, new cultural forms were generated. The new analytical mechanics was one of them. The political shifts that contributed to the creation of this fecund environment for change were introduced in chapter 3 and will be explored further in chapter 7. In this chapter, the intellectual and cultural shifts attendant to these political changes will be explored. To begin, let us consider the figure whose name I have chosen to denote this decisive cultural juncture because of his singular influence within it: Nicolas Malebranche.

Bernoulli's consequential visit with French mathematicians in 1691 started with a visit with Malebranche, and this is one of many illustrations of Malebranche's centrality within the larger intellectual changes in France in the decades around 1700. Bernoulli's itinerary, which moved him from Malebranche into the wider community of French mathematicians, marked one thread of his influence, for by 1691 Malebranche had become the point around which a vibrant circle of forward-looking mathematical innovators was revolving. André Robinet, the historian who has most extensively studied this history, calls by the term "the Malebranche circle" this French mathematical coterie, and it is this circle to which the French owe the reception of the Leibnizian calculus in the 1690s. Varignon's new calculus-based science of motion was also a direct product of this characteristic French milieu.

But Malebranche was more than just a facilitator of mathematical innovation. As the most publicly renowned member of the Oratorian religious order in France, and the philosophical beacon for many committed to its mission, Malebranche gave a public face to advanced mathematics that also joined it with religion, morality, and modern living. Malebranche came to personify the widely influential Oratorian conception of Christianity reformed through modern scientific, and especially mathematical, thought, and he served as the imagined leader of the order and the key proponent of its perceived mission to spread modern philosophy, mathematics, and religion as a package throughout France and the rest of Europe. Malebranche also added another cultural constellation of importance to this mix: French Cartesianism. Varignon was not, as the older scholarship had it, a French Cartesian in the sense of an ardent anti-Newtonian reader of the *Principia*, nor was his analytical mechanics a Continental "Cartesianization" of the anti-Cartesian science of Newton's *Principia*. But he was a Malebranchian Cartesian in the sense of being a mathematically oriented student of empirical phenomena who sought the underlying order of nature in its mathematical relations and rationality. Malebranche was crucial in establishing this precise strand of French Cartesianism in the late seventeenth century, and to understand the widespread fin-de-siècle French spirit that he came to personify, the spirit that incubated analytical mechanics, let us consider each aspect of his identity and influence in turn.

Malebranche, Mathematics, and the Oratorians

The Congregation of the Oratory, to which Malebranche was allied from the late 1660s until the end of his life, was founded in 1611 by Cardinal Pierre de Bérulle as part of a larger French movement toward Catholic reform.[1] It was conceived (in France at least; the Italian history of the order is different) as a teaching and service order, one organized around teacher-scholars oriented toward a worldly conception of religious service. Oratorians established new colleges at which students were taught virtuous, Christian living through modern approaches to pedagogy. As a teaching order devoted to Christian reform within lay society, the congregation also developed a program of conduct that was noteworthy for its flexibility and modernism. Like other religious orders, the Oratory was staffed by ordained priests, but it also institutionalized a practice of allowing nonordained *confrères* to remain in the order and to teach in its schools. The Oratorians also cultivated an admission policy that was noteworthy for its openness and inclusiveness. The Oratorians' motto was "enter who can, and leave who wants," a principle that they made real in their refusal to compel vows of obedience from members. This openness at both ends of the religious vocation was emblematic of the order's ethos overall. It will be remembered that the abbé Bignon established his clerical credentials within the Oratorian order before moving on to a career in royal administration, studying under Oratorian professors at the Collège d'Harcourt in Paris and then achieving ordination after theological study at the Oratorian Seminary of Saint Magloire. His career path was emblematic of many in Old Regime France who used Oratorian education as path into a religious identity that harmonized with secular professional vocations.

Oratorian teaching overall was strikingly progressive for the time, offering instruction in French rather than Latin, and instructing through intellectual dialogue between teacher and pupil in a way that eschewed the formal, scholastic disputation characteristic of other colleges, including those of the Jesuits. The Oratorian colleges further replaced the classical curriculum of Greek and Latin classics with a more modern program that included subjects such as natural science, mathematics, history, and literature. To advance their pedagogical goals, the Oratorians also developed their own textbooks, structuring the knowledge they presented in a manner appropriate to their teaching mission. Since these were vernacular texts, Oratorian textbooks also circulated easily in the wider public sphere, giving a larger audience virtual access to Oratorian instruction. Many Oratorian colleges, including the two that were proximate to Paris, also offered public lecture courses, and one, Juilly, sponsored a literary academy in which poetry was read and prizes awarded. In this way, many more people than those formally associated with the schools were exposed to Oratorian thinking.

The order struggled throughout the *ancien régime* to attract students and acolytes, and overall there were never more than a few hundred formal adherents to the order at any point in its history. The wider influence of the order was nevertheless immense, especially through the prominence of its teacher-scholars who served as exemplars of how Christianity, modern thought, and public service could be successfully wed.

The most famous and influential Oratorian teacher of the period was Bernard Lamy. He entered the congregation in 1654, while studying philosophy at Paris and Saumur.[2] After receiving his ordination in 1664 he became Professor at the Oratorian College of Angers. Lamy became notorious in 1675 when his Cartesian teachings were censured by royal and ecclesiastical authorities, forcing him into exile. He reemerged at the seminary of Grenoble and then at the college of Rouen, where he was protected for the remainder of his life. Despite his persecution, he also published a number of influential works, including a widely discussed refutation of Spinoza's *Ethics*, and an influential treatise on rhetoric, *L'Art de parler*, which embodied Oratorian principles. His *Entretiens sur les sciences* was also widely read and admired, not least by the Baron de Montesquieu and Jean-Jacques Rousseau, who reported in his *Confessions* that he kept a copy of the *Entretiens* on his nightstand. In its urbane, conversational approach to learning, the book exemplified Oratorian thinking and pedagogy as a whole.[3]

Lamy was replaced at Angers by Father Claude Jaquemet, a savant who shared with his Oratorian brother an affection for modern thought, especially Cartesianism. His interests, however, lay more with mathematics, and he thus avoided the metaphysical and theological controversies of his predecessor by making Angers a center of modern mathematical instruction. Numerous French mathematicians cut their teeth under Jaquemet's supervision, and some, like Father Charles Reyneau and Malebranche's secretary Louis Carré, remained affiliated with the Oratory as a result of its support for advanced mathematical work. Christianity leavened with mathematical reasoning exemplified Oratorianism, and Lamy, Jaquemet, Reyneau, and Carré represented the order generally in their devotion to modern mathematical science. Given this orientation, Oratorians, in contrast to the members of other religious orders such as the Jesuits, also enjoyed close, institutional ties with the French scientific academies, including the Royal Academy at Paris. Several Oratorians became royal academicians, and they were especially well represented among the academic mathematicians. The links between academic mathematics and the order as whole were also strong.[4]

Malebranche epitomized the Oratory in this respect, even though he neither served as an Oratorian priest nor taught in one of its colleges. Instead, he used the order's liberal disciplinary regime to make the Maison de l'Oratoire in Paris his home and material base for an independent intellectual life. His great masterpiece, *De la recherche de la verité*, which first began to appear in 1674, was produced within the Oratory in this way.[5] The book was in no way an official product of the order, yet through its conformity with its general mission the book became a philosophical touchstone illustrating what French Oratorian Christianity was all about. Supported by the Oratory and its intellectual and educational networks, the book also became a widely read and influential text in France and abroad.

La Recherche is a massive, Baroque treatise that eventually comprised six long books and an appendix of seventeen elucidations when its many revised editions stopped appearing in 1712. At its core, however, the book is a philosophical manual for better living through reasoned Christianity and modern philosophy, the very focus of Oratorian pedagogy. "Error is the cause of men's misery," Malebranche writes in book 1, chapter 1. "It is the sinister principle that has produced evil in the world, . . . we may hope for sound and genuine happiness only by seriously laboring to avoid it."[6] Thus framed, Malebranche's treatise amounts to a meticulous discussion of human error followed by a presentation of the proper method for avoiding it.

Like Descartes, whose connection to Malebranche and Oratorianism will be discussed in the next section, the author begins by extending our doubt to everything that does not appear to our mind in a clear and distinct way. This leads to an entire book (the first) showing all the ways that embodied sensation deceives us. A second book documents how our imagination creates deceptions, and the third book opens the door to truth by discussing "pure understanding." But here, too, error is possible when the mind is not trained to see itself correctly. Only in the sixth and final book is the method for achieving an error-free truth fully revealed.

Reaching this goal requires a discipline that depends upon a full appreciation of the manifold proclivities toward human error dissected in the previous books. Humans are finite beings, Malebranche argues in Book VI, yet they partake of the infinite through their relationship to God. Restricting humans to their appropriate finite boundaries constitutes the most important discipline in Malebranche's philosophy, while constructing bridges between finite human minds and the infinity of God is the book's central scientific project. "The soul cannot perfectly know the infinite," the author declares, and the failure to adhere to this dictum is the source of many errors. In physics, humans are lead astray when they think that their finite minds can ever directly and completely account for the infinitude of creation. Malebranche's famous skepticism about physical causation follows from this conviction. He believes that finite minds are not capable of knowing directly what happens when, for example, two bodies collide with one another. Indeed, we cannot even know what bodies are in their own terms since they too are infinite creations of God. "It is a ludicrous thing," he writes, "to see people deny the infinite divisibility of matter. . . . The proofs are conclusive if ever any were," yet savants "invent some frivolous distinction against the demonstrations . . . and embrace the opposite position. They [then] defend it with the bombast and absurdity the imagination can always supply." Humans will only be saved from such delusions, Malebranche contends, by recognizing the fundamental weakness of their mental faculties, a frailty rooted in human finitude within an infinite creation.[7]

Skepticism of this sort anchored much of Malebranchian physics and metaphysics. It was especially crucial to the Malebranchian strand of Cartesian science, which resisted dogmatic explanatory theorization and supported judiciously reasoned empirical arguments anchored in mathematical explanation. For if finite minds can never know the infinite directly and completely, they can at least approach such knowledge through rigorous discipline. Here mathematical philosophy enters as a resource for humankind. Our finite minds can never fully comprehend the actual operation of God's infinite creation, but we can conceive the finite dimension of these operations and their finite relations. "No finite mind can comprehend [the] magnitudes [of motion] in themselves and taken absolutely," Malebranche avers. But because "extension, duration, and speed" are "finite magnitudes expressed by finite ideas," their "commensurable relations can be known exactly."[8] By approaching the infinitude of creation via the finite relations that are empirically manifest within it, limited human minds can obtain knowledge of nature by reducing phenomena to a calculation of their measurable, finite components. Or as Malebranche sums up, "Mathematicians proceed properly in the search after truth, . . . especially those who avail themselves of the algebra and analysis revived and perfected by Viète and Descartes."[9]

Book VI, "Method," synthesizes this agenda and articulates it as Malebranche's philosophical and scientific program. Seeking to "render the mind as perfect as it can naturally be," Malebranche lays down "the rules that [the mind] must observe in the search after truth."[10] Descartes's conception of rational certainty is foundational since Malebranche makes it his central rule that "we must give full consent only to those propositions that appear so evidently true that we cannot withhold our consent without feeling inner pain and the secret reproaches of reason."[11] By adhering only to knowledge of this sort, we eliminate error and its miseries. But we do not by this rule alone maximize our understanding. Building scientific knowledge requires a different discipline, and here mathematics enters as a constructive method for building such a science. As Malebranche states: "Truths, then, are only relations, and knowledge of truths the knowledge of relations."[12] The French word for relations here is *rapports*.

By truth, Malebranche means truth as attainable by humans, for God is nothing less than that being which can see all the infinite relations of the world at once. Humans can never attain such knowledge, but by using the divine reason provided to them, they can construct a calculus of finite relations, a tool that makes possible a passage between human finitude and God's infinite knowledge. Algebraic analysis, especially the expansion of it that Cartesian mathematics provides, is the supreme tool for making this ascent toward knowledge. "Truth is nothing else but a real relation," Malebranche argues, and what mathematical analysis provides is a rational science for understanding these *rapports* in general terms.

Consider the following example. We cannot know the nature of a body in itself, or the character of its motion in absolute terms. But we can understand the finite relations exhibited by natural bodies in motion, such as extension, duration, and magnitude. These finite relations are reducible to simple, measurable terms—namely degrees of quantity and measures of more or less. In the case of extension, notions of bigger and smaller are accessible and comprehensible to us, and in the case of duration and magnitude, one can think of relative changes of speed over time. Absolute understanding is discarded in such a conception, replaced by a relative view of things conceived in terms of the differential *rapports* accessible to finite human observers. "Relations between ideas are the only ones that the mind can know infallibly," Malebranche writes, and thus by reasoning in terms of these relations, the relations between these relations and other relations, and so on, we can build a scientific understanding of nature.[13]

The power of mathematics resides precisely in its ability to do this kind of relational work, and for this reason "arithmetic provides," for Malebranche, "the means of expressing all the simple and complex relations" that are necessary. Or as he states elsewhere: "This is what is perfectly achieved through arithmetic and algebra, for these sciences show how to simplify ideas in such a way and to consider them so methodically that though the mind might have but little scope, it can with the help of these sciences discover very complex truths that at first seem incomprehensible."[14] Malebranche here is updating Viète's claim that algebra possesses a universal reach that can bring all of mathematical knowledge under its purview. A complex hierarchy of mathematical practice actually informs Malebranche's precise thinking. At first geometry, the science of spatial magnitudes as they appear to our senses, helps us to extract abstract relations of magnitude from the empirical details of spatial phenomena. When we reduce physical phenomena to geometrical diagrams, says Malebranche, it "supports the mind's perception" by "representing the truths that can be discovered about the matter at hand in a perceptible and convincing fashion."[15] Having abstracted motion in this way, the science of geometry then allows us to move to a new level of discernment by "showing us how to compare lines in order to determine their relations." Geometry thus conceived is a "universal science that opens the mind, makes it attentive, and gives it skill to control the imagination and draw from it all the help it can give."[16]

Geometry is thus a powerful first step, but it is less general in its scope and thus less powerful than other, more universal mathematical conceptions. More potent than geometry is arithmetic, which "provides the means of expressing all the simple and complex relations possible between magnitudes. It shows us how to perform the calculations that deduce these relations from one another, and to discover the relations of magnitude that might be useful by means of those already known. It also shows us how to do this with skill, clarity, and a remarkable exploitation of the mind's meager capacity."[17] Arithmetic thus takes the relations of magnitude provided by geometry and then establishes a more general account of them. Algebra and analysis, which Malebranche calls with Viète "very different from arithmetic," do the same thing at an even higher level. "They simplify ideas in the simplest and easiest way conceivable. What can be done with arithmetic only in a great deal of time can be done in a minute with algebra and analysis, and without the mind becoming entangled in lengthy operations. An individual arithmetic operation discovers only one truth, while a similar algebraic operation discovers an infinity of them."[18]

Approaching ever closer to the infinite is precisely the goal of Malebranche's method, and for him algebraic analysis is the royal road to this ultimate destination. Algebra provides part of the reward through its reduction of complex relations to the simplest of characters, the alphabet. "On these lettered magnitudes can be performed all the calculations needed to deduce the most difficult and complex relations. . . . These are the simplest, easiest, and at the same time most universal calculations conceivable. They reduce to simple and general expressions of only a few letters the solutions to an infinite number of problems and often even of whole sciences."[19] Analysis takes algebra a step further. It is "the art of employing the calculations of algebra and arithmetic in discovering all we want to know about magnitudes and their relations," which is to say, all we want to know about truth itself.[20] Analysis, therefore, unifies the subsidiary sciences of relational magnitudes, combining them into one universal calculus. Properly practiced, it allows finite human minds to approach the infinitude of creation with as much precision and as little error as is humanly possible. In recent times, Malebranche concludes, two new inventions, the differential and integral calculus, have "extended analysis beyond limit, as it were": "For these new calculi have placed an infinity of mechanical figures and problems of physics under its jurisdiction," and "they have provided it with a means of expressing infinitely many small elements that we can conceive as composing the circumference of curved lines, as well as the areas of figures and the volume of bodies defined by curved lines; they have provided it with a means of answering in a simple and general way by calculating the expression of these elements, problems whose solutions are both useful and the most complex that can be stated in geometry."[21] In short, says Malebranche, the new infinitesimal calculus has made the universal science of analysis even more universal. This brings us back to 1691, and the fortuitous meeting of Johann Bernoulli and the Marquis de l'Hôpital that Malebranche helped to facilitate.

It was the differential and integral calculus referred to above that Bernoulli brought with him to France, and l'Hôpital was not the only one to receive the gift of their introduction. Malebranche learned the new mathematics from Bernoulli, and after his visit with the Swiss visitor he became a student and practitioner of the new calculus, as both the differential and integral calculus came to be called together. Others acquired the new mathematics from Malebranche, including the Oratorians, who possess a manuscript from this period containing a set of lessons in the integral calculus attributed to Bernoulli. Malebranche's secretary and Oratorian *confrère* Louis Carré was one link in the chain of dissemination since the manuscript notes appear to be in his hand.[22] He certainly learned the new calculus at this time, and when Leibniz abandoned his project of publishing a book on the integral calculus, Carré filled the gap, publishing *Method for the Measure of Surfaces . . . by the Method of the Integral Calculus* in 1700, a textbook appropriate for the Oratorian colleges.[23] A letter from the Oratorian Father Jaquemet dated April 27, 1692, also advises his colleague Reyneau at the College of Angers about the latest mathematical discussions in Paris. Varignon, l'Hôpital, and Malebranche are all mentioned (along with Barrow, Wallis, Fermat, Apollonius, and others). In one telling passage, Jaquemet notes that the only thing different about Reyneau's solution to a problem, which he had previously sent to Jaquemet, was that he had not yet "applied to it the *incommensurables* that M. Newton and

Leibniz have the honor of inventing, and which are claimed to be one of the most beautiful and useful inventions of this century."[24] If Reyneau did not know about the new infinitesimal methods of Newton and Leibniz in spring 1692, he learned about them quickly, for over the next two decades he published his own influential textbooks that attest to the rapid dissemination of the new calculus within the Oratorian order in France.[25]

The spread of the new mathematics was also given a wider philosophical impetus by the esteem granted to Malebranche's *Recherche*, and by the wider vogue for Malebranchian philosophy that the book created. Lamy's *Rhétorique, ou l'art de parler*, which was reissued in its fourth and final edition in 1699, was indicative of the new possibilities that this cultural fusion offered. In a set of chapters at the end of the third book, Lamy explored "the marvelous sympathy that exists between our soul and numbers." Here, in a discussion that echoed *La Recherche* in many respects, while also reviving Renaissance Neoplatonism, Lamy connected our aesthetic appreciation of sound to the differential relationship of our bodies and our minds. "Since every movement which is made in the sense organs . . . is tied by the Author of nature to a certain movement of the soul," wrote Lamy, "sounds can excite passions. It is this liaison that is the cause of the sympathy that we have with numbers, and which makes it true that we feel different movements in the tone of everyone who speaks. A languid tone inspires sadness, while an elevated tone inspires courage; among songs, some are gay while others are melancholy according to the passions they excite."[26]

The art of speaking for Lamy involved a differential calculus that matched appropriate spoken tone with the passions that the speech desired to create. He did not go so far as to suggest an actual mathematics of beautiful speech, but, quoting Cicero, he stressed that "a speech is agreeable when the timing of the pronunciation of its syllables follows an exact mathematical order." "In geometry," he writes, "all the exact reasons are called 'number to number reasons.' This is why the masters of the art of speaking called by the word 'number,' or *numeros*, everything that the ear conceives as well proportioned in the pronunciation of a speech. . . . *Numerosa oratio* means in Latin what we in French call harmonious discourse."[27] To connect Malebranchian mathematical analysis to the art speaking, as Lamy does here, was to make it relevant to a host of general cultural concerns in turn-of-the-seventeenth-century France. Oratorian pedagogy was one place where such ideas struck wider chords, since the congregation's conviction that proper instruction in rhetoric cultivated a more ethical and civil character was in step with more widespread beliefs. Throughout Europe, but especially in France, the equation of reasoned eloquence with *honnêteté* was also foundational to elite self-fashioning.

Viewed from this perspective, Lamy's work suggested an important role for both advanced mathematics and Malebranchian mathematical philosophy in the pursuit of these much-sought-after cultural goals. Religion also played a role, even if the alliance between Oratorian Christianity and Malebranchian mathematical philosophy is probably best seen as a naturalizing, if not secularizing, impetus within late seventeenth-century French Catholicism. The mild regimen required of Oratorian clerics made it easy for them to imbed their religious convictions within a worldly lifestyle, and when Fontenelle noted in his eulogy of the academician Renau how the academician's study of Malebranche's *Recherche* had cultivated in him a spirit of piety and restrained comportment as a participant in academic assemblies, he was connecting the secular morality of Malebranche's mathematical approach to righteous living with the Christianity preached by the Oratorians.[28]

Yet this is not to say that there was not a religious, and even a theological, element to Oratorian mathematical philosophy. The order joined with many Christians across the European confessional spectrum in condemning modern philosophers for hubristically claiming to possess a direct understanding of God's infinitude. Malebranche's *Recherche* challenged precisely this heretical conceit, for to exaggerate limited human powers when claiming to know God was, Malebranche argued, to fall into a delusional discourse of error. Devoted seventeenth-century Christians of all stripes shared this same outlook, and from this perspective Malebranche's insistence, which was an Oratorian insistence as well, that claiming to understand the infinitude of God was sinful arrogance was a deeply held, and widely shared, theological position. Such views, however, were not undermined by Malebranche's equally deep conviction that advanced analytical mathematics offered devout Christians a pious resource for limiting our proclivity for error. Oratorian mathematical philosophy and reformed Christianity were thus joined into one indivisible conception of righteous, modern thinking in Europe around 1700. The eighteenth-century Italian Catholic, humanitarian, and mathematician Maria Caetano Agnesi embodied this unity completely, combining work as an analytical mathematician, including authorship of a widely influential textbook on the differential and integral calculus, with a life of Christian service to the poor and ailing of Milan. Both vocations, and especially their combination, were inspired by Oratorian examples.[29]

Oratorian Christianity thus joined seamlessly with Malebranchian mathematical philosophy in the decades around 1700 to create a melody in tune with the wider modernizing currents of the time. Other aspects of modern French culture at the end of the seventeenth century also harmonized with these trends, and the pioneering worldly periodical *Le Mercure galant* attests to this wider influence. During these years, the journal, which played an instrumental role overall in creating elite perceptions of taste and quality through its reflection and refraction of the wider culture, began to include mathematics within its pages. In this way it also contributed to the wider vogue for analytical mathematics and Malebranchian mathematical philosophy current at the time.

Numerical games and puzzles, such as those collected in Jacques Ozanam's frequently reissued *Récréations mathématiques et physiques*, first published in 1694, were most prevalent in the journal.[30] Systems for creating and cracking numerical codes were also popular, as was the posing and solving of numerically constructed enigmas. Material such as this fit comfortably within the *Mercure*'s overall project of encouraging entertainment and pleasure, yet these diversions also encouraged other, more subtle cultural shifts. Solving mathematical puzzles, for example, put readers of the journal into the habit of thinking about language in terms of number and mathematics. They also introduced the journal's elite clientele to the numerical qualities of ordinary written and spoken speech. This supported the wider interest in rhetoric, language, and civility by providing a powerful cultural motive for linking seventeenth-century mathematical philosophy with wider ideals of elite comportment and sociability. Linkages such as these made the *Mercure* an important site where advanced mathematics and Malebranchian mathematical philosophy came together with the practices of elite sociability and *honnête* self-fashioning.

Fontenelle demonstrated his distinctive excellence at grasping the cultural trends of the moment by publishing a brief "Memoir on the Number 9" in 1685 in Bayle's *Nouvelles de la république des lettres*, an article that was indicative of the cultural fusion just noted. The text unfolded the many mathematical features of the number in question while exploring "the knot" that tied "the real nature of numbers" together with their "arbitrary nature rooted in the institution of Arabic numerals."[31] Leibniz would attract even greater interest in 1702 when he proposed a system of binary arithmetic, arguing that a numeric system comprising only 1s and 0s offered a more powerful language for capturing the deep link between number and philosophical reason (a sentiment shared by many computational linguists today). He also explored the connections among mathematics, language, and universal reason by drawing analogies between binary arithmetic and the characters of the Chinese language.[32]

Explorations such as these mirrored Malebranchian mathematical philosophy in many ways but, more important, they also resonated with the wider cultural interest in the relationships among language, nature, and universal truth. The *Mercure* ran dozens of pieces touching in one way or another on these questions, and many explicitly connected the interest in proper language with elite preoccupations regarding *honnêteté*, sociability, and worldly pleasure. The cultivation of civilized manners was in fact a central preoccupation of the *Mercure*, so when the journal devoted more than one hundred pages over several issues in 1697 to the publication of a treatise on algebra, one that introduced the rudiments of the differential calculus, a complex set of motivations illustrative of this cultural climate were at work.[33]

The publication of this short mathematical essay spoke on the one hand to the journal's interest in promoting the latest scientific work, a motivation that also supported its habit of reviewing new mathematical books and reporting other scientific activities of contemporary interest. Yet framed as a letter from a learned "M. L . . ." to the "Comtesse de M . . . ," the *Mercure*'s "Ouvrage concernant l'Algebre" also confirmed the new connection described above between mathematics and elite self-fashioning in 1690s France. This was a concern that the journal was always eager to exploit, and reviewing the royal academician Thomas Fantet de Lagny's *Nouveaux élémens de mathématiques et d'algebre* in February 1698 (was he the "M[onsieur] L . . ." of the previous citation?), the journal acknowledged the current vogue for mathematics, calling it "*très à la mode* at the moment, especially with women."[34]

The reference to women here, and in the address to a countess in the letter on algebra, was especially important since the journal offered a space that emphasized male-female dialogue in the formation of elite culture.[35] Simple pleasure was also at play, for elite society in late seventeenth-century France aspired most of all to combine well-mannered civility with the free pursuit of urbane entertainment. Fin-de-siècle French analytical mathematics was shaped by this dynamic. The eighteenth-century historian of mathematics Montucla notes, for example, that in the wake of the 1696 publication of l'Hôpital's *Analyse des infiniment petits*, the "frivolous people of Paris" created a piece of "musical theater" (*un vaudeville et un air*) titled *Infiniment petits* that turned the vogue for the new mathematics into comedy. "I have done everything possible to try to find the words," Montucla wrote, but all he could report was that the songs "made jokes about the frail health of the Marquis de l'Hôpital, and about the strong distaste of his wife, the marquise, for the new geometry."[36]

When it staged a set of worldly, heterosexual exchanges with advanced mathematics at their center, the *Mercure* was making algebraic analysis a participant in the pleasurable, mixed-sex sociability that many saw as central to *mondain* taste and refinement. Fontenelle's *Entretiens sur la pluralité des mondes habités* had demonstrated how elite values could be cultivated through mixed-sex scientific dialogue, and the publishing practices of the *Mercure* were devoted to exactly the same agenda. The analytical mathematics that the journal published in the 1690s simply confirmed that this sort of mathematical learning could also, and should, participate in these culture-defining exchanges. The Oratorian Louis Carré also illustrated the importance of this new constellation by beginning a career as a popular mathematical tutor in the 1690s, one especially sought out by elite women. Carré's "secret little empire" of disciples, as Fontenelle called them, included both worldly women and the religious, and his success in the 1690s illustrates the ties that bound advanced mathematics and Malebranchian philosophy to urbane society as a whole. The same links were reinforced in countless other ways during the same period, and together they reveal the striking—peculiar, even—affinity between the newest work in analytical mathematics, as framed by Malebranchian philosophy, and the most important cultural dynamics of the day.[37]

Malebranche and French Cartesianism

In his *éloge* for Father Reyneau, Fontenelle stressed the Oratorian's modernism with reference to Malebranche, analytical mathematics, and the many other features of fin-de-siècle French culture just discussed. But he also connected it with the legacy of another thinker whose influence in shaping the "Malebranchian moment" was equally important. When teaching at Toulon, Pezenas, and Angers, Fontenelle explained, Reyneau professed only "the new philosophy," and this orientation generated friction with those professors who were still "attached to the ancient scholasticism." He later reduced the "new philosophy" to a single label, "Cartesianism," adding that "he could not have been a Cartesian, or, if one wants, a modern Philosopher, without also being a bit of a mathematician [*géomètre*]."[38] Fontenelle's equation of Cartesianism with modern philosophy, and both with innovative mathematics, and his use of both to encapsulate Reyneau's intellectual sensibilities, was indicative of the entanglement of each in the Malebranchian moment in France.

Cartesianism was integral to Malebranche's philosophy, and to the wider Oratorian movement more generally, yet the term meant many things in turn-of-the-seventeenth-century France. Four aspects of Descartes's legacy were most important in making the Cartesianism that was influential in the decades around 1700. First was the influence of his analytical, algebraic treatment of geometry, an outcome that made Descartes into a historical figure responsible for pioneering the new approach to mathematics that had modernized, and thus surpassed, the traditional geometry of the ancients. Second was the influence of Descartes's precise system of natural philosophy, especially his vortical cosmology and its models for explaining terrestrial gravity and the motions of the planets. To be a modern Cartesian in this sense was to be a defender of rationalist mechanical explanations and an enemy of the "occult natural philosophy" of the Scholastics. A third meaning of Cartesianism derived from Descartes's particular manner of reasoning. To be a "modern Cartesian" in this third sense meant defending the value of skeptical doubt as the starting point for any program of knowledge building. It also meant an urge to deploy rational skepticism as a solvent undermining ill-reasoned fantasies and dogmatism. This understanding allowed many to identify themselves with the label "Cartesian" even when they disagreed with the explicit propositions and theories that Descartes defended. Fourth and finally, the label "Cartesian" was also used loosely as an epithet to condemn any and all overly zealous advocates for modern reason. This broad-brush use of the term fit with one definition of the philosopher current at the time, namely the one that made him "a man who by practicing freethinking [*libertinage d'esprit*] puts himself above the ordinary obligations and duties of civil life." The philosopher, according to this antimodernist understanding, was a subversive, and from this perspective a Cartesian was nothing more than an intellectual rebel and troublemaker.

Each of these four aspects has its own history, but in turn-of-the-seventeenth-century France the semantic field joining them was so completely entangled as to make any attempt to differentiate them impossible. The permeability of the meanings also allowed the label "Cartesian" to circulate simultaneously as a creed, a herald, and an epithet without any one of these working to crowd out or nullify the others.[39] The label was invoked everywhere as a mark of both pride and disdain. Its use also carried real political punch, as the career of Bernard Lamy showed. Yet Cartesianism remained many things to many people, and while it is certainly accurate to talk about the widespread, even overwhelming, influence of Cartesianism in France around 1700, it would be grossly misleading to say that all French men and women who identified with Cartesianism (and French *Cartésiennes* were indeed some of its most important advocates) possessed the same intellectual outlook.

The best way to trace the history and influence of seventeenth-century

French Cartesianism, therefore, is to focus on the contestation itself. The religious unease and controversy of the period, especially that activated by the revocation of the Edict of Nantes granting religious toleration to Protestants in 1685, were one source of Cartesian controversy. Descartes's writings were placed on the index of banned books in the early years of Louis XIV's reign (1663), and in addition to his prosecution of Lamy in 1675, the Sun King shut down Pierre-Sylvain Régis's popular Cartesian lecture course in 1680, forcing the philosopher into exile. The king in 1685 further sanctioned a ban on teaching Descartes at the University of Paris. Each of these actions stemmed from Louis XIV's pursuit of an increasingly reactionary religious policy, and these agendas supported the rise, whether justified or not, of the idea that Cartesianism stood for everything that this religious policy was intended to root out. In 1706, the Society of Jesus joined with the crown in this effort, prohibiting a number of explicitly Cartesian propositions from being taught at Jesuit colleges. Louis XIV supported all of these anti-Cartesian initiatives.[40]

As was always the case, persecution did not stifle the movement so much as channel it into safer and more welcoming abodes, and attitudes about "Cartesianism" in this period are best placed along a spectrum. At one end were the confident mechanical materialists who made a certain, mathematical demonstration of rational causality a reason to discard God's presence and action from the world altogether. At the other end stood a host of equally vigorous Cartesians who proceeded by trying to hold mathematical rationalism together with Christian religion and metaphysics. The intensity of the late seventeenth-century religious struggles in France put special pressure on this latter group, and an important strand of French Cartesianism emerged in this context, one crucial to Varignon and his analytical mechanics. It worked from the premise that Cartesian philosophy and natural science could exist in a space isolated from, and irrelevant to, metaphysico-theological contestation. Cartesian mathematics was most easily divorced in this way, and many began to use the label "Cartesian" after 1650 to define their mathematical commitments alone. Harder to divorce was Cartesian physics, since it depended on the matter theory and philosophical dualism central to the wider metaphysical and theological disputes. Yet one of the important developments in this period was precisely the emergence of an avowedly nonmetaphysical and nontheological conception of Cartesian physics in France, one that Varignon adopted in his own scientific work.[41]

Most important in articulating this position was Jacques Rohault, whose popular "Cartesian Wednesdays" and widely read textbooks did much to disseminate Cartesian physics within French society as a whole.[42] Bignon was one of many who acquired his understanding of Cartesian science through his attendance at these public sessions. For those who learned Cartesian natural philosophy from Rohault, science was not grounded in strong claims about philosophical certainty, or the theological and metaphysical disputes inevitably attached to such thinking. Cartesianism was grounded in mathematically demonstrable explanations of the contingencies of empirical phenomena. Physics for Rohault only makes progress once it accepts that it can never achieve absolute certainty. As he wrote in the preface to his *Traité de physique*, first published in 1671, and subsequently reissued and translated into several languages in a way that made it arguably the most influential physics textbook in Europe in the second half of the seventeenth century, the problem with traditional natural philosophy was its "overly metaphysical manner" and its tendency to get hung up on "abstract questions."[43] Physics should avoid vain disputations, Rohault asserted, by reducing physical questions to empirical particulars, which can then be examined by experiments. Experiments, however, also needed to be joined with philosophical reasoning, he claimed, because "to wholly discard reason and yield all up to sense" is to "contract our knowledge into a very narrow compass."[44] Modern natural philosophy accordingly eschewed contentious and ultimately distracting metaphysical conundrums and searched instead for rational accounts of empirical phenomena confirmed by experiments.

From this perspective, Cartesian natural philosophy became for Rohault the best empirico-mathematical account of nature available. What he found in Descartes's writings was less a philosophically certain metaphysical system than a set of empirical and mechanical models for explaining natural phenomena, models that had an exceedingly high degree of rational plausibility. Rohault ultimately adopted Descartes's conception of matter, his arguments against the vacuum, his abandonment of the Aristotelian qualities, his belief in the rationality of point-contact mechanisms, and his system of the vortices, and Rohault made these principles the basis of his teaching. But he did so while also avoiding the metaphysical and theological conundrums that each of these positions provoked. He adopted his stance, moreover, in an assertive way by arguing that such metaphysical questions were not relevant to *physique* as he understood it, and were in fact detrimental to its modern advancement. Such a stance agreed with certain strands of Descartes's own pronouncements about method, even if it disagreed with others, and his particular orientation was especially important for the future development of Cartesian physics, because it left room for newer and more empirically powerful explanations to replace those that Descartes himself had suggested. It also detached physical explanation from any claim that physics depended upon an underlying metaphysics.

Régis learned his Cartesianism from Rohault, and he, too, defended this brand of rationalist empirical Cartesian science in his public lectures and writings. Fontenelle did the same, as did Varignon, especially when he advanced a set of "new conjectures about gravity" in 1690 that employed Descartes's vortical mechanisms. Even more indicative of this stance was his abandonment of these theories soon after, when devastating empirical objections were raised against them.[45] When Cartesians began to enter the Royal Academy in the 1690s, they brought this brand of Cartesianism with them while evincing no attachment whatsoever to the metaphysical controversies that were still swirling around topics such as the nature of matter and its causal determinants. The institutional imperatives of the Royal Academy, which insisted upon a clear detachment from theological and philosophical disputes, only reinforced the divorce between science and metaphysics suggested by Rohault. The Academy also professed a resolute independence that proscribed any attachment to particular philosophical systems. Such an ethic was central to the wider identity of the independent man of letters in the period, and by adhering to it the Royal Academy confirmed its place within the Republic of Letters. Thanks to the brand of Cartesianism personified by Rohault, Régis, Varignon, and Fontenelle, royal academicians could sustain this crucial independence while also remaining self-proclaimed Cartesians as well.

In this way, many royal academicians declared themselves Cartesians during these years while intending no polemical or partisan intent in the declaration whatsoever. Persuaded that scientific reasoning and Cartesianism were one and the same, and that Cartesian physics offered the best account of the physical world available, they defended its principles because they considered them to be the most empirically grounded, experimentally justified, and philosophically well reasoned available. Similarly, by committing themselves more to the imperatives of Cartesian critical reasoning than to any of the precise scientific or philosophical doctrines that Descartes maintained, they remained open to revising their theories when solidly reasoned alternatives were offered. They likewise accepted many alternative views without any regard for the precise relation to Descartes and his writings found within them. Cartesians of this sort were anything but a philosophical sect, therefore. They employed the self-identifying label "Cartesian" only because they adhered strongly to Descartes's manner of philosophizing and believed (for now at least) that his physical theories were the most plausible ones available.

Methodological and empirical Cartesianism of this sort united savants in France and elsewhere, including many who were uncomfortable employing the self-identifying label "Cartesian." Leibniz, for example, spent much of his life challenging basic doctrines of Descartes's metaphysics and physics, and he would never have invoked the label "Cartesian" to describe his own philosophy. Yet in philosophical style and outlook he was at one with many French Cartesians, and they were in turn receptive to his views. In a similar fashion, the Cartesian Jansenist Antoine Arnauld fought many metaphysico-theological battles with Cartesians such as Malebranche, while Malebranche in turn fought many battles over Cartesian metaphysics and physics with Cartesians such as Régis and Fontenelle. Yet amid these Cartesian civil wars, a unifying strand of method joined each antagonist in a way that solidified rather than fractured the Cartesian unity of the whole.

The uniformity of methodological Cartesianism was further cemented in 1683 when Arnauld and his fellow Jansenist Pierre Nicole published the fifth and final edition of their jointly authored *Logique, ou l'art de penser*. This "*Port-Royal Logic*," as it was more commonly called, turned Descartes's method of reasoned doubt and rational demonstration into a paradigm for correct thinking as a whole.[46] The book became the most influential philosophy textbook of its day, and in 1692, when the Irish savant William Molyneux cited it together with Locke's *Essay concerning Human Understanding* and Malebranche's *De la recherche de la verité* as one of the three works most responsible for "liberating logic from the shackles of scholasticism," he was demonstrating how Cartesian method had become a vehicle for unifying even the most disparate systems of theological metaphysics and philosophy.[47]

Viewed from Molyneux's perspective, Malebranche was both an emblematic and an idiosyncratic French Cartesian. He was idiosyncratic because he developed a singular interpretation of Cartesian metaphysics with direct theological implications and defended it vigorously (as a cleric no less) throughout his life. He was at the same time emblematic, however, because he wrote in such a way that made these precise metaphysico-theological commitments compatible with his status as an exemplar of Cartesian scientific reasoning as a whole. As a result, "Malebranchianism," which also became a label of self-identification (and derision) in the seventeenth century, carried with it the same double business that the more general label "Cartesianism" carried.[48]

For many, Malebranchianism referred to his controversial doctrine of "occasional causes," or his conception of causality that denied any real action to physical bodies while locating all, actual causal change in the "occasion" of God's direct intervention in the world. His so-called doctrine of the "vision of all things in God," which held that our ideas and mental perceptions are experienced directly through God, was similarly controversial. Many found these views philosophically problematic, and because they challenged important tenets of religious orthodoxy they also got Malebranche into theological hot water. He was forced to defend his doctrines on several occasions before the ecclesiastical authorities, and several of his works were placed on the Roman Catholic Index. One, his *Traité sur la nature de la grace*, remained there until the institution was disbanded in 1966.

Defending theologically laden philosophical positions such as these was central to Malebranche's own self-conception as an intellectual. The irony, however, is that these vigorously fought public positions often exerted less influence on those who adopted Malebranchianism as a credo than other aspects of his thought. Fontenelle, for example, became a strong advocate of Malebranchianism, but in 1685 he published a paradigmatically Cartesian essay of "doubts" about the system of occasional causes in which he argued that the impenetrability of bodies offered a foundation for real impacts between them, and thus real physical causes.[49] Such a nonmetaphysical approach to Malebranche's theologically laden metaphysics was typical of him, and other like-minded Cartesians approached Malebranche's thought in the same way. Régis's *Système de philosophie*, which appeared in 1690 and marked the philosopher's return from exile, further challenged Malebranche's philosophy on empirical, physical grounds. Like Fontenelle, he thought that real causes could be demonstrated physically, and he challenged the Malebranchian doctrine of vision in God by offering a scientific rebuttal of Malebranche's account of why the moon varies in size as it rises above the horizon.[50]

Leibniz offers a more complicated case since he was, like the Oratorian father, deeply invested in theological metaphysics. Yet while he believed that his system of the "preestablished harmony" offered a better account of the mind/body relation than Malebranche's theory of occasional causes, his major disagreement with him was on a question of physics. Malebranche sustained throughout his life the Cartesian definition of matter as pure extension, while Leibniz argued that matter must contain a substance beyond extension and that Malebranche's mechanics, especially his laws of impact, were erroneous. His debate with Malebranche on this point, which occurred in both published works and private correspondence, was important, for it revived the German's earlier challenge to Cartesian physics and kept it alive in France into the eighteenth century, when it resurfaced as the so-called *vis viva* controversy. More relevant to the context of late seventeenth-century France, however, is the evidence that even metaphysico-theological thinkers such as

Leibniz could find themselves debating Malebranche in ways that isolated physics from metaphysics and theology.[51]

For someone like Fontenelle, who like Varignon had no investment at all in theologically laden metaphysics, such a stance was even easier to adopt. It was in fact through a hybrid distillation of Malebranche's overall philosophy, one that detached theology and metaphysics from it while emphasizing the Cartesian mathematical reasoning that was its anchor, that he became a Malebranchian. Especially in his academic *éloges*, the Academy secretary continually referenced Malebranche as one of the founding figures of the modern complex of mathematical scientific philosophy that he perpetually defended. His "Éloge du père Malebranche," delivered in 1715, illustrates the point.[52] Narrating, as he did in all of his *éloges*, the trajectory of Malebranche's intellectual development, Fontenelle made the Oratorian's encounter with Descartes's *Traité de l'homme* in a Paris bookstore at the age of twenty-four a conversion experience that forever changed his thinking. Malebranche's *Recherche*, and the various metaphysico-theological disputes that it triggered, then ensued. Yet in Fontenelle's telling, the thread guiding Malebranche through all of this contentious work was the application of Cartesian reason. He also made a point of distancing those struggles from Malebranche's work as a scientist, writing that while he was a metaphysician and a theologian, "In these qualities he was a stranger to the Academy of Sciences, which would recklessly overstep its boundaries were it to treat theology, and which abstains completely from metaphysics since it has little tangible utility."[53] Fontenelle completed his eulogy by stressing the value of Malebranche's work in geometry and physics, work that served as "steps toward his metaphysics and theology . . . and all his most sublime speculations."[54] In sum, he distanced Malebranche the mathematical scientist from Malebranche the controversial theologian and metaphysician, while nevertheless making the first the source for all that was good and true in the second. In this way, he defined a scientific Malebranchianism that was distinct from and superior to the other aspects of his thought.

Fontenelle especially celebrated this brand of Malebranchianism when narrating the lives of academic mathematicians, a group that was influenced more than most by the Oratorian father. Fontenelle was also a member of this small but influential mathematics club, and his "Éloge de M. le marquis de l'Hôpital," delivered in 1704, was illustrative of his deeply held proclivities. It explained how Malebranche's *Recherche* had served l'Hôpital as "an excellent guide to the sciences," a book that the marquis "took counsel from, used practically, and developed a friendship with that lasted all his life."[55] At one level, the positive link that Fontenelle routinely drew between academic mathematics and Malebranchianism was simply a result of biography since few mathematicians ascended to prominence in France after 1690 without some connection to those in the circle of the Oratorian father. This very fact, however, speaks to the wider influence that Malebranche had on the mathematical thinking of the day. Mathematicians, especially those with connections to the Oratorian order, found in Malebranche's *Recherche* not a reservoir of Baroque metaphysico-theology, but a model of Cartesian mathematical philosophy. Fontenelle and his mathematical colleagues read the book this way, and in doing so they helped to contribute to a wider vogue for Malebranchianism defined as a special mathematical form of Cartesian natural philosophy, one with a strong attachment to advanced analytical mathematics.

The deeply rooted cultural vogue for this form of Malebranchianism in France in the decades around 1700 provided a crucial context for the genesis of analytical mechanics, and if forced to choose a single intellectual influence most important in shaping Varignon's innovative new science, fin-de-siècle mathematical Malebranchianism might win the prize. Whatever the singularity of its influence, however, it was certainly as crucial as any other influence in pushing forward these changes. Yet it was also because of the way that this peculiar Malebranchian culture of France around 1700 supported advanced mathematical and scientific thinking that this science came together when and where it did. Having thus documented the institutional and intellectual roots of the new science of motion, let us turn now to the final part of this book, which will examine in detail the actual genesis and reception of this new science, along with the controversies it provoked and the changes that it initiated.

PART III

Making Analytical Mechanics in the New Académie Royale des Sciences, 1692–1715

In July 1698, Varignon presented his paper "General Rule for All Sorts of Movements of Whatever Speed Varied Freely" to the members of the Académie Royale des Sciences.[1] Two months later, he delivered a second academic *mémoire* further advancing this research, and early the next year he delivered two more.[2] These papers laid the foundation for Varignon's new science of motion, which he subsequently developed in detail over the next decade. In 1698, this work also stood as a synthesis of the various strands of Varignon's scientific work during his first decade as a royal academician. The 1698 paper referenced a mathematical paper on logarithmic curves that Varignon had given to the Royal Academy in January 1693. The *mémoire* also deployed the new analytical approaches to geometry made possible by the differential and integral calculus, which Varignon had been developing since at least 1692. Varignon also began working on the mechanics of moving bodies during the same years, and Fontenelle would later point to Varignon's 1692 papers on the mathematical laws of falling bodies as a key starting point for his new science of motion.[3]

Boiled down to its essential innovation, Varignon's new science was novel in its explicit use of the new and controversial infinitesimal calculus to create a new and more powerful science of mechanics, one that possessed (or so Varignon claimed) unmatched capacities for capturing in general terms the motion of moving bodies with a new level of mathematical precision and economy. Yet if Varignon's innovations were anticipated by a decade of previous work in mathematics and mechanics, they were also shaped by the particular environment in which he worked, namely the French Académie Royale des Sciences. Varignon was the sole author of the papers that proved most decisive, but the influence of figures such as l'Hôpital, Malebranche, Fontenelle, and other advocates for the new infinitesimal mathematics in the Royal Academy were crucial influences upon his work. Also important was the new climate for academic mathematical research provoked by the vigorous managerial reforms initiated by the abbé Bignon in September 1691. Varignon had been a vigorous academician and a regular contributor to academic meetings even before the ascent of the Pontchartrain ministry, and in 1692 he became a welcome participant in Bignon's reform program. Preliminary work that in retrospect points toward the new science of motion appeared, for example, in the monthly Academy reports of 1692–93, and among the academic mathematicians few were as active as Varignon in contributing papers to these volumes. Overall, Varignon enthusiastically embraced academic science pursued à la Bignon, and the result was a series of papers, many recorded verbatim in the academic registers, that simultaneously marked Varignon as an exemplary academician while also laying the foundations for his later work in analytical mechanics.

Part 1 of this book looked in detail at the nature of these political and institutional developments, and part 2 looked at the intellectual influences that were equally significant. Both came together through Varignon to produce the new science of analytical mechanics after 1698. The chapters in this third and final section complete this book by exploring in detail the multifaceted history of Varignon's academic work. As a group, they try to bring together the institutional history emphasized in part 1 with the intellectual history stressed in part 2 in order to create an integrated history that treats academic politics and mathematical science as two dimensions of one inseparable historical whole.

The chapters proceed chronologically, combining institutional and intellectual analysis in each. Chapter 7 narrates the genesis of analytical mechanics in the decade before 1698. Chapter 8 looks at the contentious birth of analytical mechanics amid the comprehensive reform of the Royal Academy begun in February 1699. Chapter 9 continues the story by looking at the explosion of the initial debates about analytical mechanics into an open public controversy—*l'affaire des infiniment petits*, as it was called—a scandal that served as a crucible in which analytical mechanics was forged as a peculiar French science along with the new public Academy as new kind of royal institution for science. Chapter 10 completes the story by examining the steps that secured the consensual acceptance of analytical mechanics in France by 1710. This *pax analytica*, as I will call it, allowed analytical mechanics to become an unchallenged fixture of French academic science ever after. This settlement was not, I argue, the result of either a mathematical resolution of the intellectual debates or a reasoned agreement about the legitimacy of calculus-based mechanics as a science. No one before 1800 was able to remove the uncertainties that made the infinitesimal calculus so contentious, and debates about it erupted throughout the eighteenth century.[4] The victory of the French analysts by 1715, therefore, is best understood as a contingent and negotiated historical settlement.

A coda concludes the book by charting the emergence of a different understanding of Newton's *Principia* in France around 1710 and the initiation of a new set of scientific debates as a result of this new understanding, debates that would develop after 1720 into the midcentury "Newton Wars" that would fundamentally change the image of Newton and his *Principia* in France. The book ends by reflecting on the legacy of the bifurcated reception of the *Principia* in France—one strand mathematical and connected to calculus-based mechanics, and the other physical and connected to cosmology and natural philosophy—with respect to the practice of Newtonian mathematical physics in Enlightenment France, and our understanding of it today.

CHAPTER 7 *The Beginnings of Analytical Mechanics, 1692–98*

Accidentally, and viewed in retrospect, the summer of 1691 turned out to be a decisive period of genesis for analytical mechanics, a time when the crucial dynamics that would later converge in the creation of this new science first began to stir. In July, the royal minister Louvois died, setting in motion the ascent of the Pontchartrain ministry and its reform agendas for the Royal Academy. At the same moment, Johann Bernoulli was working with l'Hôpital and Malebranche, and through them others in France, in the dissemination of the new infinitesimal calculus. Between this time and the production of Varignon's breakthrough paper in July 1698, the intellectual syntheses that made this new science possible occurred. At the same time, and through the same work, Varignon also became a newly important academician in the newly administered Royal Academy. He also acquired new intellectual allies and rivals as a result of his efforts in both directions. His new science, announced in 1698, was anything but uncontroversial and grew directly from the tumult, both intellectual and institutional, that characterized these years; the tempest that it triggered, which was even more turbulent and transformative, continued into the eighteenth century. To elucidate this contentious beginning, this chapter will follow the steps that led to Varignon's crucial innovations of 1698, tracing them back to his activities inside the Royal Academy during the preceding decade.

A New Mathematics in a New Academy: The Initial Challenges, 1692–93

The first explicitly academic use of the new infinitesimal calculus by a French royal academician occurred in June 1693 in a *mémoire* authored by l'Hôpital for the monthly Academy reports that Bignon had ordered published starting in January 1692.[1] The paper cited the Leibnizian origins of the calculus explicitly, and then used the new method to solve a challenge problem posed by Bernoulli in the *Acta Eruditorum*. The paper also marked the marquis's debut as an academician, for he had only been appointed to the company a few weeks earlier. His arrival added another *géomètre*, as the French then called what we would now call a mathematician, to an Academy that had been, and would largely continue to be, dominated by chemists, botanists, anatomists, and the applied mathematicians who congregated at the Royal Observatory. The twenty monthly reports that the Academy issued before the initiative was terminated offer an interesting cross-section through which to situate l'Hôpital and his mathematics with respect to the wider culture of French academic science in 1693.

Except for Michel Rolle, who offered two papers on algebraic problem solving in the January and March issues of 1692 (his only contributions), l'Hôpital was alone among his academic mathematical peers in offering nothing but theoretical mathematical papers to the reports. This was true of all of his academic mathematical work as well, and in this way he continued, albeit in a newly innovative analytical way, the great French academic tradition of mathematical problem solving embodied in the work of the institution's founders, Roberval and Frénicle de Bessy. Consonant with his status as a blue-blooded aristocrat of the bluest sort, his work also exemplified the thoroughly liberal pursuit of mathematical science appropriate to a man of his station. Through correspondence with Bernoulli, Leibniz, Huygens (until his death in 1695), and other distinguished gentlemanly mathematicians, l'Hôpital also honed his mathematical skills through the pursuit of problems and projects shared within this lettered network. Especially important to it were the learned journals that exemplified this ethos, especially the *Acta Eruditorum* with respect to mathematics. In fact, after he became a royal academician, very little changed in l'Hôpital's life. He simply added to his network of gentlemanly mathematical correspondents another coterie of gentlemen interlocutors along with a new space for sharing his mathematical acumen with his esteemed peers.

Varignon was already working with the new infinitesimal calculus when l'Hôpital joined him in the Academy, and his correspondence with Johann Bernoulli, which began in September 1692, illustrates the dynamic that was created by l'Hôpital's arrival. In the very first letter, Bernoulli wrote to Varignon about the work of the abbé Catelan, and about l'Hôpital's rebuttal of it in the *Journal des savants*. In 1691, Catelan published a book on universal geometry that launched an attack on the infinitesimal calculus, and this had provoked l'Hôpital to publish under the name of Mr. G*** a rebuttal that appeared in the Parisian journal in 1692.[2] Catelan responded to l'Hôpital in the same journal two issues later, and l'Hôpital replied in the same, defending more strongly his original claim.[3] Bernoulli invoked the exchange in his initial letter, soliciting Varignon's approval for l'Hôpital's position. Because Varignon was in agreement the exchange opened a space for him to report back to Bernoulli about his work with l'Hôpital on the infinitesimal calculus.[4] As he wrote in June 1693, "I speak of your work [*problèmes*] to all the mathematicians that I know here, and I announce it openly to the members of the Academy so as to have the occasion to talk there about you. And when M. de l'Hôpital was not yet there, I used a session to throw the curve of a boat hull on the wall and to show how to determine the tangents to it."[5] In this way, Varignon worked to advocate for the calculus in the Academy at the same time as he worked with l'Hôpital and Bernoulli to perfect his own understanding of it.

L'Hôpital's mathematical contributions to the monthly academy reports revealed the continuation through him of the venerable French academic tradition of liberal, mathematical problem solving, along with the peculiar nature of this tradition when compared with the mainstream of French academic science as a whole. Of the almost ninety reports published between January 1692 and December 1693, roughly half were drawn from each of the two major domains of science represented in the company: *physique* and *mathématique*. This seeming parity, however, is deceiving given the way that work in astronomy and applied mechanics would be counted as mathematical according to the classifications of the day. Another publication from this period, which was also promoted by Bignon, illustrates the point. For in addition to the monthly reports, the Academy also published in 1693 a collection of old work that had been presented to the Academy but never published.[6] Varignon described the three folios to Bernoulli as containing "several pieces of geometry and arithmetic done by deceased academicians" (Roberval was featured prominently), "pieces in astronomy and selections from the work done at the Royal observatory," and "translations of several works in ancient Greek mechanics."[7] The latter was the work of de la Hire, who also played a key role in the production of this volume as a whole. Overall, this compilation illustrates well the dominant themes of French academic mathematics in the early 1690s.[8]

The overwhelming majority of the mathematical papers published were works of physico-mathematics, with astronomy and its related sciences being by far the dominant theme among the mathematical contributions. And as Varignon noted to Bernoulli, overall "there is more *physique* than *mathématique* in the *mémoires*."[9] Rolle's two reports offered in 1692 and l'Hôpital's three in 1693, were by and large the only work published that can be called theoretical mathematics plain and simple, and they illustrate how Rolle and l'Hôpital stood alone in the early 1690s as mathematicians pursuing abstract mathematical work. Varignon and de la Hire offered contributions in mathematical problem solving, but their theoretical pieces were the exception to what was otherwise a large body of work (and they were each among the most active contributors) that combined abstract mathematical problem solving with mixed physico-mathematics. De la Hire, as his translation projects suggest, was interested in the study of ancient geometric problems, and he contributed work of this sort on occasion. But he was even more active with Cassini and Sedileau in presenting the work of those at the Royal Observatory. Varignon's papers likewise focused largely on questions of mechanics. His reports ranged from empirical presentations to works in mechanical theory—a January 1692 report, for example, on the action of water in a vessel and a March 1692 *mémoire* on the nature of hardness in bodies. But he also pursued geometric problem solving such as in his July 1692 determination of the area of the heart-shaped space formed by a semi-ellipse turned around one of its diameters.[10]

Some of Varignon's monthly report contributions also foreshadow his later analytical mechanics, such as his *mémoire* contributed in December 1692, "Rules of Movement in General," which Fontenelle singled out in 1699 as a fundamental step toward Varignon's new science of motion.[11] Varignon's May 1693 contribution, "Rules of Accelerated Motion Following Every Proportion Imaginable" also points in this direction, as does another paper published the next month, alongside l'Hôpital's debut work, that used the Leibnizian calculus to make an "Application of the General Rule of Accelerated Motion to All Motions."[12] Varignon's May paper actually preceded l'Hôpital in deploying for the first time the differential calculus in a published academic paper, but unlike the marquis's work, Varignon's did not single out this usage or make a significant point of it. He also continued to shift between calculus-based explorations regarding mathematical mechanics and other, more overtly empirical and physico-mechanical interests.

However diverse Varignon's work from 1692 to 1693 was, what is evident in it is the new availability of the infinitesimal calculus in France, along with Varignon's use of this new mathematics in his continuing academic work in mechanics. The arrival of l'Hôpital in the Academy and his immediate use of and advocacy for the new Leibnizian algorithms were a key impetus for Varignon, as was his correspondence with Bernoulli about the new mathematics.[13] With l'Hôpital and Bernoulli as his allies and teachers, Varignon soon began to intensify his explorations of the new possibilities for mechanics that this new mathematics offered. He also joined with l'Hôpital in the pursuit of calculus-based mathematical research inside the Academy, and became his ally when this work began to provoke controversies. An episode from these early years illustrates well the new dynamics that were starting to get under way.

The incident was prompted by a mathematical challenge problem issued by Vincenzo Viviani in 1692.[14] When he issued his challenge, Viviani was basking in the twilight of what had been a remarkably long and distinguished scientific career. Born in 1622, he was launched into scientific prominence at the age of seventeen when he became Galileo's assistant at Arcetri, the Florentine villa-cum-prison that Galileo occupied after his condemnation by the Roman Catholic Church in 1633. Viviani's talents as a mathematician blossomed under Galileo's tutelage, and when his mentor died in 1642, the twenty-year-old protégé became a leader of the self-proclaimed *Galileisti* who served as the guardians of, and advocates for, the master's legacy. Viviani was also a founding member of the Florentine Accademia del Cimento, a scientific academy that served as one model for the courtly conception of royal science that Louis XIV and Colbert instituted in France in 1666. Viviani was in fact invited to Paris in 1665, along with the artist and architect Gianlorenzo Bernini—they shared an interest in the science of waters—in anticipation of an official appointment to the Sun King's court. Neither man was ultimately lured to Paris, but Viviani was nevertheless awarded an annual pension from Louis XIV, an act that made him a de facto foreign associate of the French Royal Academy even though no such formal role was created at the founding. When such a title was created by the 1699 regulations, Viviani was in the first group to be named to this office, a reflection of his esteem in the eyes of the French scientific establishment.

Viviani's *aenigma*, which provoked controversy after 1692, illustrated well the way that courtly values, royal politics, and mathematical debate could still become entangled at the end of the seventeenth century. His puzzle was in fact a classic piece of gentlemanly mathematical gamesmanship, one reflective of Viviani's courtly mathematical identity and the epistemologies attendant to it. Using as the medium for his challenge not a public journal, but the diplomatic correspondence that linked Grand Duke Ferdinando II de Medici, Viviani's primary patron, with other European sovereigns, Viviani proposed a geometrical challenge problem, or *aenigma* as he called it, targeted at the savants gathered at the courts of these glorious princes.[15] Eager to activate, while at the same time not overheat, the codes of honor dear to such *honnête* mathematicians, Viviani likened his *aenigma* to a sort of civil duel, one that would provoke noble competition in pursuit of "beautiful new geometrical truths" while avoiding the "loathsome" practice of actual intellectual combat.

He also conveyed his own estimable character in issuing such a challenge by framing it in a typically artful way. He did not use his own name in the paper proposing his challenge, but that of D. Pio Lisci Pusillo Geometra, a fictitious anagram that clever readers would have decoded as *Postremo Galilei Discipulo* (Last Disciple of Galileo). This was Viviani's moniker in learned circles,[16] and his manner of delivery also brought princely authority into the mix by invoking the name and reputation of the famous court mathematician and philosopher of the Medici. Leibniz, in service to the Duke of Brunswick-Lüneberg at the time, acknowledged this link by addressing his response directly to Grand Duke Ferdinando, writing, "I truly confess that I would not easily have assented to this investigation had your advice not persuaded me."[17] To secure other audiences for his solution, Leibniz also published his solution in the *Acta Eruditorum*.

Along with these characteristic social pulls, Viviani's *aenigma* further drew upon links tying courtly mathematicians to classically liberal mathematical epistemologies. To be a mathematical *Galileista*, in Viviani's mind, meant practicing geometry according to the classical epistemological canons of Euclid and Archimedes. Modern innovations, such as the geometry of indivisibles developed by Viviani's Galilean colleague Cavalieri, could be absorbed comfortably within this classical understanding of mathematical science. But overall, and especially as he aged—Viviani was over seventy when he circulated his *aenigma*—the doyen of the Galilean movement was an ardent mathematical traditionalist who emphasized the importance of classical conceptions of rigor in geometric work. He also disparaged the new "mechanical" and analytical mathematics that eschewed these methods. For him, the most beautiful and noble geometry was the kind achieved through the stringent discipline of demonstrative rigor, and this led him to celebrate the ancient mathematics of figures like Archimedes and to dismiss the innovations of the moderns.

Much of Viviani's actual mathematical work, in fact, involved attempts to reconstruct the lost work of ancient mathematicians—a marker, like de la Hire's translation work, that the humanist mode of mathematical practice was still alive and well in the late seventeenth century. The precise *aenigma* that Viviani offered in 1692 also bore all the traces of this time-honored gentlemanly approach. It centered on finding equal surface areas upon a spherical surface, a problem that both Archimedes and Galileo had wrestled with. Viviani's challenge was also directed against the new analytical approaches to such problems developed in the seventeenth century, and especially against the new differential calculus of Leibniz. Among his motivations in offering this of all challenges, in fact, was his desire to pose a problem that showed the limits of the new and suspect modern mathematics.[18]

Leibniz claimed to have solved the *aenigma* the minute he set eyes upon it, and other advocates of the new analysis, as the calculus was coming to be called, such as Johann and Jakob Bernoulli and the Marquis de l'Hôpital, also submitted solutions to Viviani's problem that deployed the new calculus. Others explicitly avoided this approach, however, including de la Hire. He valued Viviani's manner of mathematical reasoning more than he valued Leibniz's, for as he wrote to Viviani in the wake of the *aenigma* challenge: "I do not say to you, as M. Leibniz did, that I resolved your enigma the same day that I received it, . . . for even if I had given a solution such as his, which I estimate as a small thing, I would have been ashamed after seeing the diversity of ways with which you solved it with so much elegance and simplicity."[19] As his Galilean colleague Torricelli had often done,[20] Viviani produced numerous different demonstrations of the correct answer, each with its own particular character, and de la Hire's comment about this virtuosity shows how geometers often valued the particular aesthetic character of a solution more than the ease and accuracy with which the mathematician solved the problem at hand. In speaking to Viviani in this way, de la Hire also illustrated how liberal conceptions of nobility and *honnêteté* in the practice of mathematical science were still exerting a powerful influence on the mathematics that was practiced in turn-of-the-seventeenth-century France. That de la Hire was in all other respects a mathematician comfortable with modern, mechanical approaches, and a fixture, through his appointment at the Royal Observatory, of the administrative orientation to academic mathematics that was becoming ever more ascendant, only reveals the complexity of these French epistemo-social entanglements.

The *aenigma* episode ultimately served as a prelude to the epistemo-social struggles that would animate the introduction of the calculus in France over the next four years. L'Hôpital and Varignon became more aggressive after 1693 in the pursuit of calculus-based mathematical research, and their work in turn began to provoke controversy, often in terms that mirrored those of the *aenigma* challenge. Out of this controversy, analytical mechanics crystallized. To understand its emergence, let us look in detail at the rise of calculus-based mathematics inside the French Royal Academy, and the controversies that it triggered.

The Calculus and Its Discontents in the Académie Royale des Sciences, 1694–97

Situated within the context of the Academy as it had developed by 1692, and the reform efforts under way, the use of the infinitesimal calculus within the institution posed a range of different challenges. To academic mathematicians, it offered stunning new capacities for solving mathematical problems. But it also required the acquisition of new technical skills that needed to be mastered before these new powers could be utilized. The proponents of the calculus argued that through the use of analysis, reams of cumbersome geometric demonstration could be condensed into a few algebraic equations, producing an economy and elegance that vastly improved the discipline as whole. These gains, however, were bought at the price of a new burden to become expert with arcane and abstruse symbolism. Even comprehending an analytic solution to a complex geometric problem required literacy in a cryptic new mathematical language, and to become an actual practitioner of the new mathematics required even more expertise still. A wide array of complex mathematical procedures had to be comprehended before one could actually use the new methods, and in the face of these challenges, mathematicians were forced to choose: Either embrace the claims of the advocates and learn the difficult techniques that allow the calculus to do its important work, or resist the claims of novelty and continue to defend the time-honored practices of old.

This choice was made even harder by the absence of any clear epistemological ground to stand upon when choosing a position. The calculus certainly worked, and when it was used correctly it could accomplish all the innovations that its advocates championed. But there was no way to account for its success in a demonstratively rigorous way, because the method itself was built upon a patent contradiction. Judging whether an analytic solution to a geometric problem was in fact demonstratively correct was not possible from within the calculus-based algebraic articulation of it, and accordingly it was easy for skeptics to look at the arcane display of letters, symbols, and equations and see a suspect art rather than a new mathematical science. Such tensions became rampant within the French Royal Academy after 1693, and very quickly a division began to form between those Modern mathematicians who advocated for the new mathematics and those Ancient practitioners who criticized it and defended the traditional geometry of old.

The emergence of this struggle also posed problems for the Academy as a whole. If it was hard for mathematicians to sort out their position with respect to these new and highly technical innovations, nonmathematicians found the challenges even greater. How were those who spent their time with plant specimens or tinctures of phosphorous supposed to participate as academicians in the discussion of such highly technical matters? The Academy secretary, in particular, experienced new burdens as this new analytical mathematics began to be presented to the Academy. He was expected to know enough about it to record the papers and their significance in the academic registers, and Bignon's insistence that the secretary produce a better and more comprehensive academic record only intensified the burden. For the elderly Latinist du Hamel, keeping up with the new mathematical innovations became a daunting task, and one he never really overcame. One solution was Bignon's decision to assign responsibility for transcribing academic mathematical papers to Varignon in 1695, but the ultimate resolution came in 1697, when du Hamel was replaced as secretary by Fontenelle, a highly skilled practitioner of the new mathematics and an intimate of those in the Malebranche circle. We will have occasion to talk about Fontenelle's role later, but before his appointment, it should be remembered that Varignon was not only a participant in these mathematical developments, he was also an academic official responsible for recording them for the Academy. This administrative side of Varignon's work was not insignificant in the ultimate success of calculus-based mathematics in France.

A survey of the emergence of infinitesimal analysis as a French academic practice between June 1693, when l'Hôpital joined the Academy, and November 1697, when Fontenelle fully assumed the job as Academy secretary, reveals a variety of dynamics attendant to these shifts. Before Bignon's new orders regarding the transcription of academic *mémoires* into the registers, it is hard to know what was happening inside the Academy; the records offer us little access to the details of the papers read and the debates pursued. Evidence of the presentation of the calculus in the Academy from as early as April 1692 is present in a paper by Varignon, which was recorded in exceptional detail in the registers. It uses the new technique to find, as Varignon described it in his title, "a new universal quadrature of every genre and every species of the parabola imaginable."[21] More common, however, is the suggestion of its use in titles of papers recorded in the registers without any further explanation of what the paper contained.

In March 1694, for example, Varignon read a paper that "refuted the ideas of Father Guldin, John Wallis, and Bernard Nieuwentijt on the length of the Archimedean spiral, showing that it is no longer than half the length of the circumference of its inscribed circle."[22] The register offers no more access to this intervention than this description, but considering that the topic was one often used as an occasion for debating the relative merits of ancient versus modern mathematics, it is certainly possible that Varignon's refutation involved an analytical and infinitesimal line of attack. This possibility is made even more likely once one recognizes that Father Paul Guldin was a famous Jesuit critic of infinitesimal methods, and Bernard Nieuwentijt was the author of *Considerations regarding the Analysis of Infinite Quantities*, a direct attack on the legitimacy of Leibniz's calculus.[23] Did Varignon use this paper to engage in an open defense of the calculus in the Academy? The records are unfortunately silent on the matter, as is Varignon's correspondence with Bernoulli during these years.[24] The record of l'Hôpital's presentation delivered in June 1694, which du Hamel described as offering "new reflections on the tangent lines at the points of inflection for the greatest and smallest quantities," is similar in strongly suggesting a work in differential analysis while in no way confirming it directly.[25]

Even after 1695, under the new regime, which asked academicians to become more assiduous in delivering academic *mémoires* to the secretary for transcription, the academic registers do not offer us a comprehensive view into the actual mathematical debates inside the Academy. One problem was the sheer burden of transcribing all this work into the records. Many more fully transcribed papers appear after 1695, but on many occasions, as is documented in the records, *mémoires* that were submitted to the secretary for transcription were never actually transcribed. Their oral delivery is recorded so as to note the submission for Bignon, who continued to sign the registers each week. But no other record of the papers or the debates they provoked survives.

There is also a strong imbalance in the character and subject matter of the academic *mémoires* that were in fact recorded. The overwhelming majority are prose works in botany, natural history, anatomy, and other empirical disciplines, and very few diagrams or images of any sort appear. At one level, this imbalance reflects the actual membership of the company, which was skewed toward *physique* more than *mathématique* in these years. Yet mathematical papers were certainly presented, and nominally at least, half the sessions were devoted to work in this area as opposed to work in *physique*. Yet while theoretical mathematical papers were presented by Varignon, l'Hôpital, and Rolle in particular, the mathematical papers that were transcribed tended to be the narrative accounts of the work at the Royal Observatory offered by Cassini, sometimes with tables of data accompanying the text. Papers in mechanics that were long on empirical description and short on mathematical explanation or analysis also predominate.

Varignon's appointment in 1695 to help du Hamel record the mathematical papers was in theory designed to help correct this imbalance, and Varignon did contribute transcriptions, both of his papers and of those submitted by others, to the register. The growing pains of this new arrangement were in evidence at first as Varignon's transcriptions were done on different-sized paper, prohibiting their seamless inclusion into the register book. The separation between the secretary's record of the session where the paper had been delivered and Varignon's transcription of the paper after the fact also created problems. For while du Hamel would often introduce the paper he was preparing to transcribe before moving into its transcription, Varignon's transcriptions appear as awkward and isolated inserts with little explanatory set-up.

The gap between the perfunctory inscription of the paper into the registers and the actual digestion of it into the academic discussions that the registers purported to document is also illustrated by a paper that Varignon gave in September 1697, two months before Fontenelle took over as Academy secretary. The paper was based on a mechanical lever submitted to the Academy by a M. de la Garouste. Varignon's paper analyzed the mechanics of the lever, but in doing so he used differential analysis to suggest ways that the machine could be improved through a better grasp of the forces involved. He dutifully transcribed the paper into the registers, yet the signatures of Bignon and du Hamel approving the work are found on the penultimate page of the *mémoire*, after Varignon's evaluative conclusions have been stated but before he had completed his differential analysis suggesting improvements. The suggestion left by this confused transcription is that Bignon and du Hamel signed the report before fully reading Varignon's technical mathematical evaluation. This evidence points to the difficulties that the new calculus presented to those nonexperts who were not passionate in the pursuit of its methods, along with the general problem of translating its technicalities into the general program for royal academic science that was emerging during these years.

Yet despite the many intellectual and institutional obstacles thwarting the smooth and easy reception of infinitesimal analysis into the mathematical work of the French academy, its presence began to grow after 1696, so much so that it began to provoke open contestation. One reason for its increasing presence was demographic. Bignon's efforts to invigorate the Academy had involved expanding its membership, and two mathematicians appointed as part of this growth proved instrumental in giving infinitesimal analysis a new prominence. The first, Joseph Saveur, was not technically appointed to the Academy by Bignon, for evidence of his participation in academic meetings is found in 1690, and he may have begun attending meetings earlier. David

Sturdy, who has studied these matters most fully, claims that Bignon appointed Sauveur to the Academy in 1696, but evidence of his presence in the meetings before this date is abundant.[26] What is clear is that he became newly active after 1696, especially in the pursuit of mathematical work that deployed the new infinitesimal calculus.

Sauveur's background illuminates the new social currents bringing mathematicians together with both the Royal Academy and the new infinitesimal mathematics. Born into middling stock, Sauveur excelled in his studies at the Jesuit college at La Flèche (Descartes's alma mater), and this led to his admission into the circle of Bishop Bossuet, the leading cleric in Louis XIV's monarchy, and then into a seminary in Paris to train as a priest. There he met the abbé Jean-Louis de Cordemoy, a member of Bossuet's circle, who combined evangelical pursuits with interests in architectural theory, the latter resulting in a renowned architectural treatise published in 1706.[27] Cordemoy's mechanico-mathematical interests also brought him into the orbit of Jérôme de Pontchartrain as secretary of the navy, and in 1700 Cordemoy was sent on a mission to investigate the feasibility of floating logs from the Auvergne forests to the shipbuilding yards on the coast. Cordemoy drew up a hydrographic map of the area that was approved by Marshall Vauban in 1705 but never implemented.[28] When Sauveur decided not to pursue a career in the clergy, Cordemoy's connections with mathematically minded royal officials such as Pontchartrain proved helpful. By 1690, when his name began to appear in the Academy registers, Sauveur had become a professor of mathematics at the Collège Royale, a mathematical tutor to the children of many elites at Louis XIV's court, and a royal military engineer, serving particularly under Vauban as the principal mathematical examiner of new recruits. From this position he also began to cultivate his trademark mathematical specialty: the mathematics of probability as it relates to games of chance. Gambling was a fashionable diversion in court society, and the *Mercure galant*, the organ of *mondain* sociability, attests to this fascination in the mathematical material it published. Sauveur became a well-known expert in the mathematical calculations useful in such games, and as a result he attracted a lucrative clientele of courtiers eager to pay for his services. He also advanced new theories in musical theory, another dimension of the mathematical sciences that was still active in the late seventeenth century. Overall, his eclecticism marked him out as a serious mathematician with wide-ranging and innovative interests.[29]

Thomas Fantet de Lagny, the other friend of infinitesimal analysis appointed to the Academy after 1696, shared a number of traits with Sauveur in a way that marks each as exemplary of the new persona of the advanced academic mathematician emerging at the time. Lagny's father was a royal officer in the Chancellery of Grenoble, and as a youth he was pointed toward a career in the law. Yet he found himself attracted to mathematics instead, and in a career path reminiscent of Varignon's, he went to Paris in 1686 in order to pursue mathematics and try to win entry into the Royal Academy. He succeeded initially in securing a tutoring position with the elevated Noailles family, and through these contacts he made the acquaintance of l'Hôpital and Fontenelle and became a member of the new community of analysts congregated in the 1690s around Malebranche. He acquired his expertise in the infinitesimal calculus there, and these connections in turn connected him to Bignon, who appointed him to two positions in 1696: Professor of Hydrography at Rochefort, an important naval port, and member of the Académie Royale des Sciences. His appointment in 1696 was not the first time Lagny was granted access to academic meetings, however. In May 1693, the Academy registers record that "Monsieur de Lagny presented himself at the assembly" and gave "a method for squaring an infinity of sections of a sphere exactly and geometrically independently of squaring a circle." No other record of this presentation is to be found, but after 1696 Lagny would pick up right where he left off by pursuing problems in geometry in the Academy through the new infinitesimal calculus.[30]

Lagny's formal academic entrée in fact occurred at exactly the moment when controversies over the calculus were beginning to become more visible, but his participation in them was shaped by his wider position within the French state. Residing a significant portion of the year in Rochefort, Lagny sometimes sent papers to the Academy rather than presenting them orally, and in February 1696 he did just that, contributing "Two Analytical Quadratures of the Circle and of Every Sector or Segment Given." The paper was inserted at the end of the register book, which terminated in March of that year, and the transcription was signed by Lagny and du Hamel.[31] The paper deployed the differential calculus to accomplish the quadratures indicated, and on February 22, the records report that "M. de la Hire read his report on the paper by M. Lagny touching on quadrature," an indication (that is otherwise undocumented) that Bignon had asked for such an evaluation and that de la Hire was providing this service to the Academy's president. Du Hamel reported that de la Hire "found [Lagny's] manner of approaching infinitely [*sic*] very ingenious and novel." The entry is in fact garbled in that du Hamel's hand seems to trail off after writing "*infiniment*," and he leaves a space where a noun or another word should be.[32] Was he struggling to understand exactly what de la Hire and Lagny were talking about? In any case, this passage marks the first direct evidence of one academic mathematician responding to the use of the calculus by another, and that it was de la Hire who was offering the commentary is not without significance, as we will soon see.

Lagny's interventions continued a few months later when he sent another paper to the Academy, this time a work in algebraic problem solving. Bignon forwarded the paper to the members, and the registers for May 12 record that "M. de la Hire read the paper to the company, and then I [i.e., du Hamel] put it in the hands of M. Varignon, who made the copy that one sees here." A transcription indeed followed.[33] Rolle also gave several algebraic papers in 1696 that were akin to the work of Lagny here.[34] But in June, Lagny returned to Paris and began attending Academy sessions in person. The result was a new and never before seen intensity of academic presentations relating to the new infinitesimal analysis. The papers presented also intersected with questions of mechanics in ways that prefigured Varignon's work in the coming years.

The first was a paper (which was not transcribed) that Lagny read in June that considered what du Hamel called "the acceleration of movement in the fall of bodies." It was followed a week later, at the next meeting devoted to mathematics, by a paper read by Sauveur (which was transcribed) that du Hamel described as a "demonstration by lines of the rules of the differential calculus [*du calcul des différentiels*] for multiplication and division." These were followed by two Sauveur papers in subsequent weeks, one on "a rule for powers" that used the calculus and a second on July 7 that did not use the Leibnizian algorithm per se but did treat the problem in terms of the summing of indivisible quantities. The next week Lagny offered a new quadrature of the hyperbola that explicitly deployed the calculus to reach its conclusions. Finally, on July 20 Varignon gave an analytical paper treating the "ovals of M. Descartes following his methods and those of analysis." At the same session Lagny used his own approach to critique the traditional geometric quadrature of the hyperbola offered by Mercator.

Without a fuller transcription of the discussions that occurred at these sessions, it is hard to know what to make of this flurry of analytical mathematics inside the Royal Academy in the early summer of 1696. Whatever the conversation, it was provocative enough for de la Hire to intervene with the first of what would be several rebuttals to the new analytical turn among his academic colleagues. De la Hire's first protest was rather mild and nonpolemical. In August 1696 he came to the defense of Mercator's quadrature of the hyperbola, arguing that Lagny and others were going too far in indicting him for a failure to offer a truly universal solution to this problem. "Having examined his quadrature," de la Hire wrote, "I found that it was good and universal when one follows the application of his method." He also called out Lagny for suggesting otherwise, insisting that there was nothing that was not sound in Mercator's work. Nothing more came of this, at least not in these months, but the exchange makes it clear that many established academic mathematicians were beginning to feel challenged by the claims of overwhelming novelty and innovation being thrown around by the new practitioners of infinitesimal analysts.

Amid this emerging struggle, Varignon contributed a paper that brought Newton's *Principia* into the discussion in ways indicative of its place in France at this time. His paper was a reaction to an obscure 1691 book by an equally obscure author, Johann Caspar Eisenschmidt, on the shape of the earth.[35] Unbeknownst to Varignon, the book that he was reacting to was one of the first to engage in what would become a titanic battle waged for over half a century, and one that would become a great trial of strength regarding the validity of the Newtonian cosmological system.[36] The question was whether the earth was oblong, and thus shaped like a lemon, or oblate, and thus shaped like a grapefruit. Newton and Huygens had both defended the oblate thesis on mathematical hydrodynamic grounds, and Eisenschmidt was defending their view against the new claims, soon to be championed with vehemence by Cassini and the French royal astronomers, that empirical evidence obtained through geodesy supported the oblong thesis instead. Unlike Varignon's academic descendants, who would fight this battle empirically and geodesically in the years around 1740, Varignon's interest was entirely mathematical. He sided comfortably with Newton and Huygens, and he invoked, as others would do later, Picard's early measurements in South America showing varied pendulum motion at the equator as evidence in support of the *sphéroid applati* theory. But this was not really his concern. He instead intervened in the mathematical calculations that Newton and Huygens had used to reach their conclusions. As he wrote, "Regarding the precise and accurate nature of the curvature of the earth, M. Newton only gave it by approximate calculations, and M. Huygens held that it was very difficult to find the result by ordinary analysis." "The difficulty is in fact great if you use ordinary analytical mathematics alone," Varignon continued, "and it is insurmountable with geometry unaided by analytical calculation. But the calculus of M. Leibniz allows us to reach the goal easily and without struggle, and here is how." What followed was a mathematical derivation of the oblate spheroid earth using hydrodynamic theory and the Leibnizian calculus to reach its conclusions.[37]

No better illustration of the initial French reception of Newton's *Principia* as a work of mathematics, and not as a work of physics, exists. Yet if Varignon was reading Newton as a mathematician, he was also reading him as a deficient one as well. His point in his paper was that Newton's mathematical approach to the shape of the earth, while accurate enough, could be vastly improved were it to be developed instead using the Leibnizian calculus. His academic colleagues Lagny, Sauveur, and l'Hôpital were allied with him in this conviction, and as these figures raised the intensity of their advocacy for analytical mathematics, they began to find newly vigorous opposing voices rising against them. This in turn led the advocates for the new analysis to become even more strident in its defense. By the summer of 1696, a major debate was beginning to percolate, both in the French Royal Academy and throughout the lettered networks that constituted the professional mathematical community of the day. No one was more vociferous in fighting this battle than Johann Bernoulli. The new contests in play are well illustrated in the episode of the "Brachistochrone Problem," a challenge that Bernoulli issued to mathematicians in the summer of 1696, and one that was conceived as a Trojan horse designed to show the value of the new calculus.

Bernoulli posed the challenge via the *Acta Eruditorum*, and the debate that ensued illustrates well the initial formation and consolidation of the "infinitesimalist" mathematical community in France and across Europe. It also illustrates the character of their opponents. Stated simply, the "Brachistochrone Problem" asks what curve a falling projectile will take if it moves through a space in the least possible time.[38] Finding the solution requires a complex set of mathematical manipulations, and the genius of the challenge rested in the way that it was most easily and elegantly solved through recourse to the new infinitesimal calculus. In posing this of all problems, Bernoulli was therefore testing the potency of his fellow mathematicians while also demonstrating his own mathematical firepower. He was also offering a public demonstration of why the new analysis, of which he was a champion, was so powerful. The *Acta Eruditorum* announced the problem to its readers in June 1696. At the same moment, Lagny, Varignon, and Sauveur were becoming more active in their advocacy for the calculus inside the Royal Academy. Bernoulli sent personal copies of the challenge to Varignon, l'Hôpital, Newton, and Leibniz during the same summer.[39] He asked for correct answers to be submitted before Easter 1697. Throughout the subsequent months, mathematicians across Europe attempted to find the "curve of most rapid descent," triggering by consequence the first widespread public engagement with the new calculus among mathematicians.

Leibniz was the first to submit a correct solution, and in France, l'Hôpital was able to solve the problem, but Varignon struggled with it before giving up. This outcome illustrates why Varignon's mathematical thinking has often been disparaged when it is compared to the work of the other mathematicians with whom he collaborated. More interesting, however, were the efforts of others in France to find the correct curve. Both Varignon and l'Hôpital encouraged others in France to attempt a solution, and Sauveur came close. In December 1696, l'Hôpital sent Sauveur's solution to Bernoulli, bragging that he "thoroughly understands infinitesimal analysis, as you will see from this piece, for he applies it here advantageously in the solution of the problem."[40] In fact, Sauveur's solution was not complete and contained errors, which l'Hôpital, Bernoulli, and Leibniz corrected over the next few months. He in fact never succeeded in fully solving the problem, but his place within the infinitesimal community in France is revealed by his efforts. Others in France, including de la Hire, also wrestled with the problem, but no one else got as close as Sauveur. In de la Hire's case, he persuaded himself that he had demonstrated the validity of a wrong answer, earning the ridicule of Bernoulli as a result. His errors were compounded in Bernoulli's mind by de la Hire's obstinate refusal to embrace the new infinitesimal methods in his work.[41]

The emerging rift between the defenders and antagonists of the new calculus that the brachistochrone contest exposed was in fact one of its more significant outcomes. To taunt his rivals in this emerging dispute, Bernoulli developed what he called "a beautiful synthetic demonstration," which is to say rigorous and geometric, of the correct answer, which he offered as proof that the more simple and elegant solution provided by the differential calculus was valid. Sending this demonstration to Varignon in July 1697, months after the correct answers had been published, he wrote, "I have proven in the manner of the ancients that only the cycloid can be the curve of most rapid descent. It should serve to convince M. de la Hire, who not having a taste for our new analysis and decrying it as fallacious, believes that he has found by three different means that the curve is a cubic parabola."[42] Bernoulli also noted that Leibniz's German colleague Tschirnhaus needed to receive the same demonstration for the same reason, and four months earlier he had sent the demonstration to l'Hôpital writing that "it should convince those like M. Nieuwentijt, who do not understand our calculus yet nevertheless endeavor to mock it."[43] Overall, what these responses to the brachistochrone problem reveal was the intense strife that was developing over the validity and legitimacy of the new infinitesimal calculus.[44]

The specific references to de la Hire reflect the way that he had become by this time the single most assertive antagonist against the new calculus inside the Royal Academy. As early as February 1695, l'Hôpital wrote to Leibniz about his academic colleague's opposition to the new calculus, and in February 1697, in the middle of the period set by Bernoulli for the brachistochrone challenge, he presented an explicit critique of the new analysis to the members of the Academy. Titled "Remark on the usage that is made of certain assumptions in the method of the *infiniment petits*," de la Hire's paper was a calm and judicious, but also direct and assertive, indictment of the Leibnizian calculus. The author conceded that mathematicians had made "an infinity of very interesting discoveries" by treating curves "as polygon figures with sides so small that they have no sensible difference between them." However, he continued, "Since it is often difficult to avoid falling into error using this method, we should take precaution in using it as I will show in the example here." He then offered a case in which using the new calculus produced an erroneous solution. Overall, he concluded, "I could give other examples like this, but this one seems so sensible that I do not doubt that those who give it attention will conclude that it is best to assure oneself through ordinary geometry before accepting the conclusions arrived at through the method of infinitesimals."[45]

Bernoulli's synthetic geometric solution to the brachistochrone problem at one level did exactly as de la Hire suggested, but Bernoulli's intention was not to correct the calculus but to show its validity when practiced correctly. Inside the Royal Academy, de la Hire also used this approach to challenge the work of the analysts. One strand of resistance was noted already when we saw de la Hire rise to defend the traditional geometric work of Mercator against the claims of the analysts that they had superseded him. A similar episode occurred in January 1697, ten days after Fontenelle had been named the new Academy secretary. Varignon and de la Hire gave papers at this session on the "conchoid of the circle," an intricate mechanical curve produced, like the cycloid, by fixing a point on a circle and then rolling it on a plane in a way that generates a locus of points intersecting with it and another fixed point. Varignon's paper was not transcribed, but de la Hire's was, and to judge from it Varignon had used the calculus to claim a new universality in the measurement of these curves when compared with the traditional geometrical solutions offered half a century prior by Roberval. "It is not possible to say," asserted de la Hire, likely in reference to Varignon's claim in his unrecorded paper, "that M. Roberval only gave the quadrature of the 'Limaçon' [another related mechanical curve] and that he had not considered the conchoid of the circle and others. . . . It is easy to see that his quadrature agrees and is applicable with every sort of conchoidal space both in whole and in part, and no matter whether it is formed inside or outside the circle."[46] In short, Varignon's claim to have moved beyond Roberval's traditional geometry through new and more universal analytic solutions was, de la Hire argued, exaggerated and specious.

In cases such as these, de la Hire defended the value of traditional solutions against claims that the new mathematics made them obsolete. On June 1, 1697, as a kind of post mortem on the brachistochrone challenge decided two months earlier, Varignon further provoked de la Hire by offering "a new demonstration of the isochronous movements of bodies along a cycloid" that again used the method of infinitesimals to make its arguments. The correct solution to the curve of most rapid descent was the cycloid, so Varignon was pouring salt into the wounds of the brachistochrone struggle by investigating this of all curves. De la Hire was sufficiently stung because at the next meeting he offered a geometric demonstration of the time that it takes a body to fall when moving along the cycloid. But after he was finished, Varignon rose at the same meeting to offer "two new demonstrations of what I showed on June 1." "These are less simple than the first," he explained, "but they offer a lot more than what M. Huygens offered on the matter, and one should search in this manner for the most economical [*aisée*] and intelligible demonstrations of this proposition so important for the measure of time." At the next meeting, on June 15, Varignon gave some new remarks on the same question, using the calculus again to offer what he claimed to be new insights. De la Hire ceased his response, yet the encounter reveals the nature of the antagonisms that the new calculus was generating.

On other occasions, de la Hire took a different tack, offering, like Bernoulli though with very different intentions, synthetic geometric solutions that he argued were just as good, and even better because more rigorous and secure, than the new analytic solutions being offered by his opponents. An interesting case of this occurred in July and August 1697 as the Academy was approaching its fall vacation. On July 20, Sauveur brought to the Academy a problem sent to him by a M. Grégoire, royal engineer of manufactures in Nivernois. How should we construct barrels, Grégoire asked, in order to maximize the number of barrels that can fit into a finite space? Though it was not stated, the finite space involved could be likened to a warehouse or the hold of a ship, and since the problem did not specify the precise shape that the barrels needed to take, it was in effect asking mathematicians to offer solutions to a problem of commercial-profit maximization. As every modern student of differential calculus knows, this is a problem tailor made for the new analysis, and in addition to bringing this problem to the Academy, Sauveur brought a proposed solution with him that used the calculus in its articulation.[47] Two weeks later

Varignon brought two other solutions to the Academy that also used the calculus, but at the same meeting de la Hire brought a solution pursued entirely through traditional geometry. No argument between the academicians was recorded in the registers, and there is no evidence that any occurred. But clearly de la Hire was once again defending the value of traditional geometry in the face of those arguing for new and modern approaches.[48] All of this work was transcribed, and the records accordingly offer an accidental illustration of the arguments being sustained. For while de la Hire needed many pages of geometric demonstration to achieve his solution, Varignon and Sauveur accomplished theirs through a page or two of intricate algebraic explanation.

Toward Analytical Mechanics, 1697–98

On January 9, 1697 Bignon attended the Academy meeting to "present on behalf of M. de Pontchartrain M. de Fontenelle as the secretary of the company." The records note that Bignon spoke of the "age and infirmities of M. du Hamel, which no longer allow him to continue his duties," and that in selecting M. de Fontenelle, Pontchartrain had not removed du Hamel from the Academy but granted him emeritus status with all the privileges and gratifications of his current appointment.[49] We will look in more detail at this appointment in the next chapter given its relevance to the 1699 reform of the Academy, but in terms of the discussion here what needs to be noted is the arrival of Fontenelle in the Academy at exactly the moment when the calculus battles were beginning to heat up, along with his status as a trained and sympathetic analyst.

At first, nothing changed with the appointment, and the Academy records indicate that du Hamel continued to perform the job of the secretary for many months after Fontenelle's arrival. Perhaps he was helping to teach Fontenelle the job? Whatever the reason, Fontenelle does not appear in the Academy records at all between January 9 and November 13 1697 except for a brief mention of him offering a report evaluating the mathematical works of a certain Erhard Wegelius, professor of Mathematics at the University of Jena, whose "machine that he had made called *Pancosme, ou Monde Universel*" had been sent to Pontchartrain by a certain Count d'Auaux and then to the Academy for evaluation.[50] Wegelius had also created a system for coordinating all the calendars of the Christian peoples of the world, and on March 16, the registers report that "M. de Fontenelle Secretary responded to M. Bignon regarding the two *mémoires* of M. Wegelius according to the spirit of the company." No transcription of the report was made.[51]

Yet while Fontenelle quietly settled into his new role as secretary of the Royal Academy, a debate of great personal interest to him, and of importance to the Academy he was now in charge of representing, had erupted over the use of the infinitesimal calculus. In August 1697, as he approached the academic vacation that would mark the complete transition of Fontenelle into du Hamel's role, Varignon noted the heat of the debate, writing to Bernoulli that

> L'Hôpital has already departed for the country, and as a result I find myself as the sole person charged with defending infinitesimals. I am a true martyr, for I have already suffered assaults in their name launched by certain old style mathematicians. Saddened to see the young reach and even surpass them by these means, they do everything they can to decry them without ever agreeing to write down their complaints. It is true, however, that since they saw the solution that M. le Marquis de l'Hôpital gave to your problem of the curve of most rapid descent, they do not talk as much or as haughtily as before.[52]

The Academy registers reveal no record of this contestation, but if we imagine ourselves as time travelers returning to watch these debates, who would we see on each side of the struggle? Varignon, l'Hôpital, Lagny, and Sauveur were clearly Moderns, but Sauveur, like Varignon, pursued a variety of scientific projects beyond advanced mathematics, and he may not have been a particular advocate for this mathematics as opposed to others. Lagny was, like l'Hôpital, a geometric problem solver of the newly modern and analytic sort, but he was also in Rochefort much of the year and may not have been present enough inside the Academy to make a difference. Michel Rolle would later go on to be a leading antagonist against the new calculus, but for whatever reason he was a nonpresence in these early debates. Other than giving a few papers on algebra in early 1696, he presented very little work of any sort during these years, and the registers indicate that he was often absent from the academic sessions when he was not presenting.

Rolle's opposition to the calculus after 1700 will be central to the story to come, and for that reason it can be noted here that in 1697 he was a twelve-year veteran of the Academy who had begun as an *élève* with a modest 400 livre pension and risen to a more elevated status thanks to Bignon's favor after the publication of his treatises on algebra in 1690 and 1691. Rolle's volumes were grounded in a traditionally synthetic form of algebra, and they exemplified his modernism in their focus on making this new form of mathematics systematic and rigorous. But his traditionalism is also evident in the books' aversion to the new Cartesian analysis.[53]

Rolle also appeared at moments to be an ambitious man with a desire to climb the ladder of prestige. His professional ardor is manifest in the contribution he made to the very first monthly report of the Academy published in January 1692, only three months after Bignon had issued the call for papers. It focused on the algebraic rules for extracting roots.[54] Rolle published another report in the monthly *mémoires* two months later, and while these were his only contributions,[55] they indicate an eagerness to align with the new academic regime being instituted by the Pontchartrains. Rolle also engaged in a controversy with Bernoulli in the pages of the *Journal des savants* in 1693, and while this controversy did not involve the use of the calculus per se, it did center on Bernoulli's claim to have found a perfectly adequate solution to Rolle's challenge problem using Cartesian analysis as opposed to Rolle's preferred algebraic method. Rolle's insistence that Bernoulli's solution lacked sufficient methodological development to earn the prize was also at the source of their dispute. Such skepticism about the epistemological validity of seemingly efficacious analytical solutions would be manifest in Rolle's later critiques of infinitesimal analysis as well.[56]

Yet besides the battle between Rolle and Bernoulli, there was little overt contestation among academic mathematicians about mathematics in spaces outside the Academy before 1699. An exception is the exchange in Latin between Leibniz and Nieuwentijt in the *Acta Eruditorum* in 1695.[57] But this debate illustrates the general pattern in its character as a restricted conversation limited only to the learned mathematicians who read the Latin mathematical discourse coming out of Leipzig. Leibniz published a nonpolemical defense of his calculus in 1694 in the *Journal des savants* arguing for the potency of what he called his "*calculus differentialis*" in the solution of problems within "*analyse ordinaire*," which is to say Cartesian analytical geometry. The article also offered a brief summary of the development of the calculus so far, noting the Bernoulli brothers as "the first who have given public proof of the great success of the new method in solving Physico-Mathematical problems." The article also noted "the taste for it which M. le Marquis de l'Hopital has shown" through the "beautiful mathematical specimens" (*beaux échantillons*) he has made with it. Leibniz also noted that Huygens had "recognized and approved the significance" of the new mathematics, and he "rendered justice to Newton" by explaining that he too had invented a similar calculus, even if he stressed that "I believe ours opens more doors."[58] The *Journal des savants* was full of Francophone mathematical argumentation in the 1690s, not least Bernoulli's arguments with Rolle, but after the argument between Catelan and l'Hôpital in 1692, none of the material expressly concerned the calculus debates until the major public battles of the eighteenth century.

The appearance in the *Mercure galant* of a short introduction to a work titled "Work concerning Algebra" in April 1697, just after the brachistochrone solutions were published, was an exception to this public silence. The work referenced the method of infinitesimal analysis, noting that it had "its advantages and its drawbacks" and that it had "attracted partisans and adversaries of various sorts."[59] The *Mercure* also reviewed in February 1698 Lagny's *Nouveaux élémens de mathématiques et d'algebre*, a work that dealt with the calculus directly. The review also implied a reference to the "M de L" who had offered lessons to the "Comtesse de M" in the "Work concerning Algebra" published the year before. It is also worth remembering that the journal noted with respect to the mathematical work of Lagny that it was "*très à la mode* at the moment, especially with women."[60] The arrival of the *Mercure galant* as a participant in the French mathematical debates of the 1690s was an important harbinger of things to come, one indicating that the dynamic of the Malebranchian moment in France had begun to exert its influence. Given Fontenelle's long-standing connections with the *Mercure*'s editor, Donneau de Visé, which will be explored in detail in the next chapter, it is in fact likely that he played a role in bringing about Lagny's appearance here. Louis Carré, Malebranche's amanuensis in the 1690s, and the mathematical tutor whom Fontenelle described as having "*une petite empire*" of female mathematical students, also entered the Academy in March 1697 as Varignon's *élève*.[61] The dynamics of academic debate were changing, and Fontenelle was at the center of the shifts, as he would be for the next half century.

But besides de la Hire, who else was a visible public critic of the new calculus in France the 1690s? Given the nature of his work and his later outspoken antagonism to the new mathematics, it is hard to imagine that Michel Rolle ever had any sympathy for it. It is also likely that the abbé Jean Gallois was a skeptic even if no evidence indicating as much is found before 1700. Gallois, it will be remembered, personified the bookish, humanistic conception of mathematics favored by Colbert in some of his first academic appointments, and Fontenelle connected Gallois's later overt hostility toward the calculus to his antiquarian orientation. "A taste for Antiquity, which is so hard to contain within reasonable limits, rendered him little disposed to the geometry of the infinite," Fontenelle wrote in his 1707 eulogy. "In general, he was not a friend of the new, and he possessed a kind of ostracism that led him to position himself against everything that provoked a free state of thinking. The geometry of the infinite had these defects."[62] In short, Fontenelle implies, Gallois was a cultural conservative who found in the sparkling innovations of the new analysis a cause for alarm. No record of him intervening in the 1697 debate exists, however, but his intellectual tendencies were manifest in the three reports that he contributed to the monthly academic *mémoires* of 1692 and 1693, all of which were bookish and lettered in nature.[63]

Other academic mathematicians, notably the distinguished astronomer Cassini, left no trace in the archive of their position in this debate, a silence that in this instance probably indicates a lack of interest in the struggle or its lines of contestation. The mathematical class in the Academy in 1697 still included no more than a handful of men, so it is likely that the debate about the calculus during this year was also a limited affair, one that agitated only the small number of advocates for and critics of the new mathematics in the company. The reform of the Royal Academy, which began in principle with the appointment of Fontenelle as the Academy secretary changed this situation, and these developments will be dealt with in the next chapter.

Yet if the French public sphere remained fairly quiet regarding the infinitesimal calculus before 1699, one fundamental work did appear that was crucial to all that followed: l'Hôpital's *Analyse des infiniment petits*, the work that Bernoulli later claimed to have written himself and delivered to the marquis as part of their patronage contract.[64] Whoever the principal author was, the text was edited in collaboration with Varignon and then published in 1696 under l'Hôpital's name. The treatise was also published by the royal printer in a way that suggests a connection to the Royal Academy despite the absence of any records indicating discussion or approval of the book by the company. The treatise also included an anonymous preface, later attributed to Fontenelle, a text that Guicciardini calls "the manifesto of the Leibnizian mathematical community on the Continent."[65] L'Hôpital's treatise appeared in print before the brachistochrone contest had begun, and while the preface briefly mentions the dispute between Leibniz and Nieuwentijt, it more accurately reflects the state of French mathematical discourse immediately before overt polemics about the new calculus would erupt.

Fontenelle's preface was not a neutral text, however, for he articulated with great eloquence the conception of the new analysis held by the inner circle of French analysts in 1696. The opening paragraphs connect the differential calculus to be explained in the treatise explicitly with the wider discourse of Malebranchian mathematical philosophy then becoming prominent. "Ordinary analysis only treats finite magnitudes," Fontenelle began, "while this one penetrates all the way to infinity itself. One can even say that this analysis extends itself beyond the infinite, for it does not restrict itself to infinitely small differences but discovers the relations [*rapports*] between these differences and other differences, and between those and still others of the third order, fourth order and so on without ever finding a stopping point. In fact, it not only captures the infinite, but the infinite of the infinite, or an infinity of infinities."[66]

These statements powerfully echo the philosophical support for the new analysis offered in Malebranche's *Recherche*, and this discourse was reinforced by a history of the calculus of Fontenelle's own invention that made it, like Malebranche's philosophy, one more step in the advance of modern thought beyond that of the ancients. "The work that the Ancients did on these matters, especially the work of Archimedes, is certainly worthy of admiration," Fontenelle declared. But the work of the "Moderns," building on these prior achievements, has simply surpassed what the "Ancients" could ever even have imagined. "In a word, they achieved what their excellent minds were capable of at that time, and we have done the same in ours. These achievements are the result of the natural equality of minds and the necessary succession of discoveries."[67] This scientific progress was also natural and good in Fontenelle's estimation, and if the new calculus had pushed mathematics beyond the limits that the ancients had achieved, then this was as it should be and was something to be praised rather than worried about.

The real enemy, Fontenelle continued, was not progressive innovation but slavish devotion to tradition. "It is surprising," he noted, "how great men, including men as great as the Ancients themselves, retain an almost superstitious admiration for [the] works [of the past]. They content themselves with merely reading, re-reading and commenting upon the books of the past without making any other use of their reason. . . . In this manner, many people work, they write, and the number of books is multiplied. Yet nothing advances. The work of centuries is reduced to filling the world with respectful commentaries and literal translations of works that are no better than these copies."[68] This passage, the most overtly polemical in the preface, drew its heat from the wider Ancients-versus-Moderns battle to which Fontenelle was also devoting his energies in 1696. By connecting the differential calculus to the larger Modernist cause, the author was therefore encouraging the celebration of this mathematics in France in a manner resonant with wider cultural dynamics. Fontenelle would deploy similar discursive strategies on numerous occasions over the next decade, as we will see in subsequent chapters.

Newton's *Principia* in France at the Dawn of Analytical Mechanics

Fontenelle's account of Newton's work in the *Principia* in his preface is also indicative of the French understanding of that treatise at the moment when analytical mechanics first came to life. In his academic presentations about the shape of the earth, Varignon had grouped the *Principia* together with the other important works of mathematical mechanics, such as those by Huygens, which had made important innovations through the use of traditional geometric methods. But having given these past authors their due, he had also positioned his own calculus-based approach as a step beyond such work because of its economy and universality. Fontenelle approached the *Principia* in a similar way in his preface while also revealing the particular complexities that the *Principia* posed for such an understanding.

He spoke without hesitation about the *Principia* as an important milestone in the overall development of infinitesimal analysis. Writing at a time when the calculus priority dispute was nowhere in sight, he stated that "justice is due to M. Newton, as M. Leibniz himself admits, since he too has found something like the differential calculus as is revealed in his excellent book *Philosophiae naturalis principia mathematica* which he gave us in 1687, and which is almost all about this calculus." Clarifying, however, Fontenelle stressed that "the characteristics of M. Leibniz's [calculus] makes his much easier and more expeditious, as well as marvelously helpful in many different areas."[69] This assessment, echoing Leibniz's own evaluation offered in the *Journal des savants* two years earlier, makes the *Principia* one important moment in the development of the calculus but also a less significant achievement than Leibniz's parallel work in infinitesimal analysis. Fontenelle further offered these comments as a concluding addendum to an otherwise fully realized history of the calculus, one that traced the genealogy of the new mathematics from Descartes's analytical geometry through the analysis of Pascal, Fermat, and Barrow to its culmination in the differential and integral calculus of Leibniz and Bernoulli. Framed this way, the *Principia* represents a mathematical exception, a work whose achievements deserve recognition in the name of intellectual justice, but one that remains outside of the progressive mainstream of the important mathematical trends of the day because of its traditional geometric approach.

Varignon's interpretation of the *Principia* was no doubt similar to this one. Throughout the 1690s, when Bernoulli, Leibniz, and l'Hôpital were helping him to advance his program with the new differential and integral calculus, they also referenced Newton's achievement in the *Principia*. When they did so, his work was always treated with respect and esteem, yet they never spoke about the *Principia* as an overwhelming triumph that determined everything else being pursued in its wake. Typical was a letter from Varignon to Bernoulli in May 1696 in which he noted the similarity of Newton's work in the *Principia* with respect to a mathematical topic Varignon was discussing, but then let the reference stop by simply describing Newton's approach as "according to his manner of working."[70] This was to dismiss the *Principia*, calmly and without rancor it should be stressed, as a brilliant mathematical work but one with peculiarities. Its argument for universal gravitation and the physics of planetary attraction in the void was also ignored altogether in this assessment. More impressive was Newton's demonstration that planets obey an inverse-square-force law in their regular planetary orbits. This opened up a whole new set of possibilities for mathematical celestial mechanics. But these possibilities were held back, or so it seemed in light of the stunning advances offered by the Leibnizian calculus, by Newton's idiosyncratic geometric method (what Varignon called above "his manner of working").

Newton nevertheless remained an influential name at the center of all these developments, and his participation in the brachistochrone contest illustrates well his complex position with respect to the Continental developments of the 1690s. Bernoulli sent Newton a personal invitation to join the contest, and he also sent one to Wallis who, unbeknownst to him, had recently died. These personal invitations, which were very few in number, reveal Bernoulli's inclusion of Newton on the short list of the very best mathematicians in Europe. Others agreed with this assessment, including l'Hôpital and Leibniz, who each expressed to Huygens a longing that Newton would make public his analytical method of finding inverse tangents that the *Principia* had revealed even as it hid its details.[71] All these mathematicians knew that Newton was capable of powerful mathematical work, even if his habits of publication led him to guard many of his most important innovations.

The brachistochrone contest further confirmed these estimations, for while Newton was one of only five mathematicians to correctly solve the problem, the nature of his solution, and the manner of his participation in the contest, confirmed his awkward relationship to the European, and especially the French, mathematical community of the period. Although he received his invitation to solve the problem directly from Bernoulli, Newton was neither flattered by the invitation, nor the esteem it conveyed. Bernoulli had addressed his challenge to "the most brilliant mathematicians of the world," believing that "nothing is more attractive to such people than an honest, challenging problem," but Newton viewed such provocations differently.[72] As he wrote to the royal astronomer Flamsteed at the time, "I do not love to be printed on every occasion, much less to be dunned and teased by foreigners about mathematical things."[73] He nevertheless derived the correct answer, but he did not send it directly to Bernoulli, opting to publish it anonymously instead in the *Philosophical Transactions of the Royal Society of London*.[74] Bernoulli, like most European savants in this period, struggled to get access to the London *Philosophical Transactions*, and like most others on the Continent he also could not read the English-language articles found within. Newton's solution to the brachistochrone problem was written in Latin, yet Bernoulli still did not learn of the solution through direct access to the journal where it was published. Instead, he got word of it from Henri Basnage de Beauval, the editor of the Dutch periodical *Histoire des ouvrages des savants de l'Europe*.[75]

Basnage de Beauval was, like Pierre Bayle, whose *Nouvelles de la république des lettres* his journal ostensibly continued, a French Huguenot living a gentleman's intellectual life in exile in Holland. He descended from a distinguished Norman family of jurists and clerics, and his brother Jacques was a politically influential Protestant pastor also living in Holland. Henri's vocation as a journalist was emblematic of many exiled Huguenot intellectuals who found an esteemed place within the Republic of Letters by editing learned periodicals. His connection to the European mathematical community was forged in this way.[76] Huygens used Basnage de Beauval's journal as an outlet for some of his scientific work, and he helped l'Hôpital to do the same. When Huygens died, it was Leibniz who notified Basnage de Beauval of the news.[77] Overall, this placed the editor and his journal at the center of the increasingly Francophone correspondence networks that, for all intents and purposes, constituted the public professional field of mathematics at the time. Basnage de Beauval's journal, like its rival edited by the similarly positioned Jean Le Clerc, was a Continental print vehicle disseminating European learning into the wider public sphere, and his correspondence networks joined London, Cambridge, and Edinburgh with Paris, Basel, and Hanover.[78] What was known in Europe about Newton before 1700, besides what was found in the *Principia*, circulated through channels such as these, and when Basnage de Beauval sent Newton's paper from the *Philosophical Transactions* to Bernoulli, he was illustrating how these networks worked to facilitate (or not) the circulation of serious mathematics in the 1690s. Even though the paper was anonymous, Bernoulli had no problem sleuthing out the author, especially since Newton mentioned the personal invitation that he had received to participate in the contest. He was also glad to receive Newton's solution, and he published it among the winners, giving him full credit for his achievement.[79]

Yet the manner by which Bernoulli learned of the Englishman's work reveals a great deal about the reasons for Newton's marginality in this period. Content to pursue his science in isolation, and mostly oriented toward the Royal Society and its primarily Anglophone audience on those rare occasions when he did publish, Newton, despite his genius, was often forgotten during these years. Little was known about him or his ongoing work, and what was known was not easily obtained.

Barriers like these continually kept Cambridge University's Lucasian professor a relative stranger on the Continent, and his solution to the brachistochrone problem, which reached a wider audience only with difficulty, also illustrates why few savants worked hard to overcome these obstacles and learn more about him. As Johann Bernoulli described his reaction in a letter to Varignon: "If he permitted me to use it in this way, I would easily prove that the solution to my brachistochrone given by M. Newton is not a real solution because there is neither a demonstration nor analysis in it, and because possibly he just pulled it from a principle of mechanics. It would also not be difficult, following the argument of my brother, . . . to prove that Mr. Newton arrived at the truth only by the means of two falsehoods that cancelled each other out. But I have too much equity to impute such nonsense to Mr. Newton."[80] Bernoulli's overall esteem for Newton is evident not only in his conclusion, but also in his judgment that Newton's genius was of a more mechanical and intuitive sort, and that his mathematical work was suspect as a result. Judgments such as these offer a framework for interpreting the many moments when Bernoulli and his colleagues judged Newton not with awed expressions of enthusiasm but with hesitant and sometimes befuddled declarations of respect.

The same admixture of attitudes also characterized Varignon's appreciation for Newton's science. He certainly read and learned from the *Principia*, but by reading it through the Continental analytical lenses that were characteristic of French mathematical modernism in the 1690s, he absorbed a particular late seventeenth-century French version of Newton's actual scientific work. The *Principia* no doubt contributed to advancing Varignon's scientific work, but it also determined very little of it. Instead, it was the combination of influences that congealed in France around 1700 that ultimately engendered his new science of motion. Crucial to this mix was the changing political culture of France, and its transformation of the working life of the Académie Royale des Sciences. Having explored the changing mathematical climate in France in the 1690s, let us turn now to these political developments and the crucial institutional changes at the Academy that contributed to launching analytical mechanics in July 1698.

CHAPTER 8 Analytical Mechanics within the New Public Academy

First Steps, 1698–1700

Varignon's academic papers of July and September 1698 launched the scientific program that would become analytical mechanics. In these papers he deployed the new Leibnizian calculus to reconceive the motion of moving bodies in a newly universal and abstractly mathematicized way. Newton's two-part mathematico-empirical architecture for the *Principia* was one model for Varignon's work because it offered him an example of a mathematicized mechanics freed from determining physical or empirical constraints. Even more influential was Malebranche's mathematico-phenomenalist philosophy, which argued for the comprehension of nature through an understanding of the underlying mathematical *rapports* evident in empirical phenomena. To treat the motion of bodies through a differential calculus of the quantitative relations of their motions was to pursue mechanics in exactly these Malebranchian terms. The Oratorian father saw the Leibnizian calculus as the most advanced tool yet achieved for pursuing such work, and Varignon accordingly found in Malebranche an intellectual lens through which to see an entirely new set of possibilities. He also found in Leibniz's differential calculus a powerful tool for realizing this undeveloped potential.

Yet because the calculus was still highly controversial in France, Varignon's breakthrough in 1698 carried with it a provocative epistemological charge. Four months after Varignon presented these papers, in February 1699, a comprehensive reform of the Royal Academy was also initiated. The reformist project that had been under way for seven years under the Pontchartrain ministry was gaining momentum, and with the break offered by the peace secured by the Treaty of Ryswick, the ministerial reform program acquired new focus and clarity of purpose. The new regulations for the Academy instituted at this time fundamentally changed the working practices of the Academy in which Varignon worked. His status in the institution was also transformed as he found himself an academic *pensionnaire* at the top of a newly expanded class of disciplinarily defined, hierarchically classified, and administratively regulated "*géomètres*." He also acquired a new *élève* as part of an expansion of the membership, and a new academic seat in the Louvre, the new home for the Academy after 1699. The 1699 regulations established a promotion ladder whereby publication and other demonstrations of public scientific accomplishment could lead to elevation to higher academic rank. The regulations further gave the academicians the power to make promotion decisions themselves through internal elections, a result that introduced a new political dimension to internal academic activities that had not been there before. As a result, the mathematics battles of the late 1690s were reshaped into a new kind of academic politics. Varignon's analytical mechanics was transformed by these new arrangements and his science also played a key role in helping to shape the new Academy defined by these new rules.

As he had always been, Varignon remained an eager recipient of the academic-reform agenda. When asked to do so, he complied with the new rules by declaring in February 1699 that his "new science of motion" would be his publicly declared personal research project, another new requirement for all academicians. His papers of 1698 defined the terms of that work, and he was invited to give a sample of it at the inaugural public assembly in April 1699. Here, in front of a glittering public in the palatial setting of the Academy's new home in the Louvre, Varignon introduced a wider public to his program of calculus-based mathematical mechanics. The success of this first performance led to Varignon's invitation to present a second paper eighteen months later in the Academy's fourth public assembly. Through visible, public academic work such as this, Varignon emerged immediately as a central figure personifying the new spirit of the Academy being instituted by Bignon and the Pontchartrains.

By the end of 1700, the development and reception of Varignon's analytical mechanics were fully under way, along with the transformation of the Royal Academy into a new kind of protoprofessional public institution. What ensued after 1700 was an outcome produced by the dual pulls of this entangled transformation: a public battle that the actors called *la querelle des infiniment petits*. This bitter argument activated the new public structures of the Royal Academy in a vigorous way while also shaping the science of analytical mechanics that sat at the debate's center. The affair and its resolution will be considered in detail in the next two chapters; the goal of this one is to set the stage for that foundational struggle by examining, in turn, Varignon's precise scientific work from 1698 to 1700, and the structures of the new public Academy through which it developed.

The 1699 Reform of the Académie Royale des Sciences and the New Public Academy

What occurred in 1699 at the Académie Royale des Sciences was a politically motivated reform that completely transformed the structure and working practices of the thirty-year-old Académie Royale des Sciences while also repositioning the institution within Old Regime society. To reduce the reform to its essentials, the Pontchartrains created in 1699 a new and explicitly written constitution for the company, along with a newly articulated set of values and governing protocols. The minutes from the assembly of February 4, 1699 record that Bignon, whose role as the agent for the wider Pontchartrain reform program has already been made clear, visited the Royal Academy to read all fifty of the new regulations to the members.[1] A set of institutional reforms was then initiated, and after three weeks the company, arranged by rank, processed formally to the Pontchartrain residence to thank the minister for his favor. The next day, the Academy reconvened in its new guise and began to formulate a new set of projects and working conditions. By January 1700, the institution had been fully reconstructed. Forty-three new members were appointed to the company during the ten-month transition,[2] roughly doubling the size of the company. Acting now as the Academy's formally titled president within a new leadership structure that combined royal control with academic self-governance, Bignon reported that his visits "in the name of the company" to royal officials had been well received and that "its protection was assured."[3]

The establishment of this new academic constitution, together with the new academic practices it mandated, marks a watershed in the history of the Académie Royale des Sciences,[4] yet historians have long debated the precise significance of these changes. Fontenelle, the Academy's first historian—writing the Academy's history was his promised personal research project declared in February 1699—believed that 1699 was a decisive break. He described the reforms as "a second and more noble birth" for the institution, one that established the company "more strongly than before."[5] Nineteenth-century historians followed Fontenelle closely in this respect,[6] but Roger Hahn's authoritative 1971 history of the Academy revised this tradition.[7] Hahn argued that 1699 constitutes a time of institutionalization rather than of reform. The new regulations did not change the character of the Academy, he contends; they merely defined in print the practices that had developed informally during the previous three decades of its existence.[8]

Hahn is certainly right about the deeper origins of the 1699 reform. Regulation XX, instituted in 1699, for example, states that "experience having shown all too clearly the difficulties that arise when the Academy tries to work entirely in common, each academician will instead choose some particular project for his studies and by the account of it that he will give during the assemblies attempt to enrich the understanding of those who compose the Academy and to profit from their remarks."[9] This shift away from the collective orientation of the early academy and toward an individualized protocol of academic research and debate was indeed prepared over many years and not invented whole cloth in 1699. Similarly, Regulation XXX specifies a new set of rules governing the practice of individual publication by academicians, and this, too, grew out of a long series of struggles regarding the appropriate place of publication within the Royal Academy.[10]

Nevertheless, in other, and I argue more important, ways the 1699 reform marks a major moment of transformation, one that systematically broke the mold of the courtly Academy, which had continued to shape academic life into the 1690s, while vigorously aligning the institution in a more focused way with administrative monarchy. In regulation XXXV, for example, the Academy is directed to hold a public assembly twice a year on the first day after Easter and the first day after Saint Martin's Day. At this assembly, the regulations declare, "everyone will have access to the academy." Similarly, in regulation XL the perpetual secretary, already charged with receiving all the papers and inventions presented to the Academy, is now directed to "give to the public at the end of December each year an extract from the registers or a narrative account [*histoire raisonné*] of the most remarkable happenings at the academy." This regulation gave birth to the annual *Histoire de l'Académie royale des sciences*, which after its launch in 1700 offered readers a sampling of the scientific papers read at the Academy, together with the secretary's accessible narrative history of the year in academic science.

The annual volume of the Academy also gave academicians a publication outlet through which to accomplish the implicit "publish-or-perish" requirements inscribed in the new protocols. These imperatives had no precedent in academic life at the founding, and in many ways they mark a complete repudiation of the ethos of courtly academic science. Regulation XXVII stipulates that academicians "will take care to maintain contacts [*commerce*] with a variety of savants both in and out of France," and it further declares that preference will be given when electing new members or promoting existing ones to those individuals who have been most diligent in their public networking. The election of new members by the existing academicians was also formalized, and still other regulations required the Academy to collect and review all important books published both in and out of France (XXVII) and to engage in (and report publicly upon) scientific experiments reported elsewhere (XXIX). In order to facilitate all of this new publication activity, a final regulation (XLVI) directed the Academy to choose an official printer who will receive from the king "all of the privileges necessary to print and distribute works by the academicians that the Academy has approved."[11]

None of these regulations have any precedent in the early practices of the Academy, nor do they resonate with the courtly conception of academic science instituted at its founding. Yet, as we have seen, the impetus in this new direction was present at the founding, and a new clarity of purpose focusing the Academy in this administrative and noncourtly direction is discernible from the moment that the Pontchartrains ascended to ministerial authority. In this respect, the 1699 reform of the Académie Royale des Sciences should be understood as an institutional reform of royal academic science that was at the same time a political reform designed to realign the constituencies and ideologies of the French state.

In an insightful article about the lessons that Peter the Great took from his visit to the Paris Academy in the first decade of the eighteenth century, Michael Gordin has shown how eighteenth-century royal academies were often much more than sites for institutionalizing the pursuit of natural knowledge. They were also institutional vehicles for inculcating political values and for creating the right kind of political elite appropriate for the good governance of a modern state.[12] Like Peter the Great, French royal administrators were eager to subdue the political power of traditional dynastic and corporatist factions within the French monarchy. They therefore conceived of France's Royal Academy of Sciences as an organ for cultivating, and then publicizing, meritocratic administrative values and for aligning French society as a whole with a political vision grounded in those values. Since they found numerous sympathizers within the wider public, the resulting initiatives that these urges produced should be viewed as a product of mutual understandings, not as a state-imposed regime enacted upon a passive and pliable society.

The explicit turn toward the public, which is arguably the most important theme running through the new regulations, was a key feature of the new politics that the Pontchartrains were promoting. In his seminal study of the genesis of eighteenth-century publicity, Jürgen Habermas noted this connection, arguing that a newly self-conscious and assertive conception of administrative monarchy in the seventeenth century was a key step in the transformation from court society into modern civil society. Seeing in the court-based government cultivated by Louis XIV a high water mark in the traditional "representative publicness" characteristic of the premodern society of orders, Habermas located one source for a new kind of publicity in the emerging conception of the state as a depersonalized bureaucratic authority (i.e., administrative monarchy), a conception that challenged the embodied ranks and titles of the traditional courtly conception of monarchy. As this bureaucratic state became ascendant, a new public of "private individuals" emerged as its partner, becoming the audience for and the object of this new administrative governance. "Civil society came into existence as the corollary of a depersonalized state authority," Habermas writes, and with the linked pair state authority-public authority in mind royal administrators in France began to take an interest in both promoting and regulating the activities of this new civil-social-political sphere.[13]

Habermas uses the emergence of critical journalism to illustrate the new relationship between the state and the public he sees, but I contend that the publicity encouraged and regulated by the 1699 Academy reform was equally indicative of the same shifts. Here, to gloss Habermas, royal administrators in France began to cultivate, and at the same time strive to contain, a "new and continuous zone of critical contact," namely the space defined by the new public of private individuals on the one hand and the new public authority of the administrative monarchy on the other.[14] Habermas's genealogy of these changes is brief and schematic, and it emphasizes long-term socioeconomic trends more than local historical contingencies. For our purposes, the latter are more germane, and among the most important, both in encouraging the new public dynamics revealed by the 1699 academy reform, and in framing the precise development of analytical mechanics within this space, was the eruption of the "quarrel between the Ancients and Moderns" at precisely this moment of change.[15] The debate over analytical mechanics was also saturated with the terms of this battle, so in order to understand better this political cultural moment, and the broader sociopolitical changes it supported, let us consider briefly this struggle in relation to the changes afoot in the 1690s.

Ancients versus Moderns and the New Public Culture of Late Seventeenth-Century France

The "quarrel between the Ancients and Moderns" was by and large a dispute about the French literary canon that grew, like its counterpart in the United States at the end of the twentieth century, into a broader fin-de-siècle culture war.[16] The battle itself began in January 1687, when Charles Perrault, author of the Grand Academy plan considered by Colbert in 1666, rose before the Académie française to read a fairly lengthy narrative poem celebrating the reign of Louis XIV. The arguments contained in the poem were largely conventional by 1687, yet they generated a controversy that provoked deep cultural tensions. Politics were clearly one flashpoint, for Perrault's poem opened by asserting the equation between science and royal glory that was central to administrative monarchy. After lamenting "the dark night" during which the science of Aristotle was dominant, Perrault celebrated the telescope, that "admirable glass" that "opened the eyes of the human understanding." He then equated the new science with the unprecedented military glory of Louis XIV, making the Augustan grandeur of the Sun King's reign a product of the new administrative definition of royal power.[17] The entire Perrault family had close ties to the administrative state, and Perrault's verses celebrating administrative monarchy were not surprising. They also generated little controversy, since, as Joan De Jean argues, "Scientific progress was among the rare Modernist tenets that always met with ready acceptance."[18]

More controversial were Perrault's next arguments. After celebrating the link between modern military glory and modern scientific knowledge, he attempted to argue similarly for the superiority of modern literature. At this point the Academy exploded into an uproar. Even before Perrault had completed his reading, individuals began expressing open dissatisfaction with the argument. The poet and royal historian Nicolas Boileau was perhaps the most disconcerted. Numerous witnesses report that he simply could not contain himself during the reading and that he screamed so loudly afterword that he lost his voice. The arguments within the Academy spilled out to the larger public, and throughout the next decade the French literary world engaged in a massive argument over the relative merits of ancient versus modern literature.

Viewed in retrospect, the puzzling thing about the quarrel was not its nature or the divisions it crystallized, but its specific location in time.[19] The academician Jean Desmarets, for example, spent much of the 1670s launching similar salvos on behalf of modern literature, and in many cases his rhetoric was more violent than Perrault's. Yet while Desmarets's modernism was absorbed quietly into the annals of French literary history, Perrault's launched a major cultural conflagration. What had changed? Primarily the character of the literary public, or so De Jean argues in an incisive analysis.[20] It was already a time-honored practice in the seventeenth century for writers and artists to invoke the public as the primary audience and judge of their works. In 1637, for example, a precursor to the Ancients-versus-Moderns controversy erupted around the relative merits of Pierre Corneille's play *Le Cid*, and several anonymous pamphlets from the dispute cite the "public voice" (*la voix public*) or just "the public" tout court as the ultimate authority and judge.[21] Similarly, the French academy, responding to this controversy in one of its first displays of institutional power, invoked the authority of the literary public even as it cautioned writers about its dangers. As Jean Chapelain wrote in his official *Sentiments of the French Academy on the Tragi-Comedy "Le Cid"*: "Those who, desiring glory, give their works to the public should not find it strange that the public becomes their judge."[22]

In this pamphlet, Chapelain echoed the widespread understanding of publicity characteristic of the middle decades of the seventeenth century. By 1687, however, the ties binding publicity with cultural authority were being realigned. Crucial was the way that literary publications such as the *Mercure galant*, first published in 1672, unified the public in a new way, creating a critical self-consciousness among individuals that they existed as members of a broader public with value as a critical judge and cultural authority. In the issues of this worldly periodical, readers were continually addressed as "the public" and asked to engage in critical acts of judgment.[23] The *Mercure*'s founder and editor from 1672 to 1710, Jean Donneau de Visé, also used the journal to explicitly foster sociability generative of this new understanding of publicness. He offered his readers social games such as enigmas designed to provoke their participation. He also designed the journal as a trigger for conversations and debates of interest to his audience. He further asked that his readers interact with one another in print, encouraging letters to the editor and other contributions of individual opinion. Donneau de Visé especially liked to stage mini-literary quarrels within the pages of the *Mercure* as a way of making the journal the centerpiece of the critical debates that he hoped to initiate within elite society more generally. In this way, his journal illustrates how new currents of participatory publicity were a kind of tinder that the Ancients-versus-Moderns battle inadvertently set ablaze.

The *Mercure*'s program was tremendously successful. Donneau de Visé reported in 1679 that he was receiving five hundred to six hundred letters each month, and under the banner of his journal, salons formed throughout the period to collectively read works and contribute opinions to the journal.[24] The overall result is best summed up in a letter to the editor published in October 1678. The reader writes in praise of Donneau de Visé: "The Public is infinitely obliged to you, Sir. Without you, it would have a hard time knowing itself [*se connaître*]."[25]

In sum, from 1672 to 1710, the *Mercure galant* helped to achieve two of the important components of later, eighteenth-century publicity. First, the journal helped to develop self-consciousness among individuals scattered throughout the Francophone world that they were part of a new collective entity called the public. In many respects, this simply fed the wider public self-consciousness that participation in the Republic of Letters as a whole offered at the same time. But the *Mercure* also created a local and worldly French community within this larger space. Second, the journal began instructing this public in the practice of making collective critical judgments in the name of their new public authority.[26] Here the *Mercure* preceded a wider development within the Republic of Letters that would only blossom fully in the eighteenth century.

In important ways, the 1699 reform of the Paris Academy of Sciences was stimulated by the increased public activity fostered by Donneau de Visé and catalyzed by the Ancients-versus-Moderns battle. In no way was the Academy reconceived as a servant of this new, critical public. Quite the contrary, the institution and its royal patrons saw publicity as a vehicle for reasserting the absolutist mission of the Academy as the supreme court of royal scientific authority in France. What the Ancients-versus-Moderns quarrel revealed, however, was that this mission now depended upon negotiating critical public discussion in new places and in new ways. Ministers such as Pontchartrain and Bignon were not outsiders seeking to control an external public in the name of state imperatives. They were rather participants themselves in the new public culture that was percolating and transforming all aspects of Old Regime society, including the French state.

Jérôme de Pontchartrain, who ascended to his father's Louis's positions in 1699 after the elder Pontchartrain's appointment as chancellor and who became after this date the ministerial manager in charge of the new public academies, illustrates well the shift. He was, to use James Pritchard's vivid description, a "heavy-jowled, thick-lipped, hunchbacked" man who had "only one-eye thanks to smallpox, and wore a glass eye in it in an ill-fitting, weeping socket." He accordingly appeared as a "loathsome beast" to some, especially those traditional aristocratic conservatives who abhorred merit-based administrative monarchy and the families that promoted it.[27] The court diarist and reactionary Saint-Simon, for example, described Jérôme de Pontchartrain as "a most detestable and contemptible individual, and looked upon as such, without exception, by all France and by all foreigners who come into contact with him."[28] For those with different political orientations, however, Jérôme de Pontchartrain was a vigorous advocate for a newly modernized French state, one empowered by a strong alliance among administrative governance, science, and the public.[29]

His tenure as naval secretary supported this assessment, for he continued without interruption the previous initiatives begun by Colbert and his successor, ensuring that the navy remained the most technically sophisticated in the world.[30] A typical example of Pontchartrain's wider approach to royal administration was the creation of the Depot of Naval Maps and Charts in 1696.[31] Made up of cartographers, astronomers, engineers, and military experts, including the technocratic military administrator Vauban, Jérôme's ministerial mentor, this new office of the secretariat of the navy was instituted to further catalyze French maritime power by integrating state-of-the-art mathematical science within it. His reforms at the Royal Academy of Sciences were designed to align with administrative state science in a similarly integral way.[32]

Other illustrations of this sort could be added, for Jérôme de Pontchartrain was unquestionably an administrator with a clear vision of the valuable role that mathematical science could play within a modern administrative state.[33] Pontchartrain also maintained an active correspondence with leading *hommes des lettres* such as La Bruyère, Fontenelle, and the abbé Renaudot. These letters further show that he took a partisan interest in the literary battles of the decades around 1700 (he tended toward the Ancients' position). The royal secretary of state, not surprisingly, avoided strong polemical commitments in his correspondence, preferring instead the *honnête* sport of the debate itself. Yet his simple participation reveals the blurry line that separated the state from the public in these years.[34] Bignon enjoyed similar connections with the burgeoning public of science and literature, even if his correspondence from this period has not survived. Even more than Pontchartrain, his network attests to the deep entanglement connecting the administrative state, learned society, and the burgeoning new public in turn-of-the-seventeenth-century France.

Within the broader public sphere, the Royal Academy of Sciences was widely viewed as a positive illustration of the fruitful bond that tied knowledge with royalty in the service of the public good. Royal administrators, therefore, turned with increasing vigor after 1699 toward the public that held such views, not in order to contain or control it, but in order to solidify the consensus that they imagined they shared with it.

The Appointment of Fontenelle as Perpetual Secretary of the Academy of Sciences

Here the appointment of Bernard le Bovier de Fontenelle to become perpetual secretary of the Royal Academy of Sciences in 1697, replacing the retiring du Hamel, stands as both the perfect piece of evidence justifying this explanation of the 1699 reforms and the perfect testimony to the brilliance of Pontchartrain and Bignon as royal administrators. Fontenelle was born in 1657 to a family of Norman notables. His parents hoped he would study law, but under the influence of his uncle, Thomas Corneille, the brother of the more famous playwright Pierre, and a French academician who was charged, as we saw earlier, with defining *méchanique* in the addendum to the 1690 *Dictionnaire de l'Académie française*, he instead chose to pursue a life in letters. The decisive moment in his early life occurred in 1677 when, together with Thomas, Fontenelle joined with Donneau de Visé and the *Mercure galant*. The May 1677 issue of the *Mercure* introduced the journal's readers to their new associate:

> M. Fontenelle, who is only twenty years old, already has all of the intelligence [*esprit*] of someone forty. . . . He is from Rouen and he currently resides there, but several people of very high quality say that it is a crime to leave him in the provinces. . . . There is no point of science on which he does not reason solidly, but he does so in an easy manner that has none of the rudeness of scholars. . . . He loves knowledge [*belles connaissances*] but only to make use of it as an *honnête homme* [*pour s'en servir en honnête homme*]. He has a fine, gallant, and delicate mind.[35]

This *éloge*, which Fontenelle swore until his death he did not write himself, effectively launched the young *homme des lettres* into the literate public then becoming so important in France. A cascade of written works followed in subsequent editions of the *Mercure*, and they quickly established Fontenelle as one of the central figures within the wider Francophone reading public.

This literary participation was joined with active social participation in the burgeoning public sphere, and by the end of 1677, Fontenelle was residing primarily in Paris, and was becoming active in the salon culture of the city. He associated himself early on with the circle of *précieuses* centered at the salon of Mademoiselle de Scudéry. There he polished his mind and his manners in conversations ranging from Cartesian science to libertine poetry.[36] These practices were developed further as a member of the parallel salon hosted by Madame de la Sablière. Fontenelle was also a regular at the scientific gatherings held by figures such as abbé Pierre Bourdelot in this period. Through these sociable contacts, and his connections to Donneau de Visé and the *Mercure galant*, Fontenelle eventually developed close ties with virtually all of the Parisian luminaries of the period. He was a fixture at the rival court hosted by Madame de Rambouillet after 1690, and in 1698 Madame de Lambert launched her new Parisian salon largely as a forum for the Parisian literati now dominated by Fontenelle. Throughout these years, Fontenelle also wrote constantly, contributing poems, essays and other pieces to the *Mercure galant* and publishing independently a number of works both under his own name and anonymously.[37]

This life at the heart of the Parisian society in the 1680s and '90s forever shaped Fontenelle's intellectual direction. In 1687, he expressed his cultural allegiances publicly by siding unequivocally with Perrault and the Moderns in the pages of the *Mercure galant*.[38] But as Fontenelle's biographer Alain Niderst notes, this orientation was overdetermined by his previous background. Defining Fontenelle's position in one of the literary battles staged in the *Mercure galant* in the 1670s, Niderst writes:

> Parisians versus *Versaillais*, galant or *précieux* poets versus sublime poets, emulators of the old Corneille against Racine, Boileau and their imitators: it was already the quarrel of the Ancients and Moderns, and the young Fontenelle, who had condemned in his *Description de l'Empire de la Poésie* the sterility of imitation, had already chosen his camp. But did he ever have a choice? The great Corneille and his hatred of Racine, the young Corneille, the *Mercure galant*, and [Parisian literary society]—these had already ensnared him. Furthermore, behind all of these intercessors stood Colbert, protector of the Perrault family and of the *Mercure*, and his politics, which Fontenelle would often celebrate against the "devout imperialism" which began to impose itself with Louvois and Madame de Maintenon.[39]

Here, in one concise passage, reside all the factors central to this discussion so far. In Fontenelle, the politics of administrative monarchy, anti-courtly literary sociability, and the new public life of the *honnête* man of letters merged with the new Cartesian science of Malebranche and the analytical mathematicians into one quintessential embodiment of all that was percolating in modern French culture at the end of the seventeenth century. His ascension into the secretary's seat at the Académie Royale des Sciences in 1697 therefore marks clearly the centrality of these new alliances for this institution.

Fontenelle's claim to be a worthy member of this of all French academies was built upon a similar combination of factors. In 1686, he published the first edition of his *Entretiens sur la pluralité des mondes habités*, a work that crystallized the alliance between Cartesian philosophy, Parisian literary sociability, and the new *mondain* ethos of publicity forged by the Moderns in the quarrel that was to erupt in the following year. In the first edition of *Des mondes*, as this immediate classic came to be called, Fontenelle presented a set of five evening conversations conducted in a park between a refined savant and a curious marquise about the nature of the world system. In later editions a sixth *soirée* was added, and in each evening the savant introduces his aristocratic companion to the nature of celestial and terrestrial mechanics as taught by the disciples of Descartes. At one level, the work was just that: an effective popularization of Cartesian science. Jacques Rohault had already pioneered the practice of literal, face-to-face popularization of Cartesian science through his Parisian public courses, and in 1671 he brought out the first edition of his *Traité de physique*, a French physics textbook that explained Cartesian terrestrial and celestial physics to informed lay readers.[40] The book was the most authoritative and influential of a spate of books that appeared in the 1670s and '80s offering to teach Cartesian philosophy and science to the public.[41]

Fontenelle's *Entretiens* was a product of, and contributor to, this same Cartesian vogue. But viewed in its precise context, the character of the popularization was far more important than the details of the science that it taught. Like the *Mercure galant* that the author knew so well, Fontenelle situated his presentation of Cartesian science squarely within the *mondain* social world of elite society. The result was "disguised philosophy," as the *Mercure* itself described it: "Physics is made accessible to all ladies without exception even if they have never even heard about it."[42] The setting for the work is a garden to which two members of the elite retreat after an evening supper. The literary form is a breezy dialogue, the salon genre par excellence. An aura of eroticism also permeates the book as the savant increasingly "seduces" the marquise to his position (fig. 6). Furthermore, while Fontenelle effectively articulated many of the most important ideas central to Cartesian physical science, he did so within an idiom that was entirely *mondain*. A decade earlier, the *Mercure* had captured the essence of Fontenelle's method in its introductory *éloge*: "There is no point of science on which he does not reason solidly, but he does so in an easy manner that has none of the rudeness of scholars." In its review of *Des mondes*, the *Journal des savants* expressed a similar sentiment: "[Fontenelle]

FIGURE 6. *Bernard Picart (1673–1733),* La Marquise et le philosophe, *1727. Courtesy of O. Meredith Wilson Library Special Collections, University of Minnesota.*

does not treat his subject like a Scholastic. He enlivens it very pleasantly."[43] Pleasure, refinement, and *politesse* were crucial values in the emerging French public of letters of the 1680s and '90s, while pedantic scholasticism and boring, didactic prose were markers of a retrograde conservatism, the kind that the Moderns used to tar the intellectual spirit of the Ancients. Fontenelle's

Des mondes was fully modern in this sense, and its modernism was significant because it demonstrated brilliantly that complex science and pedantic, scholastic discourse were not synonymous. Even mathematical science, Fontenelle showed, was compatible with *mondain* sociability, and this demonstration was to prove tremendously influential in changing the public understanding of French academic mathematics after 1697.[44]

The demonstrations offered by *Des mondes* also played an important role in the decision to appoint Fontenelle as perpetual secretary of the Royal Academy of Sciences. In this period, a number of factors were conspiring to make Fontenelle a desirable ally for state administrators. An increased interest in external publicity within the administrative monarchy, coupled with an awareness that a new and powerful public capable of influencing royal administration was emerging. These trends pushed administrators like the Pontchartrains to think of new ways to connect academic science with administrative monarchy. Building bridges with this new public was crucial for all of the reasons discussed above, and Fontenelle held the promise of cementing these ties in an especially potent and decorous way.

Indeed, it is not unlikely that the precise public orientation of the 1699 academy reforms was a product of Fontenelle's work after his appointment rather than a precondition for it in 1697. The 1699 regulations demand, for example, that the Academy produce either a set of extracts or a *histoire raisonné* of the Academy's business. Under Fontenelle, both were instituted, and one wonders what role Fontenelle played in shaping the regulations that stipulated both of these works. Similarly, the new idea of holding public assemblies certainly fit with Fontenelle's vision of public intellectual life, and he helped to make these events a bedrock of eighteenth-century French sociability and intellectual life. Given this, it is possible that Fontenelle played an important role in shaping how Bignon and Pontchartrain led the Academy toward the public after 1697.[45]

Regardless of their origin, however, the results speak for themselves. In February 1699, the Academy of Sciences embarked on a bold new campaign to make a preexisting public institution public in a new way. Behind these initiatives was the minister Jérôme de Pontchartrain, interested in using the new public emerging at the time as part of his program of reinvigorated administrative monarchy. Allied with him was the abbé Bignon, knowledgeable in the ways of *gens des lettres* and skilled in the practices of cultural administration. Enrolled in this program as well was Fontenelle, already a major figure in the Republic of Letters, and someone motivated to use his influence to cement the alliance with the public from the other direction. Also present was an expanded group of academic mathematicians already engaged in a vibrant new practice of modern mathematics. Under this aegis, the Academy of Sciences closed the seventeenth century by orienting itself squarely toward the future. One product of this new orientation was analytical mechanics, a new science distilled from the rich intellectual, cultural, and political brew that was France at the turn of the seventeenth century, and a science that came to life through the changed dynamics of France's newly public Academy of Sciences.

The Initial Steps toward Analytical Mechanics, 1698–99

Reduced to their most fundamental innovation, Varignon's papers of June and September 1698 set in motion the development of analytical mechanics by applying the new differential calculus to the analysis of the mechanics of moving bodies. Varignon had been working on each aspect of this pairing for almost a decade, and at one level the marriage amounted to a seemingly natural combination of an innovative mathematics with an innovative conceptualization of bodies in motion. To be more precise, Varignon treated motion in these papers as a continuum of discrete instants, and then claimed that the velocity of any body in these instants could be treated as uniform. This conceptualization made possible a new and comprehensive application of infinitesimal mathematical analysis to the science of mechanics. It was the combination that marked the innovation, however, not the separate pieces that Varignon brought together.

Leibniz's mechanics, for example, employed a principle of continuity that similarly made continuous motion the aggregate of infinitely small actions. For as he wrote to Simon Foucher in 1692, in a letter that Varignon no doubt read in the *Journal des savants*: "Do not fear, Monsieur, the tortoise that the Pyrrhonians make faster than Achilles. You are right to say that *every magnitude can be divided to infinity. There is absolutely nothing so exceedingly small that one cannot conceive of an infinity of further divisions which are never exhausted.* But I do not see the problem that arises from this, or what need there is to actually exhaust such divisions. . . . I believe that nature can reduce bodies to the smallness that mathematicians can consider."[46] In employing Leibniz's calculus, Varignon drew directly upon this approach to continuous motion, and Malebranche did the same when he applied the concept of the infinite divisibility of any problem to his own analytical calculus of error. Varignon drew upon each of these thinkers when developing his ideas, but he pushed beyond them in applying the analytical calculus of infinite differences to the basic problem of moving bodies in mechanics. In doing so, he effected a profound reconceptualization of the science of motion.

Most important was the way that he substituted abstract mathematical analysis for the physical and metaphysical concepts that anchored not only Leibniz's natural philosophy, but also every other seventeenth-century natural philosophy. By treating motion empirically and "cinematically" as a sensate phenomenon capable of scientific representation through quantitative mathematical analysis, Varignon's approach permitted the central categories of mechanics—space, time, and velocity—to be treated as pure algebraic relations. As Michel Blay, the historian who has studied most closely the technical nature of Varignon's achievement, sums it up: "He put in place, under the label of 'general rules,' a set of powerful algorithms that effectively allowed problems of motion to be reduced essentially to problems of mathematical calculation; in other words, to quote Auguste Comte, it permitted the reduction of these problems 'to simple analytical exercises [*recherches analytiques*].'"[47]

This analytical turn in mechanics was a major innovation. Prior to it, laws of mechanics were conceived metaphysically and captured geometrically rather than algebraically. To illustrate the difference, consider Descartes's explanation of the law of falling bodies offered to Marin Mersenne in 1629, a description that is akin in both content and diagram to Varignon's own paper on "the opinion of Galileo regarding the spaces covered by falling bodies" presented to the Royal Academy in January 1692.[48] Assume that the line ABC in figure 7 traces the actual fall of the body. Using this diagram, Descartes argues:

> The triangle ABCDE shows the proportion in which the velocity increases [in free fall]. Line 1 denotes the strength of the impressed velocity at the first moment, line 2 the strength of the velocity at the second moment, etc. . . . Thus the triangle ABE is formed and represents the *increase of the velocity* in the first half of the distance in which the body travels. As the trapezium BCDE is three times greater than the triangle ABE, it follows that the weight falls three times more quickly from B to C than from A to B. That is, if it falls from A to B in three moments, it will fall from B to C in a single moment. Thus, in four moments its path will be twice as long as in three; in twelve twice as long as in nine; and so on.[49]

Several features of this explanation are important to emphasize. First, Descartes, like Varignon, works by breaking motion up into discrete units. Unlike Varignon, however, he does not assume a cinematic conception of mo-

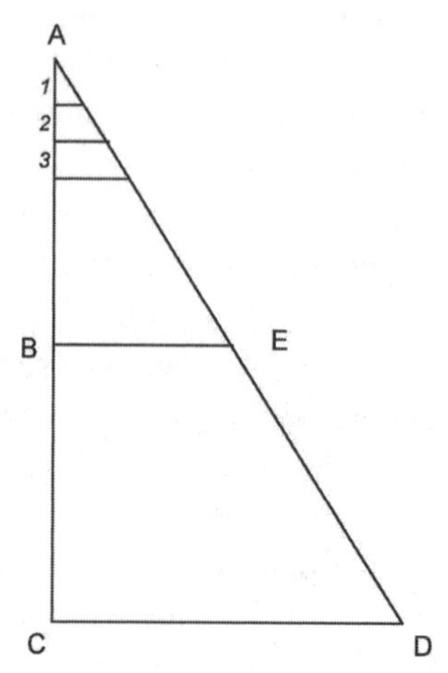

FIGURE 7. *Diagram of Descartes's account of the law of falling bodies.*

tion based upon infinitely divisible instants of uniform velocity. Rather, he uses the broader concept of "moments" and assumes that velocity is changing throughout these intervals. Second, Descartes uses geometry, not algebraic analysis, to determine the mathematical laws governing this motion. For example, while he considers magnitudes in this demonstration, he treats them geometrically by representing the increase in velocity spatially by the area of the triangle, and the constant rate of change in the velocity as a geometric ratio between figures. Neither numbers nor algebraic equations appear in his analysis. And while Descartes establishes a mathematical law in this demonstration, his final result is a geometric proportionality rather than a numerical formula or algebraic rule.

The diagram, moreover, is essential to Descartes's argument in unique ways because of his geometrical approach. Mathematical mechanics, traditionally defined, requires that motion first be translated into a comparable geometric representation. Once translated, the resulting mathematical description becomes wedded to the geometrical figure in important ways. Since magnitude, for example, is captured by a defined geometric figure, only by representing it as such can quantities be employed scientifically. Similarly, since the necessary geometric relations extant in figures determine the laws of mechanics, only by viewing motion in spatial and figural terms can a systematic science of motion be achieved. For this reason, geometrical mechanics is dependent on figural representations and their corresponding geometric relations in particularly crucial ways.

Before the 1690s, mathematical mechanics and geometrical mechanics were synonymous in Europe because no other approach to mechanics had been developed. Newton, for example, employed this same geometric approach to mechanics in the *Principia* even as he stretched the boundaries of this method in important ways.[50] Varignon's break in the 1690s, therefore, resides precisely in the development of a new algebraic approach to the science of motion that abandoned this geometric conceptualization entirely.

Mathematically, his work was a direct extension of his previous work in infinitesimal analysis. As Henk Bos has shown, mathematics in the 1690s was undergoing an important shift away from spatial and geometric conceptualizations and toward symbolic and algebraic ones. "From being a tool for the study of curves," Bos writes, "analysis developed into a separate branch of mathematics, whose subject matter was no longer the relations between geometrical quantities connected with a curve, but relations between quantities in general as expressed by formulas involving letters and numbers. . . . In the process of separation from geometry, the differential [also] underwent a corresponding change; it was stripped of its geometric connotations and came to be treated as a mere symbol, like other symbols occurring in formulas."[51]

Varignon was working at the cutting edge of these new mathematical developments, and his new science of motion depended upon them in two important and mutually reinforcing ways. First, it was his study of the new calculus that led him to his new method of conceptualizing the motion of a moving body. As with the differential calculus, Varignon's mechanics began by assuming that the continuous motion of bodies can be conceived as a series of discrete instants. He also assumed that these instants are uniform and interchangeable. In this respect, Varignon does for the movement of a body what Leibniz, Newton, Bernoulli, l'Hôpital, and the other founders of the calculus had done for the composition of a mathematical curve. Traditional mechanics, rooted in classical geometry, had not conceived of motion in this discrete, cinematic way, and Varignon's ability to see the moving body in these innovative terms grew out of his immersion in these new mathematical developments.

But if the calculus provided conceptual guidance for Varignon, the new mechanics was also conceived to take advantage of the power of the new mathematics. By conceiving of motion in terms of the conceptual categories of the calculus, Varignon was able to make differential analysis the foundation for his new science of motion. This shift was the real innovation in Varignon's work. By mapping the motion of bodies not with geometric curves, but with the symbols of mathematical analysis, Varignon was able to directly use algebraic equations as the representation of moving bodies eliminating the geometric bases that had originally guided Cartesian analysis. This permitted an unprecedented introduction of algebraic analysis into the previously geometric, not to mention physical and metaphysical, science of mechanics. For example, while Descartes and all other seventeenth-century mathematicians had represented the velocity of a moving body as the area of a geometric figure, Varignon gave it in terms of a mathematical equation, $v = dx/dt$. This exact equation, in fact, was the first general rule he derived in his initial paper of July 1698.[52] Summing up the achievement, he pronounced: "No matter how the speed of a body is presented (either accelerated, retarded, or, in a word, however you like), or how the space covered or the time employed to make the trip is revealed, it will be easy given two of these factors to find the third even if one is confronted with the most bizarre variations in speed imaginable."[53]

Given our contemporary comfort with the use of algebraic equations to represent the motion of bodies, it is hard to appreciate the innovative nature of Varignon's work. Yet the redefinition of motion in cinematic terms, and the corresponding introduction of algebraic equations as direct representations of moving bodies marked a monumental conceptual shift. A number of important features of this shift are worth emphasizing. First, Varignon's work instituted a new relationship between mathematics and physics with important consequences. One aspect of this development is illuminated by Peter Galison's study of the fundamental epistemological divisions of twentieth-century physics. In *Image and Logic*, Galison isolates a tension between two strands of scientific thinking, one that he calls the "image tradition" and another that he calls the "logic tradition." The basic distinction rests, according to Galison, in the tendency for some physicists to prefer visible, especially pictureable evidence when deciding between competing scientific claims while others prefer logical arguments, most notably mathematical explanations. What is interesting about Galison's dichotomy in relation to Varignon is that the development of analytical mechanics marks in many respects the birth of Galison's "logic" tradition in modern Western science.[54]

Galison's study concerns the material culture of modern experimental laboratory physics, so the analogy I draw here with early eighteenth-century French mechanics is necessarily a loose one. But the image–logic dichotomy as Galison draws it does offer a useful analytic for understanding Varignon's innovations in mechanics. For example, Galison defines the goal of the image tradition as wanting to produce "images of such clarity that a single picture can serve as evidence for a new entity or effect. These images are presented and defended as mimetic—they purport to preserve the form of things as they occur in the world." I would argue that the representations in geometric mechanics possess an analogous, if not identical, literalized physical relationship with the actual motion of bodies. Likewise, Galison singles out a break from images and imagistic understandings in his definition of the opposed logic tradition. This is where the analogy is especially strong. As he writes: "The logic tradition gives up, or in some cases explicitly rejects . . . image making. In its place, the logical relations between certain circumstances are determined." I would argue that this form of representation, which Galison calls "homologous" as opposed to mimetic, is akin to the representations offered by analytical mechanics. For the latter also thinks of motion in terms of an aggregate of quantifiable data points and then seeks the mathematical (logical) relations extant among these quantified aggregates in its claim to scientific theory.

Before Varignon, there was no "logically oriented" physical science according to the terms used by Galison. Mechanics, especially, was rooted inseparably in a set of pictorial representations. Traditional mathematical treatments of motion required geometric representations, because without such pictures a mathematized understanding of motion was impossible. Similarly, in geometric mechanics, the actual movement of bodies was visibly present in the pictorial representation that served as the foundation for the mathematical analysis. In the case of free fall, as noted in the example above, the acceleration is captured in the expanding angle of the triangle, and the constant rate of change is reflected in the proportionality of the figures themselves. In each case, the mathematical laws are captured pictorially in the geometric representations used to describe them. Mathematics around 1700, including the new algebraic infinitesimal analysis, was also imagistic in ways that make speaking of it in terms of Galison's logic tradition anachronistic. There simply was no distinction in this period between pure mathematics and applied or physical mathematics, and thus no distinction between a world of pure, intellectual concepts and a world of empirical, physical objects. As Lorraine Daston states: "All of [eighteenth-century] mathematics, including what they called pure mathematics, studied some*thing*. . . . Eighteenth-century mathematicians would have found the distinction between the formal apparatus of mathematics and the subject matter it treated to be an alien one. For them, mathematics, even pure mathematics, did not exist without a real interpretation."[55] It was, in fact, because of this inability to separate the logic of mathematics from its content that the infinitesimal calculus did not receive a demonstrative proof until the nineteenth century.

With analytical mechanics, however, the first steps toward a purely logical conception of physical science in Galison's sense were made, along with the first steps toward a pure logico-mathematical phenomenalism that could conceive of the potency of a mathematical description as its own epistemological justification irrespective of the interpretation of the symbols used to obtain it. By substituting algebraic analysis for geometry, and a phenomenalist mathematical description of motion for causal physics, analytical mechanics abstracted motion from its pictorial representation and turned it into a phenomenon treatable through pure mathematical analysis. Leibniz announced this innovation when he boasted that his calculus "frees mathematics from the imagination and subjects it to reason alone." Blay similarly describes this transformation when he writes that "Varignon's general rule, unlike its predecessors, does not elaborate its solutions according to a set of geometrical relations found in one or another figure, but directly according to a set of relations between symbols. . . . Progressively with Varignon, the figure . . . becomes a simple diagram. It loses its traditional value as an object of intellection and takes on instead a new value as a mere illustration."[56] This development marks the beginning of a new and, in Galison's terminology, essentially "logical" means of making scientific arguments. At the end of the eighteenth century, Lagrange would signal the ultimate triumph of this Varignonian turn by declaring proudly in the *Avertissement* of his *Mécanique analytique* that he had developed a complete system of mechanics without the use of a single diagram.[57]

A new relationship between mathematics and physics also arose as part of this shift away from pictorial literalness. With analytical mechanics, one no longer needs to create pictures of the phenomena one is treating in order to treat them mathematically. Instead, one only needs to define motion algebraically to determine the necessary relationships. The mathematics used for this approach to mechanics is also "pictureless" in a new way because the move to pure algebraic analysis, as Bos noted, detaches the mathematical symbols from any direct, mimetic attachment to their geometric referent. The shift in this direction after 1698 brought a new mathematical autonomy and abstraction into the science of motion that in many respects marks the opening of the modern, mathematized approach to physics commonplace today. As Blay sums up: "What Varignon had done was to show in an exemplary, and, at long last, inaugural way, that scientific work must aim above all at obtaining, and rigorously manipulating, *rules* and *formulas*. The field of modern mathematics was now entered into once and for all, and that of the old science of motion, with its ontological and geometric ambitions, was left behind."[58]

Varignon's work was thus strikingly innovative. "[He] definitively broke with the procedures of infinitesimal geometry adopted in the science of motion by Newton, the Bernoullis, and, to a certain extent, even Leibniz," Blay writes. "[His work] left the seventeenth century behind, announcing instead the enormous development of mathematical physics in the eighteenth and nineteenth centuries."[59] It would be wrong, however, to conclude from this that he created analytical mechanics ex nihilo through a sheer act of singular genius. Leibniz and Malebranche were obviously crucial influences, and Newton's *Principia* played a key role as well, even if the treatise was anything but the overdetermining source for his science. Overall, an idiosyncratic combination of Newton's, Leibniz's, and Malebranche's thinking forged in the peculiar cultural climate of 1690s France was most instrumental in leading Varignon in the directions he moved.

To fully understand the precise nature of Varignon's historical achievement, consider first what it did not include. One absence was Newton's physics, especially the strong argument for universal gravitation found in the *Principia*. Notably absent as well was Leibniz's actual system of metaphysical physics, a more surprising fact perhaps given the central importance of Leibnizian mathematics to Varignon's work. As early as the 1670s, Leibniz had come to believe that the evacuation of Aristotelian qualities from the modern conception of matter had gone too far. Extended matter alone could not account for natural phenomena, he argued, and he therefore began to rehabilitate the Aristotelian substantial forms as a remedy. He used the word "force" (*vis*) to describe the substance beyond mere spatial extension that matter must contain, and he further developed a complex mechanics rooted in the interaction of these forces within and among bodies.[60] In 1686, he published an application of his theory to the science of mechanics, arguing that the Cartesian measure of the force of motion (the product of a body's mass and its velocity) was erroneous. In this paper, he demonstrated that the real force of a moving body was captured by the product of the mass and the square of the velocity, and he also used this demonstration to support his larger claim that bodies must be composed of both extension and "living force," or *vis viva*. He further argued that *vis viva*, measured as mv^2, and not motion, measured as mv, was conserved in impact.[61]

Cartesians in France, including many close to Varignon, were quick to challenge Leibniz's argument, and out of these rebuttals the so-called *vis viva* debate began.[62] The debate, which amounted largely to a confusion about terms, would not be resolved until the next century, but, as Thomas Hankins and Mary Terrall argue in their accounts of the struggle, more than the measure of force itself was at stake. Rival conceptions of matter anchored the different measures that the Cartesians and Leibnizians defended, and these rival theories of matter often drove the antagonism even when the dispute itself was stalemated. Leibniz was a case in point since much of his scientific work was centered on building the philosophical and empirical explanations necessary to cement his conception of force-laden matter. Malebranche, driven by his own metaphysico-theological commitments, was equally devoted to the Cartesian notion of a forceless, purely extended theory of matter. He accordingly developed a rival mechanics and physics that conformed to his theory of matter, and debated these matters with Leibniz in public in the *Nouvelles de la république des lettres* and in his short book *Des loix de la communication des mouvements*.[63] Bernoulli made clear his support for Leibniz's position in a letter to l'Hôpital, and this triggered the marquis to defend Malebranche in a series of letters that debated the appropriate measure of force.[64]

Varignon, however, manifested no interest whatsoever in these questions, nor in the wider philosophical and metaphysical issues central to them. Typical of his approach to metaphysical physics was his response to a very long account of the infinitude of the cosmos that Bernoulli sent to him May 1698.[65] The letter was saturated with metaphysical and theological observations that appeared to Varignon to be "very true." But he reduced the entire presentation to a mathematical deduction, asserting that "all this seems to me to be a necessary consequence of the doctrine of infinities of different sorts, of which I am perfectly convinced."[66] Narrow-minded Malebranchian mathematicism was typical of Varignon, and Fontenelle described these tendencies in his quotidien routines. "[He] would pass entire days in work," the secretary wrote in his "Éloge de M. Varignon," indulging in "neither diversion nor recreation except for occasional walks when his reason forced him out of doors." As we saw earlier, he also liked to work well into the night, often experiencing with surprise "the bells announcing that it was two hours after midnight," but also accepting the notice with delight "because it meant that he could sleep for two hours before arising again at four to continue his work."[67]

An intense, single-minded focus on mathematical research alone was typical of Varignon, and his professional position in France further encouraged his narrowly specialized labors. The early courtly structures of the Royal Academy encouraged intellectual polymathy because it esteemed the liberal, gentlemanly values of the Republic of Letters more than those of the specialized technical expert. The administrative Academy, by contrast, encouraged narrow expertise, and Varignon, the son of a mechanical artisan, entered the royal company as a classic administrative hybrid who combined talents in the liberal pursuit of mathematics with an interest in developing the practical disciplines of physico-mathematics. His work accordingly situated him somewhere in between the exceedingly instrumental and practical mathematics of mathematical *élèves* and the classically liberal algebra and geometry of Rolle or l'Hôpital in the spectrum of late seventeenth-century academic mathematicians. The award of a *pensionnaire* position at the Royal Academy in 1699 supported his hybrid position, and when the new academic administrators asked him to focus on a single research project, and to publish actively in this one area, Varignon found the new discipline easy to accept. He never matched Leibniz, l'Hôpital, and Bernoulli in the realm of theoretical mathematical innovation, but he was nevertheless singularly influential within this group in bringing their insights to bear in the applied and empirical domains of mechanics. Varignon's academic position further encouraged precisely this orientation, and in this way his new science of motion was an intellectual achievement molded in powerful ways by the precise social and institutional milieu that sustained Varignon's work during its crucial years of formation.

Analytical mechanics bore the traces of this precise combination, especially in its conception of the science of motion as a mathematical endeavor applicable to, but in no way determined by, the physics and metaphysics of motion. In 1690, only two years after the *Journal des savants* reviewed Newton's *Principia*, but before Bernoulli had introduced the infinitesimal calculus into France, Varignon published his own fluid-vortex account of gravity. His study was part of a general trend, for Leibniz published his own vortical theory of celestial mechanics in 1689, claiming (mendaciously) that he had not consulted Newton's *Principia* before producing it.[68] Rohault's account of Cartesian vortical mechanics was also widely available, and Huygens published his critical analysis of the fluid-vortical explanation of *pesanteur* in 1690 as well. Bernoulli further began research into vortical mechanics in this period, a project that would occupy him for the rest of his life. After his initial explorations in mechanical physical theorizing, however, Varignon abandoned all interest in such work, evincing neither any interest in nor devotion to such research again.[69] As a result, his analytical mechanics, initiated in the years immediately after the abandonment of his mechanical theorizing, became, in ways unlike Leibniz's, Malebranche's, and Newton's, a mathematical treatment of motion detached from any particular causal physical or metaphysical explanation of it. In short, Varignon became a strong sort of mathematical phenomenalist at a moment when no such position actually existed in the scientific field of the day. His narrow single-mindedness as a mathematician, and the institutional forces that supported precisely this tendency, may offer the best framework for understanding his innovations.

Academic Mathematics in the New Public Academy after 1699

Also important in producing this outcome was the institutional environment that supported and shaped Varignon's work, namely the Académie Royale des Sciences during the most intense years of the Pontchartrain reforms. Varignon's science, when viewed in hindsight, certainly shows all the traces of the influences sketched above, and as a historical artifact it can be described as an original crystallization of all of these strands of seventeenth-century mathematical and scientific thought. But at its initiation, analytical mechanics was not conceived according to the intellectual influences that later historians would see in it. It was made through the day-to-day practice of science at the Royal Academy. It was also shaped by the debates present there without any self-conscious awareness that a new path into a new kind of mathematical physics was being charted. Varignon certainly sensed the innovative nature of his work, and the claims he made for his new science were nothing if not exuberant. Yet his claims were also directed at the field of contestation that had emerged in the Royal Academy in the 1690s and not toward some imagined historical future. What had brought Varignon to his new science of motion was his assertive use of the new differential calculus to expand the reach of mathematical science, and given the controversies surrounding the calculus in 1698, it was this issue, as opposed to all the others his new mechanics might have posed, that was the focus of debate.

The entanglement of the new mechanics with the ongoing debates regarding the validity and value of infinitesimal analysis is evident at the very moment of initiation of Varignon's new science. Varignon's breakthrough paper in July 1698 used the calculus to propose general rules, which were expressed as differential equations, for the motion of all bodies of any sort moving at whatever speed one likes. Overall, it was the exceptional generality and universality of his rules that Varignon most trumpeted. Yet underlying his claim was the necessary corollary that the rules were also applicable to all manner of particular cases.[70] From this perspective, de la Hire's paper presented on August 9 regarding the general rules governing the measure of falling bodies should be seen as an implicit response to Varignon's work even if the two papers had no direct point of contact between them. As had become his custom over the previous two years, de la Hire used his presentation to show how traditional geometry, including analytic geometry, could deal with the infinitely small moments of change without having recourse to the infinitesimal calculus. True to Varignon's claim, this required de la Hire to devote pages of geometric argumentation to show the same relation that Varignon had captured in a single differential equation, but his point was that the absence of economy in his work was more than compensated by the assurances about precise rigor that he offered.[71] Varignon's September paper also revealed the presence of this contestation in his work, for while he built upon his July paper by applying his general rules to the motion of bodies taken at whatever speed one likes along any curve whatsoever be it mechanical or geometric, he also offered a new way of showing the particular motion of falling bodies along the cycloid.[72] This precise topic had occasioned one of de la Hire's geometric interventions in opposition to the infinitesimal calculus months earlier,[73] and in his solution Varignon reduced to a few pages of differential analysis what it had taken de la Hire dozens of pages of geometry to accomplish.[74]

Varignon continued in this vein with mathematical papers presented to the Academy in November 1698 and January 1699.[75] It was at this moment that the Academy reform intervened to transform the debate in fundamental ways. One important change was demographic.[76] The reform brought over a dozen new academicians into the company, but the net result for the analyst community was a decline in membership. Malebranche entered at this moment as an *honoraire*, a classification that did not mark him as less than a full participant but as an academician with honorable status according to Old Regime notions of protocol. Malebranche was made an *honoraire* because he was an ordained priest, and l'Hôpital was likewise added to the *honoraires* because of his many aristocratic titles. From this position, both became active academicians even though their status exempted them from the disciplinary management and publication expectations enforced upon ordinary members. Varignon's *élève* Carré, admitted in 1697, was also formalized into this newly established rank as a result of the reform. This change brought new institutional rules and expectations to adhere to, but also a new career path for him with opportunities for promotion available should he perform well in his duties.

Yet countering these additions to the academic analyst community was the departure of Sauveur from his *associé géomètre* seat soon after his new appointment. The new regulation requiring Parisian residency and assiduous attendance at meetings was the reason for his departure. In a letter that was read to the Academy three weeks after the reform was initiated, Sauveur explained that his duties at court made it impossible for him to meet these demands. He therefore asked the Academy to accept his resignation. Bignon complied, making Sauveur a "veteran" who could still attend meetings at Bignon's discretion, and he then gave Sauveur's pensioned position to Lagny. Yet Lagny did not give up his positions at Rochefort as a result, a fact that reveals the continuing role of Bignon's favor in shaping academic life despite the apparent move to bureaucratic self-governance. Lagny accordingly became a member of the academic *géomètres* in 1699 even as his presence in the Academy and activity in its debates dissipated.

Also important in changing the mathematical debates within the Academy was the new disciplinary organization of the members. The first class of pensioned *géomètres* was composed of Varignon, Rolle, and Gallois. As an *honoraire*, l'Hôpital was in practice a member of this group as well, and Malebranche often joined them. But other mathematicians found themselves classified in ways that pulled them away from this identity. De la Hire is a case in point. He was made, along with Cassini, one of the first pensioned *astronomes*. His son, along with Cassini's son, was also moved out of his *élève* position and into the new *associé astronome* positions. Cassini's nephew Giacomo Filippo Maraldi, who had also been appointed as an *élève* a few years earlier, was likewise made into an *associé* in 1699, but as a *géomètre*, not an *astronome*. At one level, these disciplinary classifications carried no meaning; academicians were not constrained by them in choosing their research. Yet given the particularities of the astronomical community in France, with its separate residence, work space, and collectivist ethos, the clarification of these academicians as astronomers first and foremost did have an impact. Whether the result was a mere coincidence or a change provoked by the new institutional arrangements, de la Hire ceased being a major protagonist in the calculus debates after 1699 even though those debates increased in intensity after this date. He remained a very active academician nevertheless, and what filled the space left by his departure from questions of geometry was more astronomical work and more work on topics in empirical mechanics, a long-standing research interest for him that was newly activated by the arrival of the new and explicitly utilitarian class of *méchaniciens* added into the Academy in 1699. De la Hire's statement of personal research focus was indicative of the new arrangements. He declared a series of astronomical projects that joined with Cassini in advancing the program of the Royal Observatory. But he also joined with the new pensioned *méchaniciens* des Billettes, Jaugeon, and Father Sébastien Truchet to propose inquiries into new hydrostatic machines for pumping water and regulating its flow.[77]

Even if de la Hire's shift of focus after 1699 was not prompted by the new disciplinary organization of the Academy, other developments clearly were encouraged by this change. Most important was the emergence of Michel Rolle after 1699 as the leading antagonist of the calculus—and as Varignon's particular enemy during the initial reception of his analytical mechanics. The connection between the institutional changes at the Academy and Rolle's new intellectual commitments is illustrated in the declarations of personal research that each academician was asked to submit in February 1699.[78] In his public presentation of the new regulations, Fontenelle likened these declarations to a kind of oath that each academician swore to the company, its patrons, and the public at large.[79] Yet in recording these declarations in the academic registers, the secretary also noted the proposed deadline for completing the project offered by each academician, an indicator that these were also official acknowledgments by the academicians of the new professional discipline expected of them, along with declarations of the criteria by which they wanted their performance to be judged. Accordingly, as the academicians began to realize the programs stated in their proposals, their status in the Academy became newly tied to their visible work in precise disciplinary projects in ways that had not been the case before.

In the case of Varignon and Rolle, the most visible pensioned *géomètres* in the Academy, this new structure fueled the eruption of a classic professional turf war even if such an occurrence was a complete novelty in 1699 given the absence prior to this date of the institutionalized disciplinary structures necessary to sustain such a battle. Not all of the public declarations created this kind of internecine strife, and it was in fact the intention of the reform to foster a spirit of academic collegiality and consensus, not a climate of partisan sectarian strife. Bignon set the tone in his declaration of his personal research project, which Fontenelle noted was not even necessary because "his status suggests no other occupation than directing and caring for the Academy, and beyond this he needs to do no other work." Nevertheless, Bignon pledged to realize the publication program of the Academy, promising to use it to "render an account to the public of the manner in which [the Academy's] work is conducted, and the good intentions [*voeux*] which motivate it."[80] All of the other declarations echoed Bignon's in proposing an *honnête* and noncombative research agenda.

Yet while the declarations made by the *géomètres* were uniform in their avoidance of openly rancorous and contestatory framings, they nevertheless articulated through their precise research agendas the terms of the battle to come. Varignon promised to pursue his new theory of motion founded, as Fontenelle explained, "uniquely on a very simple principle, which he has already given in the *mémoires* of the Academy in 1692." "He will embrace everything regarding motion and push his theory to infinity," the secretary exclaimed, offering a description that would reveal, like many to come, his own personal enthusiasm for Varignon's work. "Not only the known properties and ordinary hypotheses of motion will be included in his theory, but every hypothesis imaginable including the most bizarre. For example, he will not only explain [*rendre raison*] everything regarding the fall of bodies in relation to the equal increase of speeds in the manner of Galileo and almost all the other mathematicians who have come after him, but he will also include the variation of speeds in whatever way that one would like. This should lead to a book in quarto and it should be completed by the end of next year."[81] Varignon's *élève* Carré also declared a project that aligned him with Varignon's analytical agendas. "The differential calculus, which in so little time has opened up vast and profound areas of geometry," Fontenelle wrote, "consists by considering curved lines as composed of certain infinitely small elements. . . . To treat the surfaces and solids defined by these curves, another calculus is necessary that reassembles these infinitely small parts into wholes, which is why it is called the integral calculus. . . . The integral calculus has not yet been fully studied, so M. Carré proposes to work on it. . . . The work will only require eight or ten sheets in quarto and will be finished some time after Easter."[82] Carré was in fact true to his promise, presenting a finished copy of his book on the integral calculus to the Academy for approval in August (Varignon and Malebranche were assigned to evaluate the manuscript). He then published it in early 1700 at the moment when the calculus wars inside the Academy were just beginning to grow more heated.[83]

Framing the platform of the opposition was Gallois, who proposed as his project a recovery of the work of the mathematical ancients. "Almost every *géomètre* today applies himself to the new methods of geometry," Fontenelle explained in a description tinged with sarcasm. "But since M. l'Abbe Gallois judges it important to avoid abandoning altogether the methods of the ancients, he will make the illustrious work of the ancient mathematicians his contribution to the Academy." His precise project was a new translation of Pappus of Alexandria and other antique mathematicians, and Fontenelle expected to see the first volumes realized early next year.[84]

Gallois would become an ardent opponent of Varignon during the next decade, and his other and even more strident enemy Rolle also revealed his cards in his personal research proposal. Fontenelle expressed his distaste for Rolle and his work in his terse description of it, stating only that, "he will reduce to a rigorous theory the best methods that algebra has developed to date, and he will form from that theory other methods which are necessary for perfecting geometry." Rolle's *élève* Du Torar was given similarly short shrift by the secretary, for as the secretary explained: "Experience having shown that the simplest equations in algebra have the most usage in geometry, [M. Du Torar] proposes to give the general rules for abridging the solution of equations up to the fifth degree. He should be done in about two years."[85] The subtext informing these descriptions was revealed a few months later when Rolle presented the first of his papers in pursuit of his project "Remarks on Algebra and Its Usage." In it, he stressed the need for rigorous method in algebraic work lest errors arise without being perceived. Especially worrisome, he stressed, was the inability to distinguish sound from defective method "when dissolving the unknowns common to several equals" in the algebraic solution of geometric problems. This was a gesture toward the infinitesimal method, which was being used to perform exactly this operation. Fontenelle articulated his judgment of Rolle's argument in his description of the meeting on May 9 when Rolle presented his paper. "M. Rolle began to offer some remarks of his own fashioning on the methods of Algebra," the secretary wrote, "which although widely held appear to him to be very defective."[86] Du Hamel had never broken with a tone of judicious neutrality in his academic record keeping, and the salty commentary flowing from Fontenelle's pen in this instance was both a marker of the changes afoot and a harbinger of things to come.

In the months after the reform was enacted, the pattern evident here of new calculus-based mathematical work being presented at the Academy alongside papers by Rolle that pursued algebra without the use of such methods became normal. Also established was the pattern of Fontenelle using his commentary, restricted at first to the academic registers, to shape the reception of this work whenever possible. In the case of Rolle, who presented papers related to his work in algebra in May, July, and December 1699, and then again in March, May, and November 1700, Fontenelle often simply tolerated the pensioned algebraist by leaving his ideas buried without comment in the papers that he dutifully transcribed into the registers. An exception was Rolle's December 1699 paper that directly addressed the evaporation of unknowns within the differential calculus. Rolle admitted that the calculus could be used effectively in certain cases, but he also worried openly about the absence of a clear and secure method for its reliable use. Without such a clear rule of operation, Rolle believed it was better to use other more transparently valid approaches no matter what the inconvenience. Fontenelle recorded Rolle's paper in the register, but at the end he also noted that "since the *géomètres* of the Academy were surprised to hear M. Rolle advance the idea that the equalities containing two unknowns do not express a geometric line, he promised to bring examples of this to the next meeting."[87] Rolle complied on December 12, bringing three examples, and no further commentary was offered.[88] But the case illustrates again how academic debates were unfolding in new ways given the presence of the newly assertive secretary backed by assertive administrative authorities in the assemblies.

The *géomètres* who had demanded these examples were likely Varignon and l'Hôpital, and during the same years each competed with Rolle to fill the academic sessions with their own calculus-based work. Varignon gave papers in March, July, and August 1699, and then again in January, March, May, August, and November 1700; l'Hôpital gave papers in June 1699 and in January 1700 and 1701. His June 1699 paper was also revealing of the place of Newton's *Principia* within these early discussions. L'Hôpital's problem concerned the shape of a solid in a fluid that exerts the least resistance when rotated on its axis. His point of departure was the work of Newton's disciple Fatio de Duillier, who had developed a solution that drew on Newton's work in Book II of the *Principia*. L'Hôpital used the calculus to find "the most natural and simple solution" to the problem, and to "clarify" Newton's work in ways that are not found in the *Principia*. But he also made clear that his solution "was no different from Newton's" even if his had wider reach because it was useful in solving similar problems not contained in this precise case. In this way, l'Hôpital claimed to both confirm Newton's result while also surpassing him through the use of a more powerful and universal mathematical method.[89] This relationship to the *Principia* closely mirrored Varignon's, and as we will see in the next chapter it was through a similar understanding that Varignon both built his own analytical mechanics from the work he found in Newton's treatise while also claiming to surpass it in the creation of a wholly new calculus-based science.

Viewed solely in terms of the activity of its mathematicians, the Royal Academy in the first two years after the 1699 reform witnessed a competition between Varignon and Rolle to fill academic sessions with their distinct approaches to algebraic mathematical work. Compounding the contest was the wider struggle among all the academicians to get their work in view lest Bignon find them lacking. The 1699 reform had created an imperative for all academicians to visibly display their research, and Malebranche, l'Hôpital, Varignon, Carré, and Rolle comprised less than 10 percent of the membership as whole. Every academician was competing for limited airtime, and an example of the new dynamics that this generated was the new initiative launched by an ambitious young chemist, Etienne François Geoffroy. Appointed in January 1699 as the *élève* to the chemist Homberg, himself one of Bignon's new appointments in 1692, Geoffrey was made an *associé chimiste* a month later with the enactment of the new regulations. The son of successful pharmacist, Geoffroy had spent several years in England before 1699, where he acquired fluency in English and made the acquaintance of members of the Royal Society of London, who had made him a Fellow in 1698. Seeing a comparative advantage in his ties to English science, and taking as his animating agenda the regulation that asked academicians to develop communication networks with foreign savants, Geoffroy began preparing French translations of the English articles found in the *Philosophical Transactions of the Royal Society of London* in order to read them to the Academy. These readings became a regular feature of academic meetings after 1699, and in 1706 Geoffroy began reading his own translation of the first English edition of Newton's *Opticks* to the Academy before the appearance of the Latin edition made such translation unnecessary.[90] Other academicians began to present their own precise research in chemistry, anatomy, botany, and mechanics at the same time and for the same reasons, and the result overall was a general clamor inside the Academy to produce new work and to make it visible to the Academy and its publics.

At first, the competition between Varignon and Rolle was just one example of this general competition for attention, which marked the new life of the greatly expanded Royal Academy after 1699. But in August 1700 the battle between the two academicians became direct and personal. An important source for this newly individualized animosity was the way that the long-standing mathematical differences that had always separated these two men acquired new meaning and importance within the Academy as a result of the 1699 reform. A first provocation for Rolle was likely the decision, made no doubt by Bignon, and perhaps through consultation with Fontenelle, to include Varignon and his calculus-based mechanics among the group of four academicians selected to present at the Academy's first public assembly, held on April 29, 1699. Evidence of this provocation is found in the fact that Rolle gave his first paper questioning, indirectly at first, the use of the infinitesimal method in the evaporation of unknown quantities at the very next academic session after the inaugural public assembly had been held. Even if Rolle was not directly stung by this decision, Varignon's presence on the dais at this first public spectacle was a clear indication of his favor and status within the Academy overall. He would be asked to give a second presentation about analytical mechanics at the fourth public assembly held in November 1700, and because this paper marked the intensification of his program of analytical mechanics amid the eruption of open battles about it inside the Academy and in the wider public, it will be discussed in the next chapter. But in the context of this discussion the point to emphasize is Varignon's appearance at two of the first four public assemblies, and the evidence that this favor provides regarding the esteem in which he was held within the Academy during these years.

Documentary evidence describing the events of the very first academic public assembly held in the halls of the Louvre on April 29, 1699 is thinner than we would like. What was no doubt a major public spectacle that attracted broad public interest survives for us only in a few brief documents: the official academic records of the event, a letter by a Scottish physician named John Monro who was invited to attend the proceedings by Cassini, and a report published in the *Mercure galant*. The academy records agree with Monro's letter, reporting that the session was opened by the president, who was Bignon in 1699, who "read a short unprepared statement explaining to the assembled audience, which was very large, what an academic assembly was, and how it functioned normally since this assembly, although public, would be conducted in the usual manner."[91] Monro described Bignon's presentation as a set of instructions for conduct, namely that the audience should respond to the papers rationally, not emotionally, and that they were expected to remain silent while the speakers read prepared papers. They were also asked not to interrupt the presentations. After Bignon's introduction, four academicians spoke: two recently elevated *élèves*, who presented the work of their former mentors (Cassini's son on his father's theory of comets and Geoffroy on Homberg's recently devised "aerometer") and two *pensionnaires*: Tournefort, who spoke on the distillation of plants, and Varignon, who used the differential calculus to analyze the mechanism of a water clock. Monro noted the particular complexity and difficulty of Varignon's paper, reporting that he used a diagram to illustrate his ideas and a large pointing rod to help readers follow his argument. Afterward, he reported, Bignon quipped that it was "happy for him [Varignon] that he had an audience so learned, so natural was it for men who understood not these matters to take him for a conjuror."[92]

The report in the *Mercure galant*, although brief, offers a different perspective, and one more attuned to the French audience of interest to the royal administrators supporting this new academic public outreach. The *Mercure* briefly summarized each of the scientific papers read at the assembly, yet attuned to the interests of its readers its report offered only the briefest account of the scientific presentations while directing attention instead toward the activities of Bignon and the spectacle of the elegant crowd in attendance.[93] Varignon's work was not passed over in silence, however. As Bignon noted later, after experiencing several of Varignon's public performances, his work was "over the head of most of the audience," but he nevertheless "succeeded in making himself reasonably understood."[94] It was no different in 1699. As Monro noted, Varignon's presentation on the mechanism of water clocks was "extremely complicated," but the *Mercure* was more appreciative. "The problem had previously been treated only according to particular cases," the reporter wrote. "But M. Varignon embraced it in all its infinite generality. He did so by using infinitesimal analysis [*l'Analyse des infiniment petits*], a new method which gives to geometry a sublimity and a fecundity that it has never had, but which must be practiced with an extreme circumspection and delicacy."[95] This brief and subtly provocative account reveals the extent to which the *Mercure* and its audience were attuned to the complexities of the new calculus-based mechanics beginning to come to life in France.

Yet Bignon's jibe at Varignon that he could have been viewed as a kind of mathematical conjuror was germane as well. For when one looks at the actual paper Varignon allegedly presented at the public assembly as recorded in the Academy registers, one wonders how he ever made its contents accessible to anyone who was not already trained in the methods of differential analysis. Figure 8 shows an image of one page of Varignon's paper as it appears in the Academy registers for April 29, 1699. He cannot possibly have read this paper to the audience in its entirety and verbatim. If so, how did he present the large algebraic equation at the top left of the page? And even if he used a diagram and a pointer to help, how much of the detailed mathematical content made it into Varignon's actual presentation? Given that the *Mercure galant* accurately described the larger argument of Varignon's work, and the reasons for his use of the infinitesimal calculus in it, he clearly succeeded in conveying his main claims. But how much of the journalist's report was the result of an actual comprehension of the details of Varignon's mathematics, and how much a mere parroting of Varignon's verbatim statements to this effect in his presentation? In the very next academic session, Rolle would begin arguing for the need to preserve "the most simple, exact, and natural methods of reasoning in algebra" and to avoid "deceitful shortcuts that often lead to errors."[96] Was this a response to the public conjuring of solutions that Varignon was engaging in with the differential calculus?

Whatever the reasons for the animosity, the argument between Varignon and Rolle inside the Academy grew more bitter and personal in the year following the Academy's first public assembly. Rolle's papers on the evaporation of unknowns in algebraic equations, which provoked the salty commentary of the secretary, were presented in May and December 1699, but at the end of that year the registers were signed by Bignon, Fontenelle, and Pontchartrain with the statement that "based on the account that I gave to the King of the

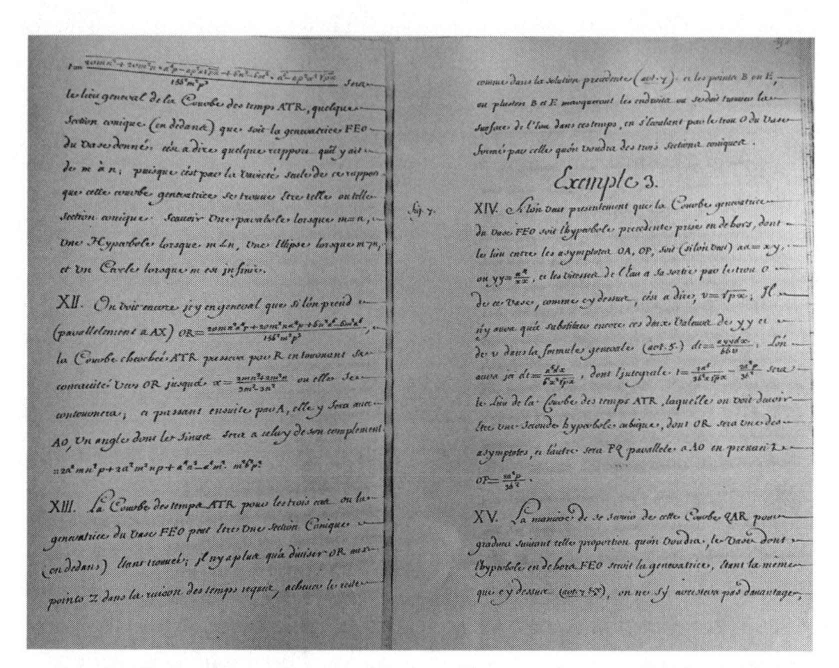

FIGURE 8. *Varignon's April 1699 public-assembly paper as transcribed in the* Registres de l'Académie royale des sciences.

contents of this present register, His Royal Highness seemed to me content with the zeal by which we have executed the new regulations."[97] The work of institutionalizing the new academy was only just beginning, however. The first public assembly marked the official move of the Academy to the halls of the Louvre, a change designed to grant more space to the greatly expanded company while also giving it a new and more public image. To set the right tone for the new meetings, a seating plan was instituted (if never revealed to the public) that effectively dissolved any formation of "disciplinary parties" around the Academy's meeting table. In this protocol, members were arrayed around the table according to disciplinary class (i.e., *méchanique*, *astronomie*, *chimie*, etc.) and then rank (*honoraire*, *pensionnaire*, *associé*, and *élève*) such that the Academy table never had two members of the same discipline sitting next to each other. The motive here was to foster a republican spirit of inquiry where each individual was individually responsible for his own views, while also disrupting factionalism and the formation of disciplinary parties.[98]

The Varignon–Rolle battle exemplified everything that this seating chart, and the overall program of administrative reform more generally, was designed to avoid. Yet fueled by their ambition and passion for their individual research agendas, the two men began to spar openly. Regulation XXVI had declared bluntly that academicians were required to maintain gentlemanly decorum, stating: "Whenever several academicians are of differing opinions, the Academy will insist that members use no term of derision [*mépris*] or ill will [*aigreur*] against one another either in their discourses or in their writings. The academy will also exhort its members to challenge the views of others in a restrained manner only."[99] Yet by August 1700, Varignon and Rolle were behaving in ways that challenged this decorum. The Academy records do not give us access to what must have been a rapid escalation of heat in the exchanges between the two academicians, but in August, Varignon put his frustrations in writing, reading a paper to the Academy titled "On the Defense of the Geometry of the Infinite against M. Rolle."[100] The remarks to which Varignon responded must have been made orally, since no preceding paper by Rolle systematically attacking infinitesimal geometry is to be found in the registers. Once positioned this way, however, Rolle responded in kind, asking at the conclusion of the reading of Varignon's paper that he be given a copy so that could prepare his own paper "showing the paralogisms into which the method of infinitesimals leads necessarily." What ensued was *l'affaire des infiniment petits*, a public *querelle* that engaged Rolle, Varignon, and a host of other individuals in an increasingly heated and public battle at the very instant that the new Academy was starting to establish its footing.

At the same moment, Varignon was also getting ready to present his second public-assembly paper, a treatment of the centripetal forces operative in planetary bodies that used the infinitesimal calculus to develop its arguments. This turn into the motion of planetary bodies marked a new intensification of Varignon's larger program in analytical mechanics, and it also brought Varignon into direct contact with Newton's work in the *Principia*. Varignon's new analytical treatment of central-force mechanics became the trigger for the escalation of the battles in France about the infinitesimal calculus, battles that grew more heated and more public during the first decade of the eighteenth century. What therefore ensued after 1700 was a threefold set of developments: the continuation, climax, and ultimate resolution of the debate about the validity of the infinitesimal calculus; the full maturation of analytical mechanics and its reception as a French academic science; and the full institutionalization of the new public Academy and its practices and protocols. Analytical mechanics became what it became in France because of all three dimensions of this historical entanglement, and it is to this ferment, and the scientific consequences that it produced, that we now turn.

CHAPTER 9 *Analytical Mechanics Goes Public "La Querelle des infiniment petits"*

Although initiated before the establishment of the new Academy regulations, and created through a synthesis of a wide array of late seventeenth-century sources, analytical mechanics was developed and established in France entirely after February 1699. In this respect, it was as much a creation of the new public Academy as it was a product of Varignon's scientific genius. This new and radically innovative science also activated the fledgling structures of the newly conceived public Academy with an intensity rivaled by no other. These public birth pangs for analytical mechanics were important on at least two levels. On the one hand, the new science challenged the institutions of public science at the outset by exposing them to the very controversy deemed most worrisome to those invested in the success of the reforms. On the other hand, it also placed Varignon and his new science of motion in a new and unprecedented relationship with the broader public at a time when important realignments were occurring with respect to the character of academic mathematics and the public identity of academic mathematicians. Together these entangled dynamics led to the establishment of a new science in France along with a new understanding of the place of its practitioners inside the Royal Academy and within French society as whole. This chapter examines this tumultuous beginning, and Varignon's second public-assembly paper, delivered on November 13, 1700, can serve as our point of reconnection with the story as it has unfolded so far.

The title of the paper was brief—"On Central Forces"—and the equally short review of it that appeared in the *Mercure galant* captures well the essence of its argument. "M. Varignon spoke about the force that directs [*fait tendre*] all the planets toward the sun," the journalist explained, "or what is called their gravity with relation to the sun [*pesanteur par rapport au soleil*]. He demonstrated geometrically that there must be such a force in order to describe the oval orbits that they trace. For the cords assumed by the Ancients, and even some of our Moderns, can no longer be sustained."[1] The review offered nothing more, but this synopsis points directly to the key aspects of the paper that made it significant. First, it notes the focus on celestial mechanics, a new topic for Varignon different from his earlier papers on the mathematical description of motion in general and its application to problems of terrestrial mechanics, especially the free fall of bodies. Second, it articulates well the focus of the paper: on the forces involved in keeping planetary bodies moving around the sun in elliptical orbits in the manner of Kepler's laws. Although the review does not state it—an absence that is very significant, as I will discuss shortly—this combination of interests points directly to Newton's *Principia*, which has as its primary focus the relation between the mathematical laws of moving bodies and their relation to the forces governing the motions of the planets. The review does not mention Newton because the speaker did not invoke this frame in presenting his work. Yet underlying Varignon's paper was a new engagement with Newton's work in the *Principia* in the further development of his analytical mechanics. Finally, while the review does not mention that he used the differential calculus to sustain the "geometrical demonstrations" of his paper, it does stress the mathematical character of the work, capturing correctly the way that it was a continuation of Varignon's calculus-based program of mathematical mechanics begun in 1698.

In 1700, amid the initial institutional upheavals attendant to the overhaul of the Royal Academy, Varignon made a turn toward celestial mechanics as the focus for his new science of motion. This turn also led him into a new engagement with Newton's work in the *Principia* as a source of insight for his work. Yet to call Varignon's work after 1700 a mere translation of Newton's science into the idioms of the Leibnizian calculus, as the traditional scholarship has done, is to ignore the complexity of his thinking and the turbulent environment shaping his work. This chapter will proceed, therefore, by returning one last time to the question of the Newtonian origins of Varignon's new science of motion, showing how Newton's work in mathematical celestial mechanics did and did not influence the full development of analytical mechanics in France. It will then examine in detail the French developments, both intellectual and institutional, that actually brought this new science into being.

The Newtonian Sources of Analytical Mechanics Revisited

Viewed abstractly and intellectually, Varignon's science ultimately derived from an idiosyncratic fusion of Leibnizian, Malebranchian, and Newtonian thinking together with a host of wider French strands of mathematical and scientific thought. Nevertheless, the analytical mechanics that he developed in the eighteenth century was crucially dependent on Newton's *Principia* in ways that cannot be quickly glossed over. In his public-assembly paper of November 1700, Varignon proposed to treat the motion of bodies according to what he called "central forces," or to use his more phenomenalist language, those forces that compel a body "toward a point C considered as a center, no matter how this motion is produced."[2] The category of "central forces" had not appeared in Varignon's earlier papers because those had set out to establish a set of general mathematical laws for the motion of bodies, applicable to all situations. With the introduction of the new category of central force, however, this generalized mathematical abstraction was lost. His analytical mechanics also became more empirically and mechanically oriented in 1700, culminating in a set of papers that applied the new science of motion to the precise problem of planetary motion and celestial mechanics.[3]

Varignon's papers of 1700 therefore mark a turning point. They introduced a new and ostensibly physical category into his analytical mechanics, "central forces," that on the one hand remained mathematical in its actual deployment but on the other suggested physical action and the application of mathematics to actual motions in nature. In making this change, Varignon also injected a dynamic, physical component into a mathematical science that did not previously have one. Indeed, prior to 1700, Varignon had continually celebrated the way that his mechanics could treat any motion whatsoever, including those that appeared bizarre and unnatural. A noticeable change is therefore present when he began to focus in a singular way after 1700 on the motions of planetary bodies in ellipses and the centripetal forces that determined such movements. The origins of this turn were clearly Newtonian.

To state the point bluntly, after 1700 Varignon's analytical mechanics became increasingly preoccupied with the concepts and problems of Newton's *Principia*. Consequently, at least from this date forward, if not earlier, the question of his Newtonianism looms large. In the papers of 1700, he adopted Newton's concept of the "accelerative quantity of the centripetal force" from Definition VII of the *Principia*, and he used it to develop his own, analytical approach to the motion of planetary bodies governed by these forces. His solutions, moreover, obtained results similar to those demonstrated by Newton in the *Principia*. When he achieved these parallels, Varignon did not hesitate to alert his readers to the match between his work and Newton's. After a demonstration in one of his papers of 1700, he declared: "It is useful to notice that without any new hypothesis, the two preceding rules have immediately given us Proposition 39 of Book I of M. Newton's *De Phil. Nat. Princ. Math.*"[4]

Other papers from this period contain similar declarations, and Varignon never hid his debt to Newton. Instead, as he summed up his achievement on another occasion: "[I have developed] a very simple formula for central forces, centrifugal as well as centripetal, which are the foundation of the excellent work of M. Newton."[5]

In this way, Varignon developed a thoroughgoing mathematical approach to celestial and terrestrial mechanics that drew profoundly on Newton's work in the *Principia* while believing that his own, more general approach made his work anything but a copy.[6] The wider intellectual community that received his work also joined Varignon in adopting this anxiety-free stance. "M. Newton and M. Leibniz were the first and only to study the different *pesanteurs* of the planets toward the sun," Fontenelle wrote in his public presentation of the papers of 1700. "However, M. Varignon has moved beyond them. . . . [His] theory gives us the solution to all the problems that one can imagine regarding the *pesanteur* of the planets toward the sun and concerning the inequalities of its action at different points in the curves traced by planetary orbits."[7] Newton claimed to have done much the same thing in the *Principia*, yet nowhere in the archive of French mathematics in the years around 1700 does one find any anxiety about this influence, or any sense that Varignon's work was ever conceived as anything other than an original scientific achievement. Likewise, the category "Newtonian" never appeared as a descriptive label for Varignon's mechanics even though the authority of Newton's *Principia* became pervasive in the discourse about it.

How was this possible? As we have seen, the *Principia* did not exert an overwhelming scientific influence on the development of analytical mechanics, and Newton was not a direct scientific influence on French mathematicians at the time in the way that Leibniz and Malebranche were. But that said, his influence upon Varignon nevertheless demands special consideration. In many of his papers, the French academician appears to do nothing more than analytically derive solutions that Newton has already demonstrated geometrically in the *Principia*. Yet nowhere does Varignon express any sense that the science that results is in some essential way Newtonian. Quite the contrary, to read Varignon is to follow someone convinced that he is discovering these principles for the first time, and someone who finds his agreement with Newton's work to be just so much icing on the cake of his own invention. Was Varignon simply naïve in adopting this attitude? Or was he a Newtonian, disingenuous or otherwise, in spite of himself, one hiding in analytical clothing?

The question is doubly important because historians have tended to treat Varignon's work in Newtonian terms. Analytical mechanics, these historians argue, constitutes little more than the translation of the discoveries of Newton's *Principia* into the language of Continental infinitesimal analysis. We saw earlier E. J. Aiton's argument that Varignon was in essence a Newtonian, but how can this analysis be reconciled with the fact that neither Varignon nor anyone else in France associated analytical mechanics with Newton's name until after 1730? This book is an attempt to answer that question by showing the depth of the French sources that actually produced Varignon's work, along with the differences between it and Newton's work in the *Principia*. Given the power and influence of this older interpretive tradition, however, the Newtonian character of analytical mechanics after 1700 demands special scrutiny.

To look more closely at the issue, consider Varignon's derivation of Proposition I.39 from Book I of Newton's *Principia*. The proposition involves the general mathematical laws governing falling bodies acted upon by central forces. Typical of Newton's style in the *Principia*, his demonstration first offers a diagram and then derives the necessary relations geometrically using the given figure. A proportionality between the force applied, the velocity of the body, and the areas covered in given times is then demonstrated, and the solution is expressed as a set of geometric ratios. Varignon proceeds differently. He conceives of the motions in terms of discrete instants expressed by the differentials dx and dt, and uses the symbol f to represent the force. He expresses the change in motion captured this way by the differential ddx and not by the changes in a geometric figure. Similarly, while Newton demonstrates the change in motion through a geometric ratio equal to the area of the square that represents the change in time, Varignon expresses this relation as $ddx = fdt^2$. Each formulation states the same conclusion and expresses the same understanding, but whereas Newton concludes with a set of corollaries that geometrically express each of the essential relationships between force, velocity, time, and distance, Varignon writes the expression $\int fdx = 1/2\, v^2$ and declares that his formula accounts for every feature of Newton's demonstration in one simple mathematical statement.[8]

At one level, Varignon's equation is the same. The same physical law is derived, and only the principles of mathematics have been employed in each. Even the obvious difference between Newton's geometry and Varignon's calculus is not as real as it might appear, because Newton assumed his own differential calculus to derive the ratios central to this demonstration even if he masked its usage within the classical geometric veneer of the demonstration overall.[9]

Nevertheless, there is a crucial set of differences between the two approaches. First, by expressing the mechanical principle as a differential equation, Varignon has conceived of the mechanics in a much more abstract and purely mathematical way. One product of this move is his justified claim to be offering greater generality than Newton. Whereas Newton needed separate demonstrations and separate geometric expressions to capture the complete truth of the proposition, Varignon can reduce them all to a single formula. In other papers, Varignon would add still more generality, considering curvilinear as well as rectilinear motion and more complex force relationships as well. Newton, of course, does something similar, but in separate demonstrations rooted in each precise case. By unifying all of these cases into one mathematical system, Varignon's analytical approach adds a new economy and generality to this previously geometric science.

Other consequences also flow from Varignon's shift to analysis. The actual motions of observable bodies remain visible in Newton's geometric approach in ways lost under Varignon's analytical approach. With Varignon, motion is captured entirely in algebraic symbols, and as a result the practice of mechanics becomes a more overtly mathematical exercise. A deeper mathematical phenomenalism follows from this manner of proceeding. While Newton's attachment to the pictorial literalness of geometry allows his mathematics to retain its claim to being a direct, if abstract, mirror of actual physical relationships, Varignon's approach does not allow for this empirical clarity. Not only is the pictorial image of the moving bodies themselves lost in the move from geometry to algebra, but the algebraic symbolization itself further adds a new ambiguity in its indirect connection with the natural phenomena it only represents homologously, to return to Peter Galison's terminology. What, for example, is the physical referent of the differential that is so central to analytical mechanics? The short answer is that it represents an infinitesimally small instant of a continuous motion. Yet conceiving of such a thing raised a host of empirical and metaphysical conundrums. No direct empirical or metaphysical definition of it is possible, yet analytical mechanics captures the quantitative relations governing the empirical phenomenon and reduces them to a powerful, predictive formula even though the mathematics contains no direct, mimetic connection to any empirical or metaphysical referent. In this way, with the analytical turn, a move toward a more deeply constituted mathematical phenomenalism is made.

This abstract mathematical universality was Varignon's own innovation, and if he used it to pull out the broader implications of Newton's work, this did not make him a Newtonian as a result. He was, rather, a Malebranchian-Leibnizian analyst, and the consequences of his adoption of this precise approach were important in France. Not only did it allow his science to resonate powerfully with Malebranchianism, and the wider intellectual discourses it supported, it also allowed the French to distance themselves from the more suspect and even pernicious aspects of Newton's physics and metaphysics.

As the *Journal des savants* indicated in its review of the *Principia* in 1688, Newton's suggestion that material attraction might be a natural fact of nature was a claim that many found unacceptable. Such an idea ran against the grain of French thought in a number of ways. First, it smacked of the outdated thought of the Ancients and threatened to return an innate attractive force into matter that would return natural philosophy to its Scholastic ignorance. Such a mysterious "occult" force also resonated with "pagan" notions of animate, active matter that lay at the heart of any number of emerging materialist philosophies. The idea of introducing an active material principle into mechanics therefore raised the specter of irreligion as well.[10] The French conception of rigorous science was also rooted strongly in the Cartesian doctrine of *évidence*, and this offered further good reasons to reject a physics that offered no rigorous epistemological justification for its claims about attractive force or its manner of physical causation. For all these reasons, the Newtonian theory of universal gravitational attraction through empty space was an idea that virtually no Frenchman in this period took seriously.

Thanks to the new analytical mechanics, however, confronting these ideas was not necessary. Because Varignon's analytical description of central-force mechanics was divorced from these problematic philosophical and physical assumptions, his science supported the introduction of Newton's mathematical principles of natural philosophy into French science without their allied physical and metaphysical arguments. Varignon used a similar approach in an obscure and posthumously published *mémoire* composed in a geometrical manner, which he wrote, for reasons that are completely unclear, on the nature of the material transformation of the Eucharist. This was Varignon the ordained priest's one and only serious work of Catholic theology, and ironically it can help us to see how Malebranchian infinitesimal analysis can be helpful for displacing knotty and contentious physical and metaphysical conundrums when seeking a scientific understanding of complex matters.[11]

At issue in the lemmas, corollaries, and scholia that Varignon offered was the proper understanding of the Catholic theology of the Eucharist, especially its claim for a miraculous material transformation of the communion wafer from ordinary bread into the divine body of the savior. How was such a metamorphosis to be understood scientifically?[12] Varignon displaced that question by offering the Malebranchian argument that whatever the actual physical and metaphysical nature of the change was, the miracle could be understood analytically as a process of infinitely many infinitely small moments of transformation. The miracle of transubstantiation was therefore scientifically comprehensible through the application of infinitesimal analysis, and analogously the actual physics and metaphysics of planetary motion was likewise not relevant to the mathematical account of it offered by infinitesimal analysis. In each case, Malebranchian mathematical analysis allowed for a complete scientific accounting of the material process of change while it avoided all claims about, or arguments over, physical and metaphysical causality. Such was the great virtue of using mathematical analysis to approach questions of physical causation, even if the absence of any direct contestation about such matters made it unnecessary to defend this philosophy in the decades around 1700.[13]

The fact that Newton often seemed to be an advocate for this same kind of mathematical phenomenalism only made Varignon's precise approach all the easier to adopt. In his *Principia*, Newton stated frequently that his mathematical principles of natural philosophy should be understood as a purely mathematical approach to physical science, and while a full reading of the treatise makes it clear that he was anything but a straightforward mathematical phenomenalist in the manner of Malebranche or Varignon, his statements to that effect in the *Principia* help to explain how Varignon could have drawn support from his treatise for his own purely mathematical approach to mechanics.[14] Varignon, to be sure, pushed mathematized mechanics much farther than Newton, creating a more complete divorce between mathematics and physics than the one suggested by the *Principia*. In realizing this agenda, he also created a science that had more differences with Newton's than similarities. The mathematics, for one, was completely different, and since Leibnizian analysis opened up possibilities that Newtonian geometry did not, this distinction was not trivial. It in fact allowed the central force that Newton called "gravitational attraction" to become in Varignon's science something inconceivable to Newton: not a fact of nature, or even a mathematical-physical principle, but an abstract, quantitative mathematical *rapport*. Similarly, while Newton built his *Principia* upon a two-part mathematico-empirical foundation, Varignon's work collapsed these two dimensions into one fully mathematical approach. The *Principia* suggested such a possibility in certain respects, but it also suggested many other different outcomes. As a result, it is a distortion to attribute Varignon's understanding of central-force mechanics to Newton alone, or to call Varignon's work Newtonian as a result.

Viewed in retrospect, what Varignon created in the first decade of the eighteenth century was a science that drew out and exaggerated those aspects of Newton's *Principia* that fit with contemporary French assumptions while ignoring and effacing others that were either less interesting or less palatable to them. By exploiting Newton's achievement exclusively in these abstract mathematical terms, Varignon also initiated a new science—analytical mechanics—that would remain dominant in France throughout the rest of the century. By unintentionally allowing this mathematical approach to displace the important empirical, experimental, and physical side of Newton's work, including the physical and empirical argument for universal gravitation, however, Varignon also pointed French science in a particular, and eventually distinctive, direction.

Modern Newtonians accept a deeply mathematized view of mechanics rooted in the practice of differential analysis. They also accept the physical reality of material forces, at least as empirically measurable facts of nature, and the empirical fact of gravitational attraction. Following Varignon, the French pioneered the first while avoiding altogether the second. In Britain and Holland, by contrast, a very different balance was struck, and it was left to Varignon's successors, in France and elsewhere, to renegotiate the future alignment of these different conceptions of physics and mechanics. The eighteenth-century Enlightenment would pick up this project around 1730 and carry it to its modernizing conclusion. In the decades around 1700, however, these negotiations were not yet necessary. For what France witnessed was not a debate about the proper relationship between mathematics and physics, or a discussion about Newton or Newtonianism—the latter term, in fact, did not even exist in the lexicon until after 1710—but a struggle over analytical mechanics and its place within French science.

Central to this struggle was the debate already under way about the validity of the infinitesimal calculus, a debate that Varignon's new science of motion energized and reconfigured. Also crucial was the expansion and reorientation of the Academy after 1699, a change that brought new protagonists and antagonists into the fray while creating new arenas and motivation for argument. While it could have posed all the knotty questions regarding the relationship among empirical facts, physical causation, and mathematical analysis and description that later students of Newton's celestial mechanics wrestled with, Varignon's public-assembly paper on central forces became instead the occasion for an intensified battle inside the Academy over the validity of the infinitesimal calculus.

In fact, as he had done after Varignon's first public-assembly paper in 1699, Rolle used one of the next academic sessions in November to read a new critique of the calculus directed at Varignon. In it he asserted that Varignon had misrepresented the similarity between his work and Newton's. "M. Newton is not as supportive as is alleged of transcendental geometry and infinitesimals," Rolle asserted.[15] This was to frame Newton as a mathematician and to describe the Newtonianism (or lack thereof) of Varignon's mechanics in terms of its mathematical claims when compared to those found in the *Principia*. Earlier we saw l'Hôpital relating to the *Principia* in exactly this mathematical way, and this was in fact the general pattern in France during these years. Therefore, as the battle over the calculus intensified, Newton's *Principia* was sucked into the whirlwind as a mathematical resource to draw upon. In this way, all the other aspects of the treatise, which the French were already largely ignoring anyway, were pushed even further to the margins while the spotlight fell with ever greater intensity on the character of the book as a treatise in mathematical mechanics.

Accentuating this focus was the intensification of the battle over the legitimacy of the calculus, which erupted as a full-fledged academic battle in 1701. Pontchartrain signed the register closing the year 1700 with the declaration that "I have read this register by order of the King, who sees with pleasure the laudable efforts of Mssrs. the Academicians in the continuation of their particular and common work."[16] Within weeks of this declaration, however, activities inside the Academy began to strain to the breaking point the ethos of *honnête* collegiality expected of the royal company. Given its new orientation toward the public, the contestation also escaped quickly beyond the halls of the Louvre, producing a spectacular public contest that the actors described as *la querelle des infiniment petits*. Varignon and his analytical mechanics were at ground zero of this struggle, and out of it his new science was brought fully into the world. It is to that *querelle* and its outcomes that we now turn.

La Querelle des infiniment petits

Varignon described the titanic battle that erupted in 1701 as an academic struggle between two parties: the Modern infinitesimalists and the Old Style mathematicians.[17] The parties themselves had begun to form a decade earlier as soon as royal academicians began using the calculus in their work. The 1699 reform did not create the parties, therefore; it solidified them into professionally positioned contesting factions. It also offered them a new institutional reason to openly contest one another. In addition to affirming the status of Varignon as one of the three senior *géomètres* in the Academy, and securing the position of his protégé Carré as an officially classed *élève*, the

1699 expansion added two new members to the Moderns' party: Malebranche and Fontenelle. It also removed two from it, however unintentionally: Sauveur and Lagny. Bernoulli, Leibniz, and Newton also arrived as foreign associates, although only the first two played an active role in the struggle.

The Old Style party was strengthened by the appointment of Rolle and Gallois as the two other *pensionnaires géomètres* alongside Varignon, and by the admission of Father Thomas Gouye, S.J., among the new class of *honoraires*. Gouye, it will be remembered, was the first Jesuit ever to be granted formal membership in the Royal Academy, a fact that is perhaps explained by his earlier service as the mathematical tutor of Jérôme de Pontchartrain.[18] Gouye was no friend of infinitesimal analysis, a trait that he shared with many other French Jesuits, as we will soon see, To his party of anti-infinitesimalists were added the foreign associates Walter Tschirnhaus and Viviani, although the latter died in 1703 and played no role whatsoever in the struggle. Historically speaking, de la Hire was also a publically declared advocate for the anti-infinitesimalist position, but whether because of his new classification among the *astronomes* and *méchaniciens* or for some other reason (it should be noted that his opposition to the calculus was never as openly polemical and heated as Rolle's), he did not participate actively in the eighteenth-century debates.[19]

In fact, if one includes the *honoraires* l'Hôpital, Malebranche, and Gouye in the group, the officially classified academic *géomètres* were almost the sole antagonists in this battle, even though the arena for it was the Academy as a whole, along with its publics. Varignon's calculus-based mechanics and Rolle's vehement opposition to the presence of infinitesimal analysis within it sat at the very heart of the struggle, and even if *la querelle des infiniment petits* grew into something that exceeded these professional academic limits, it certainly began as an internal academic struggle between two ambitious academicians and never lost its essential academic dynamic.

New sources appear in the archive after 1699 reflecting the new self-consciousness among academicians about the changing professional climate in the Academy. A new self-awareness among academicians regarding the importance of their visible work was a consequence of the 1699 reform, and whether internally generated or administratively provoked, it shaped the calculus battle in important ways. One of these new sources, a personal diary that Claude Bourdelin II, the son of a chemist by the same name who was a founding member of the Academy, offers us particular insight into the internal dynamics of the newly reorganized company. After his appointment as an *associé anatomiste* during the reorganization and expansion of 1699, Bourdelin began recording his observations of academic meetings in a personal notebook.[20] His notes offer us an inside view of the calculus battle that is not present in the academic records. He also gives us a perspective on these struggles from a nonmathematical academician engaged in judging his peers as they began to compete with one another for attention in the new public academy.

To get a flavor for Bourdelin's diary, consider these entries from his first year as a royal academician. When the *élève* François Poupart, an anatomist below him in the academic hierarchy, gave a paper a month after his admission, Bourdelin simply noted it as a "very mediocre work."[21] Likewise, when his pensioned superior Du Verney offered a dissection of the heart of a turtle at the public assembly of November 14, 1699, Bourdelin mocked the presentation, saying that "it took him the whole night to do his dissection and he did not even show the things that he said were there."[22] When Du Verney gave the same demonstration to the Academy at a regular session a month later, Bourdelin described him as "using the whole session to amuse everyone as much as he could."[23]

Bourdelin's notes also give us insight into the background politics shaping academic life in the wake of the new regulations. When, for example, Geoffroy was elected by the members to a seat as an *associé chimiste* on December 5, Bourdelin said of the runner up, a M. Denis, that "he had bragged to almost every member of the Academy that he was the doctor in service to the family of the abbé Bignon" and that had campaigned by telling them that "nothing would give Monsieur l'abbé greater pleasure than to see him get votes in the election."[24] As an anatomist, Bourdelin generally focused his attention away from the work of the academic mathematicians, but his commentary upon their work, when offered, had a similar tone. In November, for example, Fontenelle read several propositions in geometry sent to the Academy by someone who Bourdelin described as "such a mediocre mathematician that the Academy will not dignify to respond to him."[25] Interestingly, no record of this reading or judgment is found in the academic registers.[26] On another occasion, Bourdelin also described a new hydraulic machine presented to the Academy by the *élève méchanicien* Antoine Parent as a machine with "apparently no utility at all."[27] Yet Bourdelin's resolutely private diary also offered more than critical vim and vinegar. When he saw something that he liked, he was equally effusive in his praise. After a presentation by Malebranche on his theory of colors, for example, the young anatomist was exuberant. "His system was found to be very beautiful," he gushed. "I really liked it [*Je l'ai gouté fort*]. It contains many good things."[28] In the end, the particular judgments made in Bourdelin's commentary are less important individually than for what they show us generally about his participation and understanding of the new climate of competition present among academicians in the wake of the 1699 reform. Also important is the insight he offers into the new interplay between visible performance and critical judgment, which had become essential to royal academic science after 1699.

Bourdelin's diary picks up the emerging battle over the calculus at the moment when it was starting to become a personal battle between Rolle and Varignon. He noted without further comment Varignon's public-assembly paper about water clocks in April 1699, and he likewise noted casually that Rolle read a paper soon after "exploring certain methods in algebra."[29] Bourdelin also noted without comment other mathematical papers presented by each man over the next three months, and had there been any animosity evident between Varignon in Rolle in the Academy, he likely would have mentioned it. He also adopted a blasé tone when, on July 11, Varignon read a paper to the Academy sent to him by Bernoulli that used the infinitesimal method to perform a new quadrature of the cycloid. Bourdelin recorded the event, describing Bernoulli as a "professor at Groningen," and then spent the bulk of his entry reporting the long history of previous work on the cycloid that the paper contained, mentioning Galileo in particular, before noting that Bernoulli's work was radically new.[30] Likewise, when Rolle presented "Remarks on the Different Ways Available to Perfect Algebra," a paper that the registers indicate was a nonpolemical exploration of good methods in algebra,[31] Bourdelin noted the presentation without adding any elaboration.[32]

This makes his entry for August 5, 1699 significant since it may mark the first moment when Varignon and Rolle began to openly spar inside the Academy. Varignon read a *mémoire* to the Academy at this session offering his manner of solving equations of the second and third degree. He described it as "so natural and easy that I expect everyone to accept it,"[33] and Bourdelin appeared to agree, describing it as "offering a very simple and accessible method" for solving these equations. Rolle, however, disagreed, for as Bourdelin noted, "[He] rose and raised strong objections to it."[34] This contestation appears again later that year, after the Academy's long fall holiday. At the session of December 9, 1699, Rolle read a paper titled "Remarks to Explain the Different Values of Radicals in Equations" which provoked queries from the academicians about Rolle's claim that certain algebraic terms do not express any geometric line. It also led to a demand, recorded by Fontenelle in the registers, that Rolle clarify his point with examples at the next meeting.[35] Bourdelin's diary records the next meeting, at which Rolle presented "Remarks on Geometric Lines" as a reply to these queries. The academic registers simply record the reading of the paper and Rolle's belief that it offered the clarifying examples requested by the Academy.[36] But according to Bourdelin, Varignon rose in opposition to the paper, declaring its solutions "neither new nor interesting."[37]

Things appear to have settled down in the early months of 1700. Bourdelin noted Rolle's papers on algebraic problem solving presented on February 27 and May 15, but he does not note any reaction to them.[38] He likewise noted without comment the early papers by Varignon on central-force mechanics in January, March, and May, the papers that prepared the ground for his public-assembly presentation later in November. In these, Bourdelin also noted the Newtonian references in Varignon's work discussed in the previous section, a sign that Newton's name was present in the academic discussions. No mention of any reaction to these papers in the Academy was offered, however.[39] Bourdelin's diary also reveals an interesting fact about the habits of the academic mathematicians during these years, one not revealed in the academic records. On January 20, the academic register reports the presence of l'Hôpital at the session, and that he "began to give the demonstration of a geometrical problem."[40] At the next session, on January 23, his attendance is again recorded, and the register says that he "completed his demonstration of the following physico-mathematical problem," which was then transcribed verbatim.[41] Bourdelin recorded accurately that in the paper l'Hôpital "used the infinitesimal method" to treat in a new, and he claimed better, way the centrifugal forces studied by Huygens in his *Horologium oscillatorium*. But he also noted that Varignon read the paper for l'Hôpital at each session.[42] His diary notes the same happening when l'Hôpital presented his work on the resistance of figures immersed in fluid media in June 1699.[43] Why l'Hôpital did not read the papers himself even though he was present in the room during their presentation is not at all clear. The records report the practice of *élèves* reading the papers of their senior mentors at regular and public assemblies. Does this indicate that Varignon was therefore considered l'Hôpital's underling as a *pensionnaire géomètre* underneath an *honoraire*? The active *honoraires* Malebranche and Gouye apparently read their own papers. No other evidence describing this practice exists, and whatever the reason for it, it reveals the interesting fact that even when l'Hôpital was using the infinitesimal calculus in his academic work, the Academy heard about it in Varignon's voice, likely with stresses and intonations that connected l'Hôpital's to his own.

This is as close as we can come to tracing the precise steps leading up to the full explosion of *la querelle des infiniment petits* in July 1700. For whatever prompted its preparation and precise crystallization, it was at this moment that open warfare began. Rolle launched the initial attack, offering a paper that Fontenelle described in the registers as "against the fundamental suppositions of *la Géométrie des infiniment petits*."[44] Bourdelin described it as "remarks on the first principles of geometry, which is to say against *les infiniment petits*."[45] Fontenelle promised the continuation of the paper at a future meeting, and on July 21, he recorded that Rolle finished his paper and that "M. Varignon rose to express a strong desire to respond on behalf of the principal authors of the differential calculus that were not informed of this."[46] Bourdelin said that Varignon "demanded that Rolle share with the public his difficulties with the differential calculus so that they could be responded to," but Fontenelle kept the dispute inside the Academy saying that M. Varignon would respond.[47] The secretary also manipulated the situation subtly, choosing not to transcribe Rolle's paper in the registers and instead writing, "We will see in this register through the response of M. Varignon what M. Rolle's objections are."[48] After Varignon had offered in early August what Bourdelin called "his apology for infinitesimals against M. Rolle," Fontenelle shaped that intervention as well.[49] He announced in the register for August 11 that Varignon had "finished reading his response to the difficulties of M. Rolle against the differential calculus," and then shifted into his own description of what Varignon had argued before transcribing verbatim Varignon's paper. When writing in his own voice, Fontenelle noted that, "M. Varignon had proven his points easily" using "the authority of all *géomètres* both ancient and modern." He also phrased things in such a way that when he shifted seamlessly in the next sentence to a transcription of Varignon's actual paper, a careless reader might have assumed that Fontenelle was still speaking in it.[50]

It must be remembered that the academic registers were more than mere records of the Academy's work maintained for history and posterity. In these early years after the 1699 reform, they were also surveillance documents. Bignon was still signing almost every report at the bottom of each entry, and at the end of the year, Pontchartrain received the entire register for his stamp of royal approval. Bourdelin's diary also reveals the active administrative management of the academicians during these years in ways that the Academy registers do not. To cite just one example, the registers for December 19, 1699 record that Bignon read a letter to the Academy from Pontchartrain confirming the Academy's election of Geoffroy as an *associé chimiste*.[51] Fontenelle and Bignon both signed the report at the bottom, and after the report of the next session, which was the last of the year, they each signed again, along with Pontchartrain who entered his statement of approval.[52] Bourdelin's diary adds, however, that the letter about Geoffroy's appointment read by Bignon also carried in it a complaint addressed to the Academy about its failure to follow the regulations in making this promotion. The letter cited in particular Geoffroy's minimal publication record and absence of sufficiently vigorous public activity, and according to Bourdelin it also noted that this was the second time that the Academy had behaved in this lax way. They were charged with making improvements in the future even though the appointment was made as desired. What this and other episodes found in Bourdelin's diary reveal is the way that academic life was being closely scrutinized and judged during these years.[53] Accordingly, Fontenelle's artful manipulation of the academic records regarding the initial steps into the calculus wars is of no little significance when considering the development of the debate itself. His interventions as secretary, which would only grow in number and intensity as the battle continued, were also as important as any other in shaping its outcomes.

Now under way, *la querelle des infiniment petits* went on hiatus after Varignon's August response to Rolle since the Academy departed soon after for its long fall holiday. Evidence of its rapid broadening, however, is present in the paper by the *élève méchanicien* Parent that Bourdelin recorded in his diary for July 24, 1700, the session after Rolle finished his critique of the calculus. Parent, as we will see later, was a very ambitious young academician with a taste for using controversies to serve his career agendas. It is therefore significant that Bourdelin notes him giving a paper at the very moment that the calculus battle erupted that "applied the differential calculus to the study the properties of geometric surfaces."[54] Mysteriously, no record of this paper exists in the Academy registers, but its presence in Bourdelin's diary links Parent to these controversies at their inception in ways prescient of events to come.[55]

Varignon's public-assembly paper on central forces on November 13 marked the occasion for the resumption of the battle, which carried on with increasing escalation throughout the next several years. The Oratorian Father Reyneau, a mathematician who contributed his own works to the cause of the new analysis, made a summary of the *querelle* at the time, and his analysis can serve to frame and condense the debate overall. As he wrote, "All the difficulties raised by M. Rolle reduce to these two: 1) that the differential calculus has not been demonstrated; and 2) that it leads to error."[56] Rolle himself declared as much in a summary piece published in the Academy's *mémoires* of 1701. "Geometry has always been regarded as an exact science, indeed as the source of the exactitude that is characteristic of all the different fields of mathematics," he wrote. "But it appears that this exactness no longer reigns in geometry since the new system of infinitesimals has been joined to it. To my mind, [this method] has produced no truths, but only covered up for geometrical errors."[57]

Rolle's initial charge, that the calculus was not demonstratively true, was one often leveled against the new analysis. L'Hôpital asked readers of his *Analyse des infiniment petits* (1696) to consider two quantities equal even when they differed by an infinitesimally small quantity,[58] but Rolle and other critics found such a request impossible to sustain. As he wrote: "The only way to explain the foundational suppositions of the new system is to say that the infinitely small differences such as *dx* and *dy* are less than any quantity given. One can understand this clearly according to the teachings of ancient geometry, but this is not to give a reality to this difference; on the contrary, it is to say that this difference is not a quantity. . . . In the new system [however] one attributes to them a real existence . . . that completely disagrees with their absolute nothingness."[59] To challenge the idea of the infinitesimal in this way was to undermine the very foundation of the new mathematics. Only by accepting that infinitesimal magnitudes were at once a positive quantity and a nil quantity could the calculus do its work. In the end, Rolle admitted that "despite all these difficulties, infinitesimal analysis remains a truly intriguing achievement. A multitude of new and very ingenious things have been found with it."[60] Despite its ingenuity, however, the new method was patently fallacious. Other, more rigorous methods needed to be employed instead.

Even more potent was Rolle's second claim that the new analysis produced erroneous results. To demonstrate his point, he used the new method to produce allegedly incorrect outcomes, showing in the process the dangers of practicing nonrigorous mathematics. This more narrowly mathematical argument was in many respects more provocative, because the claim of those who practiced differential analysis was that the method worked despite its epistemic mysteries. Leibniz, for example, acknowledged that the calculus raised serious philosophical difficulties, yet this did not stop him from defending the new mathematics. He advocated for it because it provided, in his opinion, a simpler and more general approach to problems solved only with difficulty using traditional geometry.[61] Varignon built most of his own claims as an innovator on very similar foundations, and Rolle's demonstrations therefore offered a direct challenge to these positions. They showed that the infinitesimal method could in fact produce errors, and because Varignon was not generally inclined toward philosophical and metaphysical disputation anyway, it was these mathematical arguments that provoked his response.

After Varignon's public-assembly presentation on central-force mechanics at the November *rentrée*, Rolle resumed his attack, giving what Bourdelin called "a second set of remarks on the principles of geometry, which is to say against infinitesimals."[62] This response, along with Varignon's public-assembly paper, entangled Newton in the struggle, and his role in this debate, which would be fixed in these early exchanges, is important to note given the influence it exerted on the *Principia*'s early French reception.

In his reply to Rolle's first critique against the calculus, which he offered in August, Varignon had pointed Rolle to Newton's work in the *Principia*, citing verbatim Latin passages from the treatise in which he claimed that Newton used the infinitesimal method in ways supportive of Varignon's own use of it, thus confirming its rigor.[63] The fact that Newton buried his infinitesimal method within the veneer of synthetic geometry only added power to his argument, because it confirmed the harmony between the new analysis and traditional geometric methods. Rolle, however, did not accept the argument. "M. Newton is not as favorable as is alleged to transcendental geometry or infinitesimals [*des infiniment petits*]," he declared. "He despises imaginary distinctions and wants to honor the space between being and non-being."[64] Varignon countered in January 1701 by defending the methods that he claimed both he and Newton shared, and by challenging Rolle "to do what I have done with central forces through some other means than the differential calculus."[65] In this way, Newton's *Mathematical Principles of Natural Philosophy* was reduced, as Fontenelle had done in his preface to l'Hôpital's *Analyse des infiniment petits*, to a work "all about the new calculus." The treatise was also made into a mathematical authority suitable for fighting the French battle over infinitesimal analysis.

So framed, the battle quickly expanded to the other academic mathematicians, becoming a full-fledged party struggle. In February, Gallois rose to read a *mémoire* on the principles of the differential calculus, one that Bourdelin described as "a criticism of infinitesimals and of M. Varignon."[66] The academy registers do not record Gallois's paper and only noted its presence when the reading was completed on February 19.[67] Papers read at the Academy by l'Hôpital and Carré that employed the new mathematics successfully supported the calculus in turn.[68] Carré also rushed his textbook on the integral calculus into print, despite lingering errors in the text, because his colleagues believed the work was necessary in the current climate.[69] Fontenelle also offered his only direct intervention in an academic debate during his forty-two-year term as Academy secretary, reading a paper on February 23 that defended the metaphysics of the differential calculus. Fontenelle noted his intervention in the academic registers,[70] but the paper was neither transcribed, nor published. Bourdelin, however, found it worthy of note, describing the presentation as "very elegant" and "well received by everybody."[71]

The calculus debate eventually came to consume all the air in the Academy, and as it grew in intensity throughout the spring and summer of 1701, the particular rivalry between Rolle and Varignon became more bitter and personal. Indeed, in May 1701, the discussion became so heated that Bignon was forced to intervene. Bourdelin recorded in his diary that Varignon asked for a commission to be established to evaluate his latest response to Rolle's position.[72] On May 27, Bignon responded, eschewing the appeal for a commission and instead applying regulation XXVI of the 1699 statutes requiring academicians to "only speak *avec ménagement*" in the assemblies and to "not employ any term of *mépris* or *aigreur* against one another." Fontenelle recorded the intervention this way: "M. Rolle read a response to the last response of M. Varignon, but since his response contained a large number of purely personal remarks that were not at all relevant to the question at hand, the President [Bignon] ruled that from now on, M. Rolle will give his objections to infinitesimals solely through demonstrations with no other discourse, and that M. Varignon will respond in the same way."[73] None of the commentary itself was recorded in the minutes, and accordingly the registers offer very little record of what must have been a very bitter set of exchanges.[74] Bignon's intervention, however, almost unique in the history of the Academy before 1740, indicates just how bitter the struggle was.

The abbé Louvois, an *honoraire* in the Academy who served as its appointed president in 1702, offers another glimpse into the contours of the conflict. Charged by Pontchartrain with the task of producing an assessment of the newly reformed Academy, Louvois was given the tricky assignment of evaluating the institution's work during this period of intense mathematical contestation.[75] It served no one's interest to present the Academy as a locus of deep intellectual strife, and Louvois accordingly set out to defuse the mathematical tensions by framing them in ways that fit with the wider agendas of the Academy. He noted, for example, the strong opposition of Gallois and Rolle to the new infinitesimal mathematics, and in the case of the first he connected his skepticism to his deep appreciation for ancient mathematics, an argument that more or less echoed the discourses offered in defense of the Ancients' position in the Ancients-versus-Moderns struggles of the time. In the case of Rolle, Louvois emphasized his esteemed mathematical reputation and his judicious caution in the face of reckless innovation. Gallois's new translation of the ancient Greek mathematician Pappus was offered in his defense, while Rolle's new book on algebra, still in preparation, was celebrated for its effort to perfect this newly emergent science. Through clever rhetoric, Louvois ultimately defended the Old Style mathematicians by defending the traditional rigor that anchored their work. Varignon, by contrast, was praised for his "vigorous work to develop a universal mechanics" and he was celebrated for the "precision of his mind, especially when dealing with the most abstract things." This was a quality that Louvois believed would "contribute to the perfecting of this science." In this telling, the innovations of the new analytical mechanics were trumpeted through their association with the widely admired values of universality and intellectual precision.[76]

Placed together, these accounts situated the calculus dispute within the deeper contest between traditional liberal erudition and modern scientific innovation within the Academy. Louvois also linked the precise mathematical struggles inside the institution to the wider struggle between ancient and modern culture, which was still raging in France. Having explained away the controversy in these terms, Louvois further worked to bury it beneath positive references to the many other achievements of the Academy. Utility, productivity, and *honnête* service to society—these were the themes Louvois emphasized. His representation further displaced the mathematical disputes by making them distractions from the real achievements of academy. The early success of the *méchaniciens* was especially noted, as were the accomplishments of Cassini's Observatory, especially its work with the Roman Catholic Church to resolve the astronomical calculation of Easter. The report also lauded the astronomical and mechanical work of de la Hire, making no mention of his prior participation in the calculus dispute. Louvois ended his report by emphasizing the service to state and society that the Academy as a whole had achieved. He asked for the honor of serving as president again, and he promised Pontchartrain that if chosen he would "make sure that the assemblies each week are as usefully conducted as you could hope for."[77]

Louvois's presentation of the calculus dispute reveals the dangers that it posed to the wider program of public academic science. It also demonstrates how such dangers could be managed through an appeal to the wider mission of the Academy. The debate itself, however, did not end with the initial struggles described by Louvois. Less than a month after Bignon's initial disciplinary intervention, Bourdelin reported that "Gallois rose to read a new critique of infinitesimals" but then declined to do so because "Bignon was not present and he feared being accused of carrying the argument too far."[78] In a letter to Bernoulli from this period, Varignon indicated that Gallois was the real force behind Rolle's attacks.[79] His judgment was confirmed in July when the abbé revived the dispute by presenting a paper that echoed many of Rolle's original critiques.[80] Another presentation from Rolle followed on the same day, and on July 9 Father Gouye revealed his cards by delivering his own critique of infinitesimal analysis.[81] This last session appears to have been particularly heated. Bourdelin noted in his diary that after Gouye's paper, "which appeared to argue that the differential calculus only substitutes the word indefinite for the word infinite," Gallois read "a very short *mémoire* against infinity" only to be followed by Varignon, "who vividly provoked the sensibilities of M. Rolle when he read his third response to the difficulties raised by him against the differential calculus."[82] Varignon read a fourth response to Rolle in August, and Carré supported his views by delivering three papers employing the calculus in the same month.[83]

With the debate showing no signs of abating, Bignon decided to make a more decisive response. On September 3, 1701, at the final assembly before the Academy departed for the long fall holiday, the Academy's chief minister officially declared the argument over. "The dispute concerning the *infiniment petits* has carried on far too long," Bignon proclaimed. He then formally silenced future discussion of the matter. In lieu of further debate, he announced that a committee would be formed and charged with deciding the question definitively. The judges would be Father Gouye, Jean-Dominque Cassini, and Philippe de la Hire. The jury was stacked heavily in favor of Old Style mathematicians, an outcome that Varignon attributed to Bignon's willingness to let Gallois and Rolle recuse jury members with whom they disagreed. No other record of the decision or its perceived justice survives, and with Bignon's declaration the debate in the Academy was calmed, at least for a time.[84]

Bignon's executive order terminated the debate for the moment, but not before it changed in profound ways the reception of Varignon's analytical mechanics. Overall, the eruption of the calculus dispute meant that Varignon's science was conceived and discussed exclusively in terms of its mathematical methods, and not in terms of the many other provocations it contained. His focus on the question of planetary motion in his papers of 1700, and his use of Newton's *Principia* in this context, opened the door, at least potentially, to a whole range of physical questions related to his concept of force, along with a host of broader epistemological questions about the use of mathematics to theorize celestial mechanics. These were questions that Newton's *Principia* would eventually provoke in France. But in 1700, thanks to the interpretation of the book as a work of mathematical mechanics alone, one whose primary contribution was found in its innovative if idiosyncratic mathematics, the treatise simply became one more resource to use when fighting the battles over the calculus.

In fact, when looking for weapons to counter Rolle's charge that his infinitesimal mathematics was fallacious, Varignon often invoked Newton's work in the *Principia* as contrary evidence. One strategy he used has already been discussed. When fighting with Rolle, he challenged his antagonist to produce his central-force mechanics without using the calculus, a ploy that made the calculus the foundation of his achievement, and one that also implied that the key element in Newton's central-force mechanics was his use of the calculus in its development. Reyneau's diary of the dispute also describes another strategy used by Varignon that also brought the *Principia* into the struggle. Seeking demonstrative proof that the calculus was methodologically sound, "M. V[arignon]," says Reyneau, "pointed M. R[olle] to the first proposition of Book I of Mr. Newton's *Principia* where the calculus is rigorously demonstrated in the manner of the ancients. . . . He then argued that it was [Rolle] who had fallen into errors and that his calculus properly practiced only gives true answers to problems."[85] Varignon, as Reyneau documents, repeated the same tactic throughout the debate, continually pointing Rolle to passages in the *Principia* to prove both the rigor and the accuracy of the infinitesimal method.

Newton, of course, did not really provide the synthetic demonstration of the calculus "in the manner of the ancients" that Varignon (and Reyneau) attributed to him, nor did he employ the calculus properly speaking in the *Principia*. In fact, he explicitly avoided its use because of concerns about rigor that were not too far away from Rolle's position. What he did offer, however, was a powerful demonstration of the mechanical insights achieved by using the infinitesimalist method, along with an argument for them couched within a traditional, Euclidean geometric framework. In short, Newton offered a rigorous foundation supporting the more aggressive innovations offered by Varignon's science, one that could be invoked as a bulwark for it.

The dynamics of the polemic itself were responsible for Varignon deploying Newton in this of all ways, but the result was the reinforcement of two particular interpretations of the *Principia* that survived to become influential in eighteenth-century France. First was the claim that the *Principia* was essentially a treatise on infinitesimal mathematics. Second was the conception of the *Principia* as a work of mathematical mechanics alone and not a new kind of mathematical natural philosophy. As Varignon framed it, Newton's *Principia* was important with respect to his analytical mechanics neither because it provided the model for his mathematicized physics nor because it offered some crucial categories for it such as central forces. It was an important touchstone because it offered a geometrically rigorous presentation of infinitesimalist mathematical mechanics, one that he could use to convince critics such as Rolle that his new and original science was epistemologically sound. It will be remembered that this understanding of the *Principia* as ultimately a treatise in the infinitesimal calculus *à la manière des Anciennes* would still be found half a century later in the commentary on the *Principia* written by the Marquise du Châtelet and Alexis Clairaut. It is also found in the same period in the histories of mathematics written by Montucla, Bailly, and Lagrange. Such were the long-term legacies of the peculiar reception of the *Principia* in turn-of-the-seventeenth-century France. Needless to say, Newton would have described things differently, but since he played no direct role in these disputes, and exerted little if any influence in shaping his own reception in this space, his work became what it became irrespective of his intentions.

By silencing academic discussion of the calculus in September 1701, Bignon created a temporary pause in the debate. But at the same moment a new jolt to the dynamic was added by the appearance in January 1701 of a new learned periodical in the French public sphere, the *Mémoires pour servir à l'histoire des sciences et des beaux arts*, a serial journal more commonly called the *Journal de Trévoux*. Before 1701, the French debate on the calculus, while extremely heated, had remained largely an intra-academic affair. Indications of a wider public discussion about the calculus were present in works such as the *Mercure galant*, and it would not be surprising to learn that the debate reached far beyond the walls of the Louvre. But the scarcity of such evidence attesting to any wider public participation in the *la querelle des infiniment petits* before 1702 suggests that the Academy was still largely in control of this debate before this date. With the appearance of the *Journal de Trévoux*, however, this containment was lost. From the outset, this new periodical injected energy into the calculus disputes, and since the journal was not only a learned periodical, but an organ edited by the Society of Jesus, the results were both curious and profoundly transformative.

Public Mathematics Complicated: The Jesuits in the Public Sphere

In substance, the *Journal de Trévoux* was a reflection of its editorial staff, a group of Parisian Jesuits attached to the society's premiere French college, Louis-le-Grand. These Parisian clerics and college professors obtained from the Duc de Maine a privilege to publish a journal within his sovereign territory at Trévoux. Since their publication arrangement exempted them from many of the regulations governing the periodical press in eighteenth-century France, the editors were able to use their freedom to produce a journal that served their own intellectual agendas. As its opening "Preface" announced, "these new *mémoires*" were to contain "extracts from all the books in science printed in France, Spain, Italy, Germany, and in the Kingdoms of the North, in Holland, in England, etc."[86] A universal outlook such as this was typical of the Jesuits, and the mission statement invoked another characteristic feature of the order: a commitment to modern scientific learning. Using the wide network of correspondents that the Jesuit editors possessed, they hoped "to frequently give the public critical manuscripts, explanations of interesting medals, new clarifications on passages of holy scripture, the latest discoveries in physics, medicine and mathematics, reports on the latest new machines, etc. such that able people in all genres of science will find material to satisfy their interests."[87] Overall, the editors evinced a deep commitment to contemporary scientific thought and strove to produce an esteemed organ of learned discussion throughout Europe.

Its opening "Preface" also defined the editorial slant that would quickly make the journal a respected peer of the other leading Francophone serials. "In the many contests that often emerge between men of letters over matters of science," the editors wrote, "the authors of these *mémoires* will never take sides. They will only give a simple exposition of the writings on each side, retracting, however, any comments that appear to be rude [*d'aigre*] or injurious. They will also maintain the same neutrality in all other aspects of their work, except on questions of religion, good morals or the state, where neutrality is never permitted." "The writers of these *mémoires* are really nothing more than historians," the editors asserted, and in order to maintain the "standards of perfect neutrality so necessary to writers of this character," they declared that nothing offensive would be permitted in the journal and that extreme praise of writers and works would also be avoided.[88] This editorial policy mirrored the one practiced at the *Journal des savants*, the *Acta Eruditorum*, and the Dutch publications of Basnage de Beauval and Le Clerc, which also aspired to be objective organs of contemporary intellectual discussion.[89] In the early decades of the eighteenth century, the journal also succeeded brilliantly in its ambitions, running a number of scientific articles that placed the journal at the center of the wider intellectual discussions of the day.

Their first issue, for example, included a *mémoire* that challenged the Cartesian conception of natural inertial motion. This provoked a judicious yet spirited debate in the journal about the scientific principles of motion, one that in turn triggered a critical intervention by the Academy of Sciences in the guise of a *mémoire* written by Philippe de la Hire.[90] The journal also published astronomical observations, reported new medical discoveries, and alerted readers to the latest maps, scientific instruments, and machines. In this way, the *Journal de Trévoux* functioned as a kind of parallel public academy in France, providing an outlet for scientific work not produced within the official Academy.

The Jesuit Father Antoine de Laval, a royal hydrographer at Toulon, illustrates well the opportunities that this outlet offered. He established himself as an important astronomer by publishing astronomical data in the *Journal de Trévoux* several times a year.[91] Since publication in the journal was open to all, and since securing publication in it was quicker and easier than trying to place a *mémoire* in the Academy's annual volume, many academicians began to use the journal as a publishing outlet. The rules of the Academy prohibited academicians from using the byline "Member of the Royal Academy of Sciences" unless an academic committee gave prior approval for their work, but this was not a major obstacle, and the journal often became a space where academicians debated scientific issues in ways that blurred the boundaries between academic and nonacademic science.[92] The journal's editors also used their publishing practices to establish a reputation for their journal as a place where independent, objective discussion was present, the kind idealized by Republicans of Letters.

Its Jesuit connections were not insignificant, however, and the editors took very seriously their promise to give up neutrality in the defense of religion. The journal accordingly served as perhaps the single most important forum for public theological discussion in France before the expulsion of the Jesuits in 1767. In these debates, the prejudices of the journal were obvious and assertive. Radical Dutch Protestantism and journalism were especially singled out for attack by the editors, and equally pernicious in their eyes were the various sins committed by Descartes and his many *sectateurs* in France. More than anything else, the *Journal de Trévoux* despised the deeply mechanistic and, in their mind, deeply atheistic world-picture championed by Descartes. In this respect, the editors continued what had become by 1701 a venerable tradition of Jesuit anti-Cartesianism in France.[93] As early as 1649, French Jesuits published a list of forbidden philosophical and theological ideas heavily laden with Cartesian tenets, and in 1682 Cartesianism was officially banned in the Jesuit colleges. Similarly, in 1690, the Jesuit Father Gabriel Daniel published his *Voyage du monde de Descartes*, an urbane satire of Descartes's cosmology. This work was an immediate popular success, reissued in many subsequent editions, and it served in some respects as the anti-Cartesian Jesuit counterpoint to Fontenelle's *Entretiens sur la pluralité des mondes habités*.[94]

Several aspects of Cartesian thought most disturbed the fathers of the Society of Jesus. Most important was Descartes's radical distinction between *res extensa* (material body) and *res cogitans* (mind), which the Jesuits saw as eliminating God from nature. French Jesuits helped to pioneer the classic argument linking Descartes's mechanistic philosophy to Spinoza's materialist God-Nature monism, and while many Cartesians denied this link, it became a standard attack in the period. This line of attack also positioned the Jesuits against Cartesian science, even when the science avoided explicitly metaphysical or theological claims. As Paul Mouy sums up in his history of Cartesian physics: "However the Jesuits acted elsewhere in Europe, in France it was [Cartesian] physics that they sought to destroy."[95] Against the mechanistic science offered by Descartes, French Jesuits championed a more naturalistic and Thomist conception that positioned them alongside other so-called Baroque scientists of the period, notably Athanasius Kircher, whose writings established a natural philosophical paradigm within which many French Jesuits worked.[96] As Father René-Joseph de Tournemine, the leading intellectual at the *Journal de Trévoux* before 1720, described the approach: "The entire universe is a vast theater which god has opened before the eyes of all men, to teach them, by this great and magnificent spectacle to love his power and wisdom."[97] Tournemine also considered the proof of God's existence from universal design "the most evident of all proofs," and his theological worldview supported his deep naturalism and organicism in matters of science. According to Tournemine, and many other French Jesuits, sane natural philosophy began with wonder at God's creation and culminated in a deeper appreciation of nature's mysteries through application of restrained reason, common sense, and empiricism.[98]

These deeply Thomist views clashed strongly with the skepticism and demonstrative rationalism of French Cartesianism, and it was from this perspective that the Jesuits launched their criticisms of it.[99] Of most immediate concern to the Society of Jesus in early eighteenth-century France were three aspects of Cartesian philosophy. First, the Jesuits continued to decry the Cartesian definition of matter as nothing other than material extension. There were explicit theological reasons for this position rooted in the nature of the Eucharist, but more important was the concern that this view of matter denied God a place in the action of nature. Their stance also dovetailed with their belief that miracles (including daily miracles such as transubstantiation) were real and active evidence of God's presence in the world. It likewise supported their refusal to accept any overly mechanistic or mathematical view of God's creation. The Jesuits were similarly hostile to the doctrine of innate ideas, invoking in its place the ancient Aristotelian cum Thomist doctrine that there is nothing in the soul that has not first been in the senses.[100] Pulling these last two strands together, the Jesuits were also deeply suspicious of

Malebranche's particular integration of Cartesian skepticism and mathematical rationalism with Catholicism. As Tournemine declared, "Doubt of the material world leads straight to absolute Pyrrhonianism and hence to atheism."[101] This stance positioned Tournemine and the *Journal de Trévoux* into a critical stance vis-a-vis Malebranche.[102]

Metaphysico-theological debates about Cartesianism thus found a new impetus with the appearance of a Jesuit periodical in the French public sphere. The term "Cartesian" was routinely invoked in the pages of the Jesuit journal as an epithet describing particularly distasteful doctrines or texts. While the reviewers also used the label in discussions of mechanics and physics, such as in the 1701 article "On the Sentiments of the Cartesians on Motion,"[103] it was more commonly used to describe metaphysics and theology, and it was especially linked to Malebranchianism. In the early issues of the *Journal de Trévoux*, the Cartesians were most often those who denied the existence of the material world, those who made God the direct source of all change in nature, and those who located the presence of God only in the mind through innate ideas.[104] In short, they were Malebranchians. These associations were also supported by Leibniz, who used the journal as a venue for his own challenges to Malebranche. He also linked these pernicious Malebranchian positions to Cartesianism and joined them with his other attacks on Cartesian physics and mechanics.[105] In this way, the intense metaphysico-theological battles over Cartesianism that began in the last decades of the seventeenth century were given new vigor in the eighteenth century by the appearance of a Jesuit periodical preoccupied with these precise concerns.[106]

Informed by these larger priorities, the *Journal de Trévoux* also entered the debate about the calculus in its inaugural volume. In its very first issue, it published anonymously a manuscript titled "General Rules for Uniform Motion."[107] The paper did not employ infinitesimal analysis per se, but it did offer an algebraic as opposed to geometric demonstration of the laws of motion. Further defining motion as "nothing else than the *rapport* between space and time, or the length of the path traveled during the duration of the movement," the article defined the problem in the Malebranchian conceptual language of analytical mechanics.[108] No major contribution to the discussion about analytical mechanics was offered, but the paper did reveal that these Jesuit journalists were in tune with the changing scientific climate of the time.

The third issue of the journal, which appeared in print in the winter of 1701 within weeks of Bignon's first intervention at the Academy, offered a more polemical contribution. It ran a brief review of an article published in the *Acta Eruditorum* by Johann Bernoulli that offered a new method for analytically determining the arc of curves.[109] The very presence of the review was a provocation, because the journal rarely devoted entire reviews to single journal articles unless the article contained overtly controversial material. Clearly this article contained such material, for the journal not only reported on it at length, it stretched well beyond the breaking point its policy of neutral objectivity.

The review began by celebrating the fecundity of geometric discoveries during the last century. Descartes's analysis, Cavalieri's method of indivisibles, Wallis's arithmetic of the infinite, Fermat's method of maximums and minimums, and the work of Hudde were all praised. But no mention of either Leibniz's or Newton's infinitesimal calculus was offered. This introduced the argument of the review, because Bernoulli's work employed the calculus, and the Jesuit reviewer set out to critique this practice. "The new analysis," he wrote using italics to emphasize his points, "or the calculus of differences that is known in France as *analyse des infiniment petits*, has pushed far beyond this earlier work. *It has penetrated as far as infinity itself and gone even beyond that, embracing not only infinity, but the infinity of infinity or an infinity of infinites*."[110] Bernoulli's work, the reviewer continued, was representative of this new turn. But, he lamented, "when one reasons using *the infinite, the infinite of the infinite, the infinite of the infinite of the infinite, and so on, without ever coming to an end*, and then applies *this infinity of infinites* to finite magnitudes, one offers as an explanation, or as the foundation for one, the dark abysses that one should be clarifying. M. Bernoulli does not give any other demonstration. He does not even describe the outlines of his new method. He only says that those who would like to find it must consider it an enigma and await the day when the explication of it will be revealed."[111] The reviewer then proceeded to sketch out Bernoulli's demonstration in detail, noting that the "illustrious M. Tschirnhaus," a foreign associate of the French Academy after 1699, had demonstrated the same rule without these problematic assumptions. He then concluded with the following assessment: "Those who are accustomed to the ancient manner of geometric reasoning suffer at the thought of abandoning it for such abstract methods. They prefer to avoid the path of *the infinite of the infinite of the infinite* since the visibility on this path is very clouded and it is easy to take a wrong turn without knowing it. It is not sufficient in geometry to find the right conclusion, one must also know evidently that one has found it well."[112]

With this article, the *Journal de Trévoux* declared itself squarely on the side of Rolle and Gallois in the calculus dispute. It also positioned the foreign associate Tschirnhaus as their ally in these struggles. Furthermore, since the journal already had a wide readership, this intervention effectively broke open the Academy's monopoly on the calculus debate and turned it into a full-fledged public debate. At one level, a combination of intellectual and institutional rivalries intrinsic to the new field of public science in France account for this contestation. Jesuits were important intellectual players in the wider field of French learning, yet while the Academy's structure effectively excluded them from the official scientific establishment, their new journal, combined with the traditional power of their colleges, gave them a platform for reasserting their influence. Intellectual motives particular to the Society of Jesus also played a role in this institutional contestation since many French Jesuits, for a set of complex reasons, were simply hostile toward infinitesimal calculus. The growth of the *Journal de Trévoux* after 1701 therefore gave critics of the calculus a powerful new public forum for their views.

Why would the Jesuits have been hostile to the calculus? The role of classical Euclidean geometry in their curriculum was one reason for their skepticism. Jesuit education prized geometric reasoning because synthetic demonstration provided a rigorous discipline for orienting correct thinking overall. The Thomism that these schools taught agreed with this epistemological orientation since Aristotle had also been an advocate for science grounded in Euclidean demonstrative logic. Jesuit science was hardly the reactionary, anti-Modern monolith that it is still too often portrayed to have been. Overall, Jesuit schools taught a modernized natural philosophy that emphasized empiricism, commonsense reasoning, and wonder in the face of nature's complexity. Nevertheless, a commitment to geometric rigor was a hallmark of Jesuit science education, a fact that explains the long line of Jesuit mathematicians, from Christopher Clavius to Paul Guldin and Gregory de Saint-Vincent, who positioned themselves against the new infinitesimalist and analytical mathematics of the seventeenth century.[113] In a recent study, Amir Alexander has even argued for the centrality of Jesuit opposition to infinitesimalist mathematics at the very political and theological heart of the order.[114]

Gregory de Saint-Vincent became something of a Jesuit mathematical hero in the struggles against the new mathematical analysis, for as a contemporary of Cavalieri, Pascal, and Barrow, he was among the pioneers of infinitesimal geometry, while also a traditionalist who developed his innovations within, rather than against, the canons of Euclidean rigor.[115] His infinitesimalist mathematics displayed a judicious balance between ancient tradition and modern innovation, and overall, French Jesuits defended the same ideal. Since they also combined their Jesuit devotion to classical geometry with a particular French hostility to the "reckless" geometric reasoning that they associated with Descartes and Malebranche, they became vigorous allies of the Ancient mathematicians in the French disputes over the calculus.

The editorial line of the *Journal de Trévoux* drew upon these wider Jesuit assumptions. On the one hand, their commitment to a complex, nonmechanical conception of nature made Jesuits on the whole sympathetic to all modes of scientific explanation that culminated in a sense of wonder about nature. They were also quick to contest any science that reduced nature to an abstract, rational system. Infinitesimal analysis did not automatically find itself indicted by this particular Jesuit orientation, but when Jesuits saw modern mathematics being used to intensify an already dangerous trend toward overly abstract and excessively mathematical reductions of nature, they tended to cry foul. As Father Tournemine once declared: "There is not in life an inclination more dangerous, or more ridiculous, than to conduct oneself by means of geometry, unless one can be entirely sure that the idea, or the principle of demonstration, is conformable to reality."[116]

Infinitesimal geometry lacked precisely this comprehensible anchor in commonsense reality, and for this reason it is not surprising that many French Jesuits were uncomfortable with it. Yet this same Jesuit outlook could also lead to sympathy with the new mathematics so long as it was construed in different terms. Interpreted one way, for example, the new infinitesimal calculus supported Jesuit scientific thinking, for as Bernoulli's reviewer had intimated, the most powerful result of the new mathematics was a more acute appreciation for the mysteries of the infinite. For the Jesuits, this mysterious quality of the new mathematics could be a source of praise. Indeed, on one occasion the *Journal de Trévoux* turned this interpretation of infinitesimal analysis into a veritable theological position, defending the study of the new mathematics because it made people more receptive to believing in miracles.[117]

A frontispiece to a compilation of Gregory de Saint-Vincent's geometry illustrates well the French Jesuit understanding of infinitesimal analysis (fig. 9). The foreground shows geometry being practiced using human tools, while *putti* in the background square the circle by directing the divine light of God's radiance alone. The message here is that human reason can approximate divine truths, but only God can achieve complete mastery of nature's mysteries, including the mysterious relation between discrete and continuous magnitudes. As with Malebranche, whose intellectual differences with the Jesuits are neatly articulated by this example, instrumental mathematical reasoning is presented as a valuable tool that can serve as an aid and bridge for finite humans to comprehend the mysteries of God's omnipotence and infinitude. But unlike Malebranche, Jesuits such as Saint-Vincent and those at the

FIGURE 9. *Frontispiece, Gregory de Saint-Vincent, S.J.,* Opus geometricum quadraturae circuli et sectionum coni: Decem libris comprehensum *(Antwerp: Meurius, 1647). Courtesy of the Koninklijke Bibliotheek, The Hague.*

Journal de Trévoux, were loath to accept any claim for mathematical analysis as an instrumental means for bringing human mastery over these mysteries. Mathematics allows us to see and approach the infinitude of God, but any claim to transcend the gap between man and God through mathematics was dangerous hubris.

Obviously, support for the calculus such as this was little more than a different kind of opposition to it in the mind of Varignon and his supporters, and overall the French Jesuits, especially those who staffed the editorial team at the *Journal de Trévoux*, were vigorous opponents of the new mathematics. The institutional power of the Jesuits in France made these challenges all the more important, as did the success of their journal, which quickly established itself as a respected scientific authority in the newly restructured public sphere of science. The fledgling public academy felt this pressure particularly strongly, and for this reason exchanges like the debate about inertial motion that placed the Academy into contest with the Jesuits were common in the first years of the eighteenth century.[118] The Jesuit interventions against the new calculus drew upon this influence, catalyzing an already intense dynamic. Their first review, in fact, called forth a series of responses that turned the journal into a public arena of mathematical argument, one linked to the parallel debate going on in the Academy even if the two discussions never overtly acknowledged each other.

The first response was a very brief article, "Diverse Problems on the Nature of Curved Lines." It appeared in the fall of 1701 at about the same time that Bignon silenced the calculus debate in the Academy.[119] The anonymous author of this original piece offered a "natural," algebraic method for describing a parabola, and his work touched on the calculus debate in only the most oblique way. The paper did deal with its central mathematical foundation, however: the algebraic description of curves. The article also offered a different perspective on these questions by connecting it with the mechanical arts. "In the practice of the arts," the writer declared, "one needs curved lines. . . . All these questions are [therefore] important for the practice of the arts as I hope to show one day."[120] Ultimately, this paper was a diversion from the calculus debate proper, but the final issue of 1701 fueled the dispute directly. Not only did the issue contain the latest round of exchanges regarding the Cartesian theory of inertial motion, a debate that implicated the Academy directly, it published a letter from Leibniz "concerning his sentiments on the differential calculus."[121] Leibniz had already contributed to several other discussions in the journal's first year, but with his letter about infinitesimal mathematics, he entered not only the calculus debate in the journal for the first time, but also the wider French mathematical debate overall.[122]

Leibniz's debut was in some respects orchestrated by the French defenders of the infinitesimal calculus. L'Hôpital had asked Leibniz to intervene, and the letter he offered, which he later believed had disappointed the marquis, was encouraged by Bernoulli, who wanted someone to respond directly to the criticisms of his mathematics in the journal. Leibniz based his defense on a pragmatic approach to the new analysis, arguing that the discoveries made by infinitesimal analysis were considerable. While he conceded that "only the newness of the method excuses its lack of beauty," he also insisted that M. de l'Hôpital's work would assure anyone that the new method was sound.[123] In fact, Leibniz charged, the reviewer for the *Journal de Trévoux* had misrepresented the approach of both l'Hôpital and Bernoulli in arguing that their method of infinites leads to an infinite regression. "As the illustrious M. de l'Hôpital says himself," Leibniz chastised, "one need not take infinity in a rigorous sense here, but only in the way one uses it in optics."[124] Leibniz then offered an example from astronomy where the notion of infinity is employed in a largely empirical and commonsensical way as a means of conceptualizing a difficult process of physical change. Afterward, he offered his conclusion: "For instead of the infinite or the infinitely small [in this example], one takes quantities as big or as small as is necessary so that the error is less than the error given. As a result, this method only differs from that of Archimedes in that our method offers a more direct means of expression and conforms better with the art of invention."[125]

Leibniz does two things in this defense that are worth noting. First, he attempts to apply a pragmatic and empirical justification for the new analysis as a means of diffusing the philosophical arguments leveled against it. Second, he attempts to subvert the appeal to ancient, rigorous geometry by erasing the differences between old and new. Other defenders of the calculus in France would take a different approach, and judging by Leibniz's assessment of l'Hôpital's reaction, his was not the approach that the marquis was looking for. He likely wanted an actual defense of infinitesimals useful for countering the attacks being leveled against them in the Academy. The Jesuits at the *Journal de Trévoux* had still other agendas, and they preferred to rebut Leibniz's defense directly. Not wanting to let him get the last word, the journal appended a disclaimer at the end of his letter suggesting that "several mathematicians who have examined with great care l'Hôpital's *Analyse des infiniment petits*, and who even profess to follow his method themselves, say that it is necessary to consider the infinite in a rigorous sense and not in the manner explained by M. Leibniz here." "It was based on their reports," they argued, "that we spoke as we did in the *Mémoires* of May and June."[126]

This disclaimer suggested that the Jesuits disagreed with Leibniz's views. Varignon, however, saw things differently. He was stunned and dismayed by what he read from Leibniz, and perhaps echoing l'Hôpital's sentiments, he wondered whether the journal had falsely attributed to Leibniz the views of another savant.[127] On November 28, he wrote a letter to Leibniz, his first, asking for a clarification. "The enemies of your calculus are declaring victory and using this letter as a brief and concise statement of your views on the matter," he lamented.[128] In January 1702, when he had not yet received a response, he wrote a letter to Bernoulli imploring him to intervene with Leibniz on his behalf. Varignon was clearly agitated, and it was Leibniz's distinction between rigorous and pragmatic notions of the infinite that triggered his frustration. The dichotomy itself was subtle, and in his letters he clarified two possible interpretations, one that would generate agreement with Leibniz and a second that would make Leibniz the enemy of everything that he was trying to defend against Rolle. What irked Varignon was the way that Leibniz's concept of the "nonrigorous infinite" was being employed by his "enemies" to enlist Leibniz into the camp of those sympathetic to Rolle without any acknowledgment of the complexity of the issues involved.

He was reassured in February, when Leibniz responded with the clarifications Varignon hoped for. Leibniz also agreed to allow Varignon to publish his letter in the *Journal des savants*, an event that occurred in March 1702.[129] Leibniz's second piece more fully laid out his views on the nature of the infinitesimal, and it made clear Leibniz's belief that the differential calculus was both rigorous in its methodology and fecund in its mathematical potency. In a letter to Bernoulli soon after the article appeared, Varignon noted with joy the discomfort that the article had created in the Academy. Father Gouye, who had requested and received Leibniz's original letter on behalf of his Jesuit colleagues at the journal, expressed anger at Leibniz for writing misleadingly in his first missive. He also wondered why he had not clarified these views more fully before. De la Hire was similarly moved to reconsider the views that the original letter had led him to defend. Varignon was thus able to indulge in a brief moment of satisfaction, but he noted that Bignon's academy commission had not yet ruled, and that his own set of responses (which he sent to Bernoulli) were not going to be published, a silencing that he attributed to the discipline that Bignon was continuing to impose on him.[130]

La Querelle des infiniment petits Goes Public

Indeed, the wider public dispute about the calculus was far from over. Between the appearance of Leibniz's original letter in the *Journal de Trévoux* in November 1701 and its subsequent clarification in the *Journal des savants* in March 1702, the *Journal de Trévoux* published several works that kept the debate alive. One was a rare supplement to the journal's first volume of 1701 that contained two short mathematical papers. The first was a further analysis of Bernoulli's *Acta Eruditorum* paper that only pushed farther the arguments of the journal's first critique. The second was a paper by a mathematician named Joseph Saurin demonstrating geometrically a principle of refraction relating to rainbows.[131] Saurin also published a paper on Huygens's theory of pendulum motion in a second supplement to the journal at the end of 1702, and the significance of his appearance in this debate will be examined shortly.[132] In the context of early 1702, however, these papers were most important for the confirmation they provided that the journal was in no way dissuaded by Leibniz to subdue its participation in the debate about infinitesimal analysis.

The confidence of the journal's staff was further revealed in the first issues of 1702. In February, as the journal began appearing monthly rather than bimonthly, de la Hire published the Royal Academy's official response to the inertial motion debate, situating the institution squarely on the side of Descartes, Galileo, and the principle of inertia. Undeterred, the journal editors responded with a rebuttal that kept the debate alive.[133] At the end of the same issue, again as a kind of appendix, they also published an original mathematical piece: "Parallel between the differential calculus and Fermat's method of *maximis et minimis*."[134] This paper did not explicitly critique the new analysis but rather staged a classic test case by showing how Fermat's more geometrically rigorous method produced the same outcomes. Demonstrations such as these either supporting the calculus by showing that it did produce accurate geometric results, or challenging it by showing how ordinary geometry, when practiced by masters such as Fermat, were often used to question the supposed innovations of the new analysis, and this paper engaged in a similar practice. In May, yet another exceptional appendix was added to the issue for that month that offered two responses to the paper from the previous September on curves used in the mechanical arts. Both offered more systematic and elaborate mathematical demonstrations of the property described in the original paper, and neither employed the calculus.[135] No sources are available that make possible either an identification of the authors of these articles, or a reconstruction of their motives in writing them, or a recounting of the deliberations of the editors in publishing these pieces in this exceptional way. But what can be concluded from this flurry of mathematical publication in 1701–2 was the eagerness of the Jesuits to make their journal a participant in the calculus struggle and their general inclination as editors against the new mathematics.

By the spring of 1702, therefore, and despite the efforts of Bignon to use the disciplinary mechanisms of the Academy to silence it, the calculus debate raged on more powerfully than ever, this time outside the walls of the Academy. The expanded public space for science created by the *Journal de Trévoux* had generated this possibility, and in the wake of Bignon's refusal to let the Academy serve as a forum for its discussion, the "quarrel of the infinitesimals" spilled out into the public sphere. Rolle was quick to seize upon this new dynamic for his own advantage. Silenced like Varignon at the Academy, he chose to publish a paper in the *Journal des savants* in April 1702 that made many of the same arguments featured in the academic debate.[136] The article in effect transferred the internal academic debate about the calculus into the official periodical of the French intellectual establishment. Appearing as it did less than a month after Leibniz's clarification of his position on infinitesimal analysis in the *Journal de Trévoux*, Rolle's *Journal des savants* article, which listed his title as a royal academician in the byline, effectively placed the Royal Academy into open public debate with Leibniz, the Society of Jesus, and the broad readership of these journals.

Shifting the venue in this manner was not a frictionless move for Rolle to have made. Rolle's ally, Gallois, was a founding editor of the journal, and his influence was certainly important in pushing Rolle in this direction. But at the end of 1701, just months before Rolle's article appeared, the editorial team at the journal had been changed. Royal administrators declared the journal ineffective, and Bignon was ordered to execute a reform here just as he had been asked to do in so many other areas of official French culture. This meant that it was Bignon who supervised the publication of Leibniz's letter about the nature of the infinitesimals in March, and then Rolle's resumption of the calculus dispute in the journal in April. An *Avertissement* in the first issue of 1702 announced the editorial change, asserting that the increased number of books published each year had made it impossible for one man alone (Gallois) to edit the journal. The text also lamented the difficulty of producing such a widely read text. "The public's taste is not easily satisfied" the *Avertissement* explained. "One [reader] wants only theology while another wants only mathematics and physics. Still others want medicine and anatomy, *bel-lettres*, antiquities, and history. To satisfy everyone, we would need to produce separate journals devoted to each subject." Stylistically, the same dilemmas pertained. "Most savants and scholars [*gens de cabinets*] care little for clever turns of phrase, while the worldly [*gens du monde*] care little about the depth of things and want only a fine and agreeable critique. Clarity can charm the latter, but they have no tolerance for the more abstract and difficult topics. How can one find the right balance to please each?" To meet these demands, the *Avertissement* explained, a new "company of *gens des lettres*" would be put in charge of editing the journal, and under Bignon's supervision they would collectively produce the content that appeared each week.[137]

This new journalistic context for the calculus dispute changed the character of the debate in important ways. While the Paris Academy had already become immersed in a bitter dispute that pitted rival mathematicians beholden to rival conceptions of the appropriate character of academic mathematics in contestation with one another, the argument at first was an intramural affair. The reform under way at the Royal Academy complicated things by injecting a new competitive emphasis on specialization and publicly declared disciplinary expertise into the older working practices of the company. It had also explicitly invited the broader public in as an auditor of these new practices even if the shift was only in its barest infancy in 1701. The tensions that these new institutional dynamics created were among the pressures driving the battle over the infinitesimals inside the Academy. To extend these struggles into the learned press, as Rolle effectively did with his article, was to bring new constituencies into the fray, and even more complexity into the struggle. The *Avertissement* in the new *Journal des savants* had noted how the wider learned world was going through its own reorganization around 1700, one that placed traditional *gens de cabinets* devoted to the older values of the Republic of Letters into a new relationship with the worldly and pleasurable sociability favored by *gens du monde*. The insertion of the Academy's newly professionalizing scientific experts into the same public space was no easy undertaking, yet after 1699 this was precisely what French academicians were asked to do. As a result, academicians not only confronted new dynamics inside the Academy as a result of the reform, they were forced to establish their new identity through public negotiations involving every constituency within this changing social and cultural milieu.

Among the things shaped by this new public dynamic was the perception of analytical mechanics that Varignon was fighting to establish. That Bignon's new management of the *Journal des savants* had led to Rolle being offered a new public platform for his views supported Varignon's perception that Bignon was silencing him in particular. The struggles that Varignon and his allies experienced in the wake of this new publicity of the academic debate illustrate well the dynamics that these new institutional arrangements produced. Assuming that Bignon was already in charge of the *Journal des savants* in early 1702, a conclusion that is warranted by the records documenting the reform, then it was through the collective decision of the editor-in-chief, Bignon, and his "*petit académie*" of supporting editors that Rolle's challenge to infinitesimal mathematics was allowed outside the Academy and in the journal. This forced the defenders of the new mathematics to come up with their own public strategy for circumventing Bignon's silencing of debate inside the Academy and his management of it in the *Journal des savants*.

Varignon reports that he convened with his allies, l'Hôpital, Malebranche, and Father Reyneau, at the Oratorian library on the rue St. Honoré in Paris soon after Rolle's article appeared. Varignon had multiple refutations ready to be published, but according to him, Bignon refused his rebuttals because they contained "personal attacks on Rolle."[138] Varignon was certainly not immune to ad hominem invectives, and his letters to Bernoulli reveal the depth of his enmity toward Rolle. But the defenders of the calculus also found themselves caught in a difficult institutional vise. The *honnête* ethic of the Republic of Letters forbid personal polemics, and the *Journal des savants*, like its peer in this respect, the *Journal de Trévoux*, attempted to institutionalize these values by prohibiting heated intellectual wars on its pages. Since Rolle's piece was an objective defense of his own mathematical position, it was deemed a constructive piece of mathematical work and accordingly published. What the defenders of the calculus needed was an equally positive and nonconfrontational piece articulating their own point of view. But as Varignon asked in frustration, "How do you expose Rolle's errors without mentioning him by name?"[139] As academicians, Varignon and l'Hôpital were further beholden to the discipline imposed by Bignon and to the rules of the Academy that made open, named, intellectual contestation a practice unworthy of their title and office. How could the new analysis be defended and scientific progress be secured while also securing the decorum demanded of royal academicians?

This was the dilemma that confronted Varignon, l'Hôpital, and their allies in 1702, and one resource at their disposal was Johann Bernoulli. In August, he sent a paper to Varignon on the integral calculus asking that it be published in the *Journal des savants*. The paper had the virtue of using the new calculus to solve a complicated mathematical problem, and since it in no way addressed Rolle or the calculus controversy directly, it could serve as the positive statement supporting the mathematics that the defenders of infinitesimal analysis needed. Unfortunately for its advocates, however, the paper was deemed unsuitable for publication according to the new publishing agendas at the recently reformed *Journal des savants*. According to Varignon, Bignon declined to publish the article, claiming that there had been too much mathematical disputation in the journal of late, and that readers were growing tired of it. He also pleaded economic hardship, claiming that the journal was not selling well, not least because of its excessive mathematical content. The journal therefore could not afford to publish Bernoulli's paper, Bignon contended. Varignon further reported that Bignon had announced that Rolle's article and the letter from Leibniz were to be the last mathematical works that would appear in the journal since he wanted all future mathematical papers to be saved for special issues devoted only to these topics. Publishing rationales such as these fit with the agendas articulated in the *Avertissement* announcing the new editorial direction of the journal, but they also worked, whether intentionally or not, against the desire of the infinitesimalists to offer a public response to Rolle's critique.

In the end, Varignon counseled Bernoulli to send the paper to Otto Mencke, editor of the *Acta Eruditorum*, arguing that, "his journal is infinitely more accommodating than ours."[140] Varignon also employed this other outlet himself, sending a paper to Bernoulli soon after Rolle's article appeared that challenged the latter's work. He asked that it be published in the *Acta Eruditorum*, either anonymously or under Bernoulli's name, insisting further that his correspondent reveal nothing about their discussions to any member of the Academy. His insistence about these precise tactics indicates that his obligations under the new academy regulations, and Bignon's disciplinary power, were influencing his thinking at this moment of intense struggle.

Varignon's piece never appeared. During the summer, the German mathematician Jakob Hermann published his own refutation of Rolle's paper in the *Acta Eruditorum*, and this satisfied Varignon.[141] More difficult was the problem of challenging Rolle within the French public sphere, especially that part of it composed of *gens du monde* and other worldly readers who paid little attention to the erudite Latin commentary coming out of Leipzig. This public mattered given the new public orientation of the Academy, and to be fully successful, the new analytical mathematics had to maintain its central position in the new public academic science initiated in 1699. At their meeting in the Oratorian library, Varignon and his allies assessed what few options they had. They lamented in particular the need for a publicly accessible presentation of their views that would overcome Bignon's refusal to publish any public displays of technical mathematical bickering. In the end, they opted to give the pen to a relatively unknown mathematician who could speak on their behalf free of the constraints imposed by the Academy regulations.

The mathematician recruited was Joseph Saurin, already something of a celebrity in 1702 even if he had not yet made his name as a mathematician.[142] Born in Grenoble in 1659, Saurin was the third son of a Calvinist minister who, like his two brothers, had followed his father into the clergy. His pastoral style, however, proved distasteful to French authorities, and when he delivered an inappropriate sermon sometime around 1680, he was forced to flee to Switzerland under threat of arrest. The religious leaders of Bern gave him a large parish near Yverdon, and he witnessed the revocation of the Edict of Nantes from his new post. After 1685, Saurin welcomed many newly exiled Huguenots to Switzerland, but he soon clashed with them as well, and in 1688, the more orthodox Calvinist theologians in Switzerland forced him to flee to Holland. Once in the Low Countries, he made the decision that would guarantee his fame in France. Writing to Bishop Bossuet, the leading religious figure in Louis XIV's monarchy, he asked to consult the bishop on matters of religious conscience. Under Bossuet's supervision, Saurin returned to France and engaged in a widely noticed conversion to Catholicism, which he completed in 1690. His conversion even included a secret rescue of his wife from Switzerland, a mission personally arranged by Bossuet and the king. When Bossuet himself presented Saurin and his wife to the royal court at Versailles in September 1690, the notoriety of the former Huguenot was secured.[143]

It was after the conclusion of this romance, as Saurin himself called it, that his participation in the history of French science began. "Free and tranquil in Paris," wrote Fontenelle in his *éloge* for Saurin, "he needed to choose an occupation." He debated between mathematics and the law, but opted for the former because he wanted to "escape the contests of theology." As Fontenelle described his thinking: "He believed that in giving himself to geometry, he would be able to inhabit a region where truth is less cloudy and where his reason, agitated for far too long, could enjoy a modicum of repose through certainty. He also had a naturally geometric spirit, and had been a geometer even at the pulpit."[144] From 1690 on, Saurin devoted himself to the serious practice of mathematics, and his work from the beginning was deeply informed by the new infinitesimal analysis. It is likely, in fact, that he joined Varignon, l'Hôpital, Fontenelle, Reyneau, Carré, and others in the orbit Malebranche soon after 1690, but since Bossuet was instrumental in also moving Sauveur from the clergy into mathematics, one wonders what role the legendary "Eagle of Meaux" played in orienting Saurin in this direction.

Saurin also implicated himself within the wider mathematical community in other ways, and these might have shaped his career choices as well. Describing Saurin's recreational habits, Fontenelle wrote that "his only diversion was to go each day to a café where *gens des lettres* of all sorts congregated."[145] This was the Cafe Gradot, where Saurin later acquired notoriety when he was drawn into one of the great scandals of the early eighteenth century.[146] Called *la querelle des rimes*, or "the affair of the couplets," the scandal sprang from the common practice of singing satirical songs at the café. Fontenelle called the sport "worthy of the three Furies had they possessed *esprit*," and many suspected the talented poet Jean-Baptiste Rousseau of initiating the practice. The songs alone did not cause the scandal, however. The controversy erupted instead when Rousseau used the Parisian court system to sue Saurin for slander in the creation of a medley against him. Saurin was arrested and sent to prison in 1711, an event that set Parisian society abuzz.

Saurin attempted to turn this energy to his own purposes, writing, in Fontenelle's words, "extremely touching letters from prison to the well-connected people who protected M. Rousseau, letters where the truth made itself felt." He also published appeals "written with the same tone" that "addressed the public as much as the judges [of the Parlement of Paris]." He likewise composed *mémoires* in which he placed his morals side by side with those of his accuser.[147] In the end, Saurin's campaign succeeded, and he was acquitted of all charges while Rousseau was banned from the kingdom in perpetuity and ordered to pay large fines.

The scandal reveals the importance of Saurin among the *gens du monde* who frequented the Gradot and other cafes like it, and since nothing other than mathematical works left Saurin's hands before the scandals, it also reveals the centrality of mathematics, especially analytical mathematics, in these influential worldly circles. Saurin's public name in this French mathematical world was first fashioned with his interventions against Rolle in 1702. Through this work, he became, along with Fontenelle, another Gradot regular, one of the key mathematical thinkers in the camp of the Moderns. Fontenelle indicated as much in the closing lines of his *éloge*, listing Saurin's closest friends as Bishop Bossuet, l'Hôpital, Malebranche, and M. de la Motte.[148] The presence of his spiritual mentor Bossuet's on the list is not surprising, nor is the presence of the analytical mathematicians l'Hôpital and Malebranche. De la Motte's presence is also fitting once one recognizes the full place of Saurin in the cultural world of his day. Antoine Houdar de la Motte was a philosophical skeptic, libertine poet, and man of letters with close ties to Fontenelle and the other Moderns. He was a regular at the Gradot, a fixture at

FIGURE 10. *Jacques Autreau (1657–1745),* Fontenelle, La Motte et Saurin chez Madame de Tencin, *also called* La maison d'Auteuil, *c. 1716. Oil on canvas. Musée du Château de Versailles, France, MV5573. Photo: Gianni Dagli Orti / The Art Archive at Art Resource, NY.*

the salon of Madame de Lambert, and, thanks to her connections, a member of the Académie française.[149] Saurin and de la Motte were intellectual compatriots, as a portrait of them together with Fontenelle and Madame de Tencin, another well-known Parisian *salonnière*, shows (fig. 10). To recognize the ties that bound this group together is to recognize the importance of Saurin and his infinitesimal mathematics in this libertine *mondain* milieu.[150]

Saurin began the public intellectual work that would launch him on his mathematical career in August 1702. Supported at every step by Varignon and l'Hôpital, he responded to Rolle's *Journal des savants* article with an article of his own published in the same journal. The paper criticized Rolle's work directly while vigorously defending the new infinitesimal method.[151] As Bignon had promised, the article was published as a special stand-alone issue, and a further set of special restrictions were imposed before the editor allowed the piece into print. According to Varignon, "even with all of his *crédit*, M. le Marquis de l'Hôpital had enormous difficulty getting the work published." First, all the personal invective had to be removed, a demand that Varignon believed "disfigured the response." Bignon also required that l'Hôpital support the publication financially. The marquis agreed to buy forty copies of the issue at a total cost of twelve livres (pocket change for him) and he paid for all the publication costs associated with printing Saurin's three diagrams. Sufficiently remunerated, Bignon allowed the article into print.[152] Rolle, it appears, was also constrained in getting a further response published. In early 1703, the *Journal des savants* printed an anonymous set of "remarks" about Rolle's original paper of 1702 that defended the infinitesimal approach while completely avoiding the controversy itself. The journal also published a solution soon after to a long-standing mathematical question along with a prize problem posed by Bernoulli, articles which proved that the journal's prohibition against mathematical work, if really there ever had been one, was anything but absolute. Indeed, it appears that Bignon was concerned primarily with silencing the overt polemics about the calculus, and this pushed Rolle elsewhere than the *Journal des savants* in his quest to respond to Saurin. He chose to publish a pamphlet, which appeared in July 1703 with his affiliation "*de l'Académie Royale des Sciences*" boldly printed on the title page. Technically, such a publication would have required prior approval by the Academy, but the registers reveal no evidence that such an examination ever occurred.

Whatever its precise publishing history, the appearance of the pamphlet sustained the controversy while elevating its polemical intensity. Varignon wrote to Bernoulli in August that "the quarrel over the infinitesimals has resumed with more heat than ever." Rolle's pamphlet, he continued,

> is full of ignorance and impudence. The first piece would have been fine had he not attacked the very methods that he pillaged himself. But this second piece is from a man who is drowning and grabbing for anything. He exercises so little discernment in what he offers that he evidently expects support only from the ignorant for whom the words infinite and infinitesimal generate fear. Since a man of the Academy would not dare to respond to him as he deserves, Mr. Saurin is going to respond without restraint. He will unmask him as an imposter who seeks only to seduce the ignorant. Indeed, since his blunders and his fictions are so enormous, I say he will expose his ignorance and his imposture so clearly that he will forever be delivered to the disdain of every savant and even to those with only mediocre qualifications.[153]

Not surprisingly, the *Journal des savants*, which offered a brief review of Rolle's pamphlet on July 31, was more judicious in its commentary. It summarized the details of the dispute, offering those who had not yet noticed the controversy an accessible introduction to the principal points of contention. And rather than diffusing the controversy, it instead emphasized its significance, writing that while the debate was largely centered on the rivalry between two mathematicians, Rolle's piece "offers a critique of the differential calculus and infinitesimal analysis to which those who practice this analysis should pay particular attention." The journal also noted the complexity of the issues in question and their inaccessibility to those unschooled in abstract algebra. Concluding, the reviewer wrote: "Since it is geometry that is in question here, and geometers are accustomed to a precise and exact manner of reasoning, one hopes that these disputes will not last much longer and that the truth will finally be recognized. Maybe the disputes of these savant men will also produce something, such as the impetus for someone to produce a new *Elements* that will demonstrate the correct method of algebraic geometry and thus push this science as far as it can go."[154]

These hopes were not realized in the short term, however. Throughout the next two years Rolle continued to publish critiques of the calculus wherever possible, and in May 1705 Varignon wrote to Bernoulli that with "so much of Rolle's audaciousness combined with so many of his errors infecting our journals, M. Saurin has lost his patience and decided to revive his old dispute with him about the method of tangents."[155] In this way "the quarrel of the infinitesimals" was rekindled with "as much heat as ever." Rolle had succeeded by this time in getting his views regularly articulated in the normal weekly issues of the *Journal des savants*, but when Saurin responded, it was in a special issue devoted entirely to this topic. One appeared on April 23, and Rolle's response was included in the regular Monday issue published on May 11. A second special issue was devoted to Saurin's reply on June 11, and in July, Rolle was given a special issue of his own to reply in turn.[156] The only explanation for all this activity is to assume that Gallois was actually pulling the editorial strings at the journal, and whatever the mechanism for it, the debate was by now an embarrassing and absurd stalemate. In August, recognizing this morass, Saurin sought out Bignon to effect a resolution.

As editor-in-chief of the *Journal des savants*, Bignon possessed the requisite authority to end the debate in the journal. Interestingly, however, the process did not play out here or in the wider public sphere. Instead, the minister turned to the Academy of Sciences, forming another commission on August 8 charged with ending the debate once and for all. The commission comprised only academicians. Father Gouye, Cassini, and de la Hire were carried over from the first calculus commission, which had never issued a ruling, and two new members joined them: Gallois and Fontenelle.[157] The latter's presence was especially noteworthy because the perpetual secretary rarely participated in the work of the Academy in this direct way, and he was, as Varignon noted with relief in a letter to Leibniz, "the only supporter of the calculus not recused."[158]

The committee presented a fairly lengthy report of its resolution at the first academy assembly of 1706. Overall, the report concluded, the disagreement was rooted in technicalities and was not substantive. Rolle, the committee stated, had misrepresented his own debt to the method of infinitesimal analysis while Saurin, for his part, had misunderstood Rolle's method, which was largely "equivalent" (*co-égale*) to his own. More important in the eyes of the committee, however, were the conduct violations that transcended the intellectual disagreements themselves. As Bignon announced in reading the report before the Academy:

> Even if Mrs. Rolle and Saurin had good reason to continue their dispute, they have too often and too flagrantly violated the rules that all sorts of laws impose on savants as well as on *honnête gens*. The public has been scandalized by their various writings where they abandon themselves to blatant invectives against one another, even neglecting the most basic rules of politeness, propriety, and good faith. For this reason, we order M. Rolle to adhere to the statutes of the Academy to which he has the honor of being a member, and as for M. Saurin we ask that he adhere to the same out of the goodness of his heart.[159]

The academy's decree of January 1706 effectively ended Rolle's campaign against the infinitesimal calculus. A fifty-word account of the decision was published in Bayle's *Nouvelles de la république des lettres*, in the very last paragraph of the "News" section in the January 1706 issue. It began by announcing that Bianchini had been named to replace the deceased Jakob Bernoulli as an academy foreign associate, and then described the resolution of "the difference of opinion between M. Rolle and M. Saurin on *infiniment petits*." The journal reported that Rolle was directed to adhere to the Academy regulations, which require that all academicians moderate their actions with circumspection, and that Saurin was called to the goodness of his heart.[160] No other public discussion of the judgment appears to have occurred.

Ten months later, Varignon wrote to Bernoulli and announced Rolle's conversion. "I have written to M. Hermann telling him that M. Rolle has finally been converted. He will tell you how he came to express these sentiments to me and to M. Fontenelle, and about how he expressed them to Father Malebranche as well, saying that the opposition to infinitesimals had been pushed far enough and that he was tired of it. I hope, both for his honor and for our repose, that this conversion is sincere."[161] Yet there is evidence that his conversion was not as complete as Varignon would have liked. As late as February 1708, Rolle was delivering papers to the Academy that tried to solve mathematical problems without invoking "the most extraordinary assumptions used in the new system of infinites."[162] Other disputes between the Old Style mathematicians and the new analysts also erupted during this period. In the late summer of 1706, Varignon's new *élève*, Nicolas Guisnée, elected after Carré had been elevated to the rank of associate, engaged in a dispute with Gallois over issues of mathematical methodology. Similarly, in August 1709, Bourdelin reported that Saurin, now a member of the Academy after an admission process that we will examine in the next chapter, rose to "challenge the problems which M. Rolle had found in his general method."[163] Other academic disputes of this sort occurred throughout the early decades of the eighteenth century, but the open, polemical exchanges about analytical mechanics characteristic of the period 1696–1706 had come to an end.[164]

This outcome was favorable overall to analytical mechanics. Born of warfare, it emerged from the conflicts of the 1700s battle hardened and secure. It was still a controversial feature of the new public science of the Academy, and not yet a fixture of it, so in order to secure analytical mechanics as a central component of French academic science, a place for it had to be crafted within the institutional ethos of the institution. Building this institutional ethos was Bignon's job, but Fontenelle was his key agent in this project, and his role in the establishment of analytical mechanics as a respected French academic science cannot be overstated. He was charged in 1697 with channeling his stupendous rhetorical gifts toward the creation of an ideology for the new Académie Royale des Sciences that served the aspirations of the crown and savants alike. Since Fontenelle was also an ardent supporter of the new mathematical science practiced by Varignon and the other Malebranchians, he wove their epistemic values seamlessly into the ideological fabric he created for the Academy as whole. The result was the creation of a *pax analytica* in France that also allowed for the solidification of the new mathematical mechanics initiated in the 1690s as a mainstay of French academic science. Documenting how this solidification occurred is the project of the next chapter.

CHAPTER 10 *Managing toward Consensus* Bignon, *Fontenelle, and the Creation of the* Pax Analytica *in France*

Amid the clamorous public struggles over the calculus, the newly reformed Royal Academy worked to establish its new working patterns and influence within the terrain of French public science as a whole. The calculus wars posed considerable challenges for those eager to redefine academic science in a newly public way, and for this reason the controversy presented as many political problems as intellectual ones to those involved. Its resolution, moreover, was as much a political settlement as an intellectual rapprochement. The eventual detente made possible the ongoing development of analytical mechanics in France, and for this reason it is important to consider the political and intellectual forces that created and sustained the new peace.

Especially crucial were the efforts of Bignon and Fontenelle, who used their managerial power to secure a settlement that served their respective conceptions of French academic science. Managing the conduct of the Academy was Bignon's direct ministerial responsibility, but Fontenelle played an even more important role given how the 1699 reform had created for him a new job as institutional public spokesperson and manager of the Academy's public image. At one level, Fontenelle served Bignon, Pontchartrain, and the king in this work, but because the political rationale for the new Academy centered on enlisting the self-interested participation of the academicians themselves, Fontenelle was largely given free rein to play his new role as he saw fit. He was also selected for the new job precisely because of the harmony between his talents and interests and the agendas of the crown. Bignon and Fontenelle also shared many of the same goals, and consequently they worked together after 1699 (though not always without friction) to solidify the structures, practices, and protocols of public academic science in France. The outcome was the establishment of the eighteenth-century Académie Royale des Sciences as a model for many of what a modern institution of state-funded science should be like. They also engineered a site where the calculus-based mathematical physics developed after 1690 could be further developed and prosper.[1]

Yet if Fontenelle's cooperation with Bignon as an academic manager played a decisive role in bringing about this general institutional outcome, it was largely due to his individual influence that analytical mechanics became at the same time a scientific pillar firmly planted into the institutional foundation of the new Academy. In fact, if a consensus was established in France after 1706 accepting the legitimacy of the infinitesimal calculus and its application to mechanics, it was the governmental management of public science by the minister and the Academy secretary that played the most decisive role in securing this outcome. Ideas, as we will see, were crucial to this result, but intellectual factors alone, and especially mathematical considerations, do not explain why French analytical mechanics became an accepted and established academic science after almost a decade of strong challenges to it. To understand this result, we must look instead at the institutional management of mathematical science at the Royal Academy after 1699, and especially at Fontenelle's role as a defender of the new science who made calculus-based science an integral part of the wider public image of the Academy overall. In the end, it was not any particular scientific result, achievement, or rational demonstration that secured a place for analytical mechanics within the institutional bedrock of French academic science. Instead, it was Fontenelle's public propaganda work on its behalf that accomplished this outcome. Accordingly, it is to this work and the outcomes it generated that this chapter now turns (fig. 11).

Fontenelle as Academic Perpetual Secretary after 1699

Fontenelle's contribution to the development and establishment of analytical mechanics in France was twofold: administrative and intellectual. His intellectual influence worked simultaneously on two levels. The first stemmed from his own expertise with infinitesimal analysis and his ardent support for the new mathematics. This made him an unwavering advocate for analytical mechanics and a direct supporter of it inside the Academy. As secretary, however, Fontenelle rarely found occasions to express his scientific views directly. Yet in the same role he held other levers of intellectual influence useful for making his opinions felt. Especially important was his control over the records of the Academy, and after 1702, when the first published volumes began to appear, the annual publications of the company. These became a yearly vehicle for Fontenelle to showcase between the covers of the official academic publication his understanding of royal academic science.

Whenever possible, Fontenelle used the publications of the Academy to

FIGURE 11. *Louis Galloche (1670–1761),* Bernard le Bovier de Fontenelle, *1723. Oil on canvas; 128 × 96 cm. Châteaux de Versailles et de Trianon, France, MV4374. Photo: Gérard Blot. © RMN-Grand Palais / Art Resource, NY.*

articulate his support for the new science. This exerted a tremendous influence, but even more influential was his tendency to weave his praise for this science into his wider academic discourse. The general academic ideology that Fontenelle began to publicly propagate after 1699 was crucial in securing the new identity and culture of the Academy overall. By insinuating his personal enthusiasm for the new analytical mathematical sciences into this general discourse, he further aligned the official voice of public academic science in France overall with the particular voices of those supportive of this precise science. The result helped enormously to secure the presence of the new science at the heart of the Academy overall.

In addition to supporting analytical mechanics intellectually, Fontenelle also shaped its establishment institutionally through his capacity to shape academic outcomes and manage the perceptions of academicians and their work. He performed this managerial work in conjunction with Bignon, and we have already seen how he began to assert himself in this regard immediately after his appointment in 1697. His control over the academic register gave him one set of powers, and since Bignon and Pontchartrain used the records to oversee the conduct of the company and its compliance with the new regulations, Fontenelle's record keeping directly influenced the image of the academicians that these administrators absorbed. As we have seen already, he also used his precise presentation of academic work in the register to frame its perception in ways that suited his own agendas. Whether Fontenelle played any role in selecting the speakers for the public assemblies is not clear, but he quickly assumed the role of emcee at these events, replacing Bignon who played this role at the first session. Public assemblies then became a further site of influence for him. He also took charge of the publication program of the Academy, which was up and running by 1700 even if the first academic volume did not appear until 1702.

A survey of Fontenelle's new responsibilities as perpetual secretary after 1699 reveals the contours of his influence. One important interface was that between the weekly practices of the academicians themselves and the inscription of their work in the academic registers. When he was appointed in 1697, he picked up from the elderly du Hamel the new imperatives toward record keeping that the Pontchartrains through Bignon had insisted upon after 1692. This included the assiduous recording of attendance at every meeting, the composition of a detailed summary of the papers presented and the business discussed, and the transcription of academic papers read at the meetings. Du Hamel's struggle to satisfy these demands is palpable in the academic registers from 1692 to 1697, but after Fontenelle took charge the weekly Academy registers became noteworthy for their clarity, order, and legibility. What is also evident is Fontenelle's urge to manage the image of the Academy even at the level of these mundane records.

After September 1697, when Fontenelle fully began his job, the weekly entries started to include most of the papers read before the body including those that were never published. They also contained reports of the experiments and other activities conducted at the institution, including those that did not get reported in the annual *histoire*. The *registres* also recorded the names of the successful and unsuccessful nominees for academic seats, and in some cases the tally of the voting. They further contained the internal reports of academicians concerning treatises, discoveries, and machines submitted for review. Finally they record the ministerial comings and goings of importance to the institution, and report on the changes of significance. In all these ways, Fontenelle's record keeping left an invaluable archive that offers historians a comprehensive window into the detailed practices of the Academy in the first half of the eighteenth century.

But the historian looking for insight into the character of academic debates, or for an appreciation of the nature and conduct of academic science as it was actually pursued at the weekly meetings in the Louvre, is continually frustrated by the *registres*. Even at the level of these private, internal records, the ideal image of consensus desired by royal officials was maintained. Indeed, the *registres* taken at face value project a picture of academic practice largely devoid of any real contestation. Certainly, the Academy had many vigorous contests, and Bourdelin's diary reveals the spirit of critical judgment and competition animating the Academy in the wake of the 1699 reform. On occasion, these battles punched through the decorous veneer of reasoned consent projected by the Academy records.[2] But even here, as we saw with the calculus wars, what is missing in the records is any detailed account of the actual debates themselves, the positions taken by academicians, their justifications, or any other evidence regarding the actual conversations that occurred during the Academy's twice-weekly assemblies.

Two conclusions are possible in the face of this evidence: either the Academy did in fact realize perfectly its stated ideal of gentlemanly decorum and restraint, or the secretary distilled this image out of a more contested reality. Obviously, the latter hypothesis describes the reality even if it would be wrong to push this to the extreme by describing the Academy as a hornet's nest of party strife. Quite the contrary, the ideal of *honnête* science was deeply internalized by all the members, and professional collegiality prevailed inside the

Academy far more often than not. Nevertheless, the Academy's practices were certainly shaped by the kind of factional, institutional politics that one finds in any serious, intellectual institution, and even if the republican ideal of reasoned, gentlemanly consent was widely and scrupulously honored, it is also clear that Fontenelle made sure to present this image whenever possible no matter what the actual reality was. When one remembers that these registers were not just historical records destined for the archive, but also surveillance documents used by royal ministers to form judgments about the academicians and their work, Fontenelle's interest in this self-conscious management becomes all the more clear.

To hold the power to shape what Bignon and Pontchartrain saw of the Academy and its work was one important resource, and we have already seen how Fontenelle used his power to manage the calculus wars in ways that served the analysts and prejudiced Rolle, Gallois, and the other Old Style mathematicians. But if Fontenelle began in 1697 to exert an influential managerial presence through his control over the weekly recording of academic activity, his influence grew exponentially in 1699 with the creation of the new publication organs of the royal company.

Producing the annual *Histoire de l'Académie royale des sciences*, which became a major assignment for Fontenelle soon after the new regulations were instituted, gave the perpetual secretary a whole new array of academic management opportunities. One dimension involved the *mémoires* that the Academy began to publish every year starting in 1702 when the inaugural volume for the year 1699 first appeared. Academicians had been writing and presenting *mémoires* to the Academy ever since the institution was founded, but the new regulations made the writing of such papers an academic obligation. Their publication in the new academic volume created expressly for this purpose was also made a new measure of successful academic performance. While many, such as Varignon, welcomed the opportunity to write and publish their work, others were not accustomed to producing specialized *mémoires* of this sort. From 1699 forward, however, the status of an academician came to depend upon producing and publishing academic *mémoires*. In this way, in a manner akin to the new conception of the self-conscious "author" that Alain Viala sees as one outcome of the new disciplinization of literature at the newly created Académie française in the 1630s, a new kind of self-conscious disciplinary scientist began to be created in France after 1699 as a result of the new administrative imperatives toward specialized research, writing, and publication instituted at the Royal Academy.[3]

Bignon was in charge of the surveillance and remonstration necessary to encourage academicians to become specialized and publicly visible scientists in this way, and the perpetual secretary was in charge of the new publication program that allowed for the realization and ratification of these ministerial goals. We have already seen Fontenelle using his interested pen to shape the transfer of oral academic discourse into the written archival account of it found in the academic *registres*. The move from these manuscript *registres* to the published *mémoires* offered another space for intervention. Studying this interface is made difficult, however, by the dearth of documents recording the publishing procedures followed in the post-1699 academy. James McClellan is certainly correct to begin his study of the eighteenth-century "Comité de Librairie" inside the Academy in 1700, and to stress the committee's importance in the production of a new kind of scientific author with a new and more specialized relation to his research. But McClellan's sources largely come from after 1730, when the procedures developed after the reform started to become so familiar that they began to appear in the academic records themselves. At first, few traces of this work were left behind, and we therefore know almost nothing about how this committee formed initially, and what sorts of procedures it followed in its early years. It makes sense to assume, as McClellan does, that the protocols established in 1700 continued into the eighteenth century,[4] but how were these structures established initially, and through what manner of controversy and contestation? Fontenelle's registers are silent on the matter, and no other academic record exists.

At the end of the academic register for 1699, which also included a significant portion of the 1698 sessions as well, a "Table of Contents" is found that itemizes each paper presented and the author and date of its presentation. No similar table is found at the end of the registers for 1700 or 1701, and the difference indicates the changes under way regarding the translation of the academic material found in the registers into print. No record of how the initial translation from manuscript to print was accomplished is to be found, but a revealing set of letters from July 1718 offers insight. They are between the Chevalier de Louville, an astronomer who had won a special exemption from the Parisian residency requirement, which allowed him to conduct his astronomical research at his manor home in the Loire valley, and the astronomer Joseph-Nicolas Delisle, who was Louville's eyes and ears in Paris. Their correspondence concerned the fate of an academic paper that Louville had delivered concerning the obliquity of the ecliptic. In the paper, Louville had challenged the findings of Cassini, the director of the Royal Observatory, and he wanted to know why his paper had not been accepted for publication in the year's *mémoires*.[5] His correspondence with Delisle offers insight into how the academic publication program was working by this date.

Delisle responded confirming that the Academy had rejected Louville's paper, and after he inquired further about the publication protocols of the institution, Delisle offered him the following explanation.[6] Members of the library committee, Delisle wrote, were appointed by the Academy's officers and were directed by the perpetual secretary to nominate the papers suitable for publication. This committee, meeting in isolation from the Academy as a whole, deliberated and chose the papers to be published. Their decision, moreover, was final, and no appeal was permitted.[7] In the face of this, Louville wondered whether he could publish his paper elsewhere, perhaps in the Jesuit *Journal de Trévoux*.[8] Delisle responded by reminding Louville that while he could publish his work this way, he could not use his academic title in the byline unless the use had been approved by the Academy. This would require the formation of another committee charged with this task, and once again their decision would be final. He also discouraged Louville from trying to provoke the Academy in this way.[9] In the face of these obstacles, Louville appears to have suppressed his work, choosing to find other ways to change the mind of Cassini and the public about the obliquity of the ecliptic.

This episode illustrates the routine institutional contestation that was likely present each time the Academy's *Histoire et mémoires* were assembled for publication. Yet nowhere do these ordinary institutional struggles appear in the academic records. Nor do they appear in the published *mémoires* themselves because the *mémoires* as a whole were published separately, in their own independently paginated volume, and were arranged in a "Table of Contents" that presented them one after another without comment. Indeed, a reader of the Academy's annual volume unfamiliar with the Academy's institutional practices might comfortably assume that these were the only works written by the academicians during the year indicated on the title page.

The raison d'être of the Academy's annual published *mémoires* was to establish the technical scientific credentials of the academicians and to facilitate the broad circulation of their specialized scientific work. This manner of publication suited that mission perfectly. Yet the annual volume itself consisted of more than the *Mémoires in Mathematics and Physics Pulled from the Registers of This Academy*, as the full title indicated. The publication of undigested technical science, which this part of the volume accomplished, pursued one role for the Academy, but royal administrators had other goals in mind as well when they launched the new reform. Also essential to its mission was the creation of a broad public understanding of royal academic science compatible with royal political agendas. To accomplish this equally crucial program, technical science had to be translated into the more accessible idioms of French public discourse more generally, and had to be shown to be compatible with its values and norms. The public assemblies illustrate one effort in this direction, and another was the decision to include a critical narrative *histoire* in the annual academic publication. This was a separately paginated stand-alone volume that offered an explanatory prelude to the technical papers published together with it. The *histoire* summarized the specialized work found in the *mémoires* while also creating a context for it by explaining the preceding developments that made this work significant and by summarizing the *mémoires*' accomplishments and innovations. Together, this annual *Histoire et mémoires* of the Académie Royale des Sciences introduced the broad Francophone public each year to the work of this important royal institution, and if Fontenelle likely played a role in selecting the academic *mémoires* chosen for official publication, he was singularly responsible for the creation of the annual *histoire* that presented the Academy and explained its work to the general public.

Writing the annual *histoire* became after 1699 arguably the most important annual duty of the Academy secretary, and Fontenelle established the genre and its protocols by narrating each year a kind of "year in academic science" with paragraphs that highlighted the important academic events of note and sections devoted to the work of each of the disciplinary classes marked out by the 1699 regulations. The *mémoires* appeared afterward, arranged in chronological order according to the date of their academic presentation. Accordingly, while the *mémoires* recorded without context the perceived high points of the Academy's weekly deliberations during the year, and while these technical academic papers could be read individually as discrete scientific works, the wider public also had Fontenelle's contextualization of them framed in terms of the priorities he invented for their clarification and assessment. And similarly, if the Academy records that contained these papers were already an idealized reduction created by the Academy secretary of the actual back and forth of weekly academic debate, Fontenelle's *histoire* offered an even more packaged presentation, one oriented toward the wider public and generated out of a mission to serve both royal political agendas and Fontenelle's personal, scientific, and institutional interests.

The contrasts between an eighteenth-century Academy *histoire* and a representative article in a popular science journal such as *Scientific American* illustrate well the particular agendas that shaped this work. Both genres share the ambition of conveying complex and sometimes abstruse science to a broad audience, and both need to account for the disagreements that are typical of actual scientific practice. But while a typical *Scientific American* article focuses on the range of opinions surrounding a particular question without shying away from the deep disagreements that may dominate discussions about it, the Academy *histoire* needed to adopt a different narrative strategy. Academic *histoires* offered an authoritative summary of the state of thinking about a given question while emphasizing the important contributions already made by previous savants and those added to it by the particular French academician in question. Dissenting views were usually suppressed in these accounts, and if they were offered an attempt was made to reconcile the differing positions, thus preserving a vision of unity in the republic of science. In this way, the *histoire* attempted to construct a harmonious image of the Academy and of science as a whole, one free of overt partisan contestation.

A crucial balance was essential to the successful realization of the *histoire*'s purpose. On the one hand, the secretary needed to find ways to sell the virtues of academic science to the public at large. This meant first of all demonstrating how the science in question was publicly useful and important. But it also meant showing that academic science was comprehensible, and therefore accessible to the public as a whole. Here the secretary was building bridges between the public and the Academy by inviting individuals to think of themselves as academic auditors, or even "virtual academicians" (the link here with the parallel effort of the public assemblies is worth noting) who could participate through reading in the public work being done on their behalf by royal academicians. Given the priorities of the wider public toward which these *histoires* were addressed, the secretary also had to make his presentations appealing on an aesthetic level. Effective clarification of complex topics was not sufficient to an audience that often read substance in terms of style. Equally important was a presentation that was pleasing to read and in agreement with cherished stylistic conventions. Here the secretary needed to build a second bridge to the public by making academic science appear as a worthy subject of *mondain* conversation.

Fontenelle had already shown the crucial balance in *Des mondes*, for as he declared in his introductory "Preface": "I have tried to bring philosophy to a point where it is neither too dry for the worldly public [*gens du monde*] nor too light [*badine*] for savants. . . . I have sought a middle ground where philosophy is accessible to everyone."[10] Much the same agenda was operative in his academic *histoires* even if the genre conventions and intellectual expectations of academic officialdom constrained his aesthetic freedom in crucial ways. Fontenelle's pedigree in the *mondain* world of literary sociability provided him with the perfect combination of talents to strike the chords effectively, and under his control, the annual *histoires* of the Academy became a widely respected organ of public science. The volumes also reinforced the public image of the Academy desired simultaneously by the secretary, the academicians, and their administrative patrons.

Fontenelle also took it upon himself to use other organs of the new public Academy to articulate and advance this same program. Most important was his invention of the annual *éloges* delivered by the secretary in honor of recently deceased academicians. The institution of the *éloges* at the Academy of Sciences constitutes perhaps Fontenelle's greatest legacy as a member of the Academy of Sciences. Moreover, the practice illustrates once again the way that the new public Academy sought to project itself as the institutional embodiment of a new public culture founded on the use of science to achieve individual honor, state service, and the public good.[11]

The academy regulations of 1699 did not stipulate that the perpetual secretary would deliver a funeral oration in honor of recently deceased academicians, and Bourdelin notes in his diary that the practice actually began at the spring public assembly of 1700 when Bignon read a eulogy for the diarist's father, a founder of the Academy who died in December 1699.[12] Nevertheless, at the public assembly of April 6, 1701, Fontenelle invoked the "rules of the Academy" in announcing that he would deliver a eulogy in honor of the academician Daniel Tauvry.[13] Thereafter, hardly a public assembly passed without the perpetual secretary devoting a significant portion of the session to a reading of one of his *éloges*. He also arranged to have the *éloges* printed in the annual publication of the Academy (they appeared after his *histoire* at the end of that volume). By the time that he retired as perpetual secretary in 1740, the practice had become an unassailable feature of academic ritual. All of Fontenelle's successors carried on the ritual of delivering these orations at the public assemblies, and the speeches, along with the written *histoire* issued each year, became the two public duties most characteristic of the job of the perpetual secretary. The *éloges* also became one the most influential public presentations of royal academic science throughout the eighteenth century.

Under Fontenelle, the *éloges* were used to define the new public nature of the man of science essential to the public culture of importance to the state. The selection of this genre for this precise work is reflective of Fontenelle's genius. The Académie française had been composing *éloges oratoires* for its deceased members from as early as 1635, invoking in the process the institution's ties to the great antique tradition of funeral oratory personified in classical writers such as Plutarch and Cicero. The Academy of Sciences had no similar tradition. The closest an academician in the sciences got to a funeral oration before 1700 was a brief obituary in the *Journal des savants*. Doubtless, the recent battle between the Ancients and the Moderns, which went through a second iteration in the late 1690s, was in the back of Fontenelle's mind as he conceived of this new program for the Academy of Sciences. What better way to project the new academy as a premiere public institution of the realm than to lay claim to an ancient and venerable practice, update it in accordance with the modern ethos important at the time, and then perfect it as a vehicle for articulating a new vision of the public scientist as both a modern royal official and a modern intellectual. This is exactly what Fontenelle did in the sixty-nine *éloges* that he composed and publicly delivered from 1701 to 1740.

Taken as a whole, these orations define a new social type—a man (for academicians were all men) who from no particular social position finds his way to science and then, through self-sacrificing devotion to its virtues, serves humanity as a whole. The plotting in each case was different, and the values asserted in each story were particular to the individual case. But the goal of every *éloge* was exactly the same. Each defined one particular exemplar of an ideal type: an individual who personified the new administrative definition of public science and the conception of public service essential to it. Furthermore, to judge by their effect on one celebrated figure, they were profoundly successful. Among the factors that led the young Marquis de Condorcet to surrender his aristocratic ties to the military and pursue a career in science was the image of the man of science articulated by Fontenelle's *éloges*.[14] Condorcet would continue Fontenelle's legacy as the perpetual secretary of the Royal Academy himself, and countless other Frenchmen (and more than a few women as well) also made the transition from traditional roles to more modern ones as a result of the stories offered by Fontenelle.

In the immediate context of 1699, however, a second dimension of the *éloges* was equally important: their aesthetic beauty. The ideal of the man of science as public servant was crucial to the administrative program of monarchy in general, but only if this vision could capture the imagination of the elite public on which its political hopes depended. This public had come into being around the pursuit of *honnête science*, a domain of knowledge that admitted no firm demarcation between truth and beauty, reason and eloquence, or science and style. As a result, *honnête gens des lettres* tended to approach Fontenelle's orations as literary documents first and foremost, and then if they succeeded on that level as documents containing an important vision of public scientific service. Unless the *éloges* could win favor at the level of style, therefore, they had no hope of winning the attention of the elite public of interest to administrators. The brilliance of using Fontenelle as the bridge between these two worlds is again revealed here. No matter how critical later thinkers became of Fontenelle's other intellectual positions, no one ever questioned his genius as a stylist. Throughout the eighteenth century, the caution issued to young writers was not to forget Fontenelle, but to refrain from trying to imitate the inimitable master. Only Fontenelle could write like Fontenelle; on this critics agreed. But no one questioned the merits of the writing itself or its claim to genius. For the Academy of Sciences, this respect translated into almost universal praise for both the letter and the spirit of his *éloges*, praise that reflected upon their author and the Academy as a whole.

Through the *histoires* and the *éloges*, then, the perpetual secretary of the Royal Academy of Sciences attempted to articulate a vision of the Academy as a public institution devoted to the common good through service to science. He also attempted to legitimize this notion of public science and public service against the challenges posed by alternative visions of the same thing. He achieved his goals by projecting a vision of the Academy as a merit-based republic where rational deliberation produced universal consent thanks only to the natural truth of science itself. He also achieved his goals when he projected the Academy as an *honnête* community where reason and honor, and truth and virtue reinforced one another. At its best, this academic ideology, which appealed to a public of elites largely committed to the same meritocratic values, effectively cemented the alliances necessary to advance the power of the administrative monarchy in France. Under Fontenelle in particular, these ties became especially strong because he moved so comfortably across the border (which was not really much of a barrier anyway) separating the state from the elite public. He also personified completely the image of the public figure forged out of their alliance.

Before delivering his *éloges* at the Academy, Fontenelle often delivered practice readings at the salon of Madame de Lambert. In these Parisian gatherings, the crucial marriage between the public Academy and Parisian society was cemented in an informal way, making the formal establishment of the same thing at the twice-yearly public assembly all the easier. As Roger Marchal, writing of this nexus from the perspective of Madame de Lambert, explains: "Thanks to Fontenelle, the salon on the rue de Richelieu found itself suddenly at the heart of literary life and quickly becoming the site of an ambitious project designed to effect the triumph of modern ideas: the feminine and *mondain* public, the academic institutions, and certain influential political spheres. It was necessary, therefore, to give [Fontenelle] a retinue and a pomp worthy of these new ambitions."[15]

The presence of other publics with other understandings of public science and the state complicated the easy realization of this dream. The 1699 reform attempted to place the Academy at the authoritative heart of French public science, but this claim was continually challenged by those wedded to other institutions. Salons, cafes, universities, the Society of Jesus, the royal *bureaux*—these institutions, together with the people that inhabited them, often had their own claim upon public science in France, and theirs did not always harmonize neatly with the public agendas of the Royal Academy. For state officials, the goal was to efface this diversity and effect a unity in the name of a reformed, if still absolutist, public authority. It was also to accomplish this mission while maintaining the ideal of free and open public reason central to the Academy's republican self-image. Here again, Fontenelle, charged with articulating these goals as the Academy's official public spokesperson, was forced to tread carefully.

To the extent that the secretary was successful in harmonizing academic science, in all its eighteenth-century manifestations, with administrative monarchy, publicity, and elite sociability, then a powerful cultural weapon was forged. In this case, the Academy and the administrative monarchy both benefited from being allied with this program. Yet the forces pulling this unity apart were powerful as well, and despite all efforts to the contrary, 1699 did not mark the unification of public science into a harmonious, absolutist whole centered on the Royal Academy. It instead created a new dynamic that placed a number of distinct if interconnected political constituencies and publics into competition with one another for control of the same thing: public knowledge and culture. All developments in French science after 1699 were profoundly shaped by this new institutional dynamic, not least analytical mechanics, which remained forever a favorite topic for the secretary as he performed his decisive discursive work. Accordingly as Fontenelle worked to establish the image and ideology of the new public Academy, he also worked directly, and sometimes self-consciously, to ensure a place for analytical mechanics in the institutional bedrock of this new institution.

Laying the Foundations of the *Pax Analytica*: Fontenelle's Public Academic Discourse of 1701–2

If one wanted to set a precise date for Fontenelle's debut as public manager of the Académie Royale des Sciences, a good choice might be April 6, 1701. The academy held its fifth public assembly on this date, and since the first volume of *Histoire de l'Académie royale des sciences*, documenting the work of the year 1699, was, according to Fontenelle, "in press" (it in fact appeared nine months later), the secretary used the occasion to read a preface to that volume that he had composed.[16] In it, he described the new academy regulations since they were published verbatim in this first volume, and he also explained their rationale. The April 1701 assembly was also the first to be held after Varignon's public-assembly presentation about his calculus-based central-force mechanics, and since it occurred at the moment when the vitriolic battle regarding the calculus was starting to erupt inside the Academy—Fontenelle's own intervention in this debate had occurred five weeks earlier—the secretary also decided to invisibly address the struggle by offering a general discourse on the "Utility of Mathematics for the Sciences," which he would publish as a kind of preface to the first academic volume. At the April assembly, he read the text as an oration addressed to the assembled crowd. Fontenelle also gave his inaugural *éloge* at the same public session, so the assembly was overwhelmingly dominated by his voice. In fact, so much of the time was devoted to Fontenelle's many presentations that there was only time left for two formal scientific presentations, an examination of a monstrous fetus given by the *élève* Littré and some chemical experiments performed by Homberg.[17]

Fontenelle was no doubt very eager by the spring of 1701 to start exploiting the new public organs made available to him by the 1699 regulations. He was also especially interested in these opportunities given the controversies that were erupting regarding the calculus, a topic that provoked both his personal and professional interests. Institutional snags, however, put a damper on his ambitions. The actual *Histoire* published in 1702 dealt only with the work of the Academy in 1699, and thus, when it appeared, it was addressing what was by then three-year-old scientific news. Fontenelle lamented these frustrating delays in the opening pages of his narrative, but publication obstacles would continue to plague the Academy's publication agendas throughout the eighteenth century. Virtually none of the annual academic volumes appeared within a year after the close of the academic year, and many took as long as five years to appear. These delays were important. Since work read at the

Academy only appeared in print much later, there was always a lag between the official written pronouncements of the Academy and the wider public debates occurring at the same time. Here one sees why periodicals like the *Journal des savants* and the *Journal de Trévoux* provided such an appealing publication alternative for some academicians in this period.

The publication lags were especially frustrating to Fontenelle in the early 1700s. Having not yet published any work, the Academy was not yet a direct participant in the public discussion of science that its reform had provoked, and implicitly authorized, in 1699. Furthermore, without the presence of the Academy's official voice, the public sphere was also embroiled by 1702 in *la querelle des infiniment petits*, a debate of crucial importance to the secretary personally, and to the institution overall. Indeed, before Fontenelle could get his own official account of either Varignon's science or the calculus debate that it catalyzed into print, a discussion of both was already reverberating throughout the public sphere. Thus, at the very beginning of his tenure as perpetual secretary, Fontenelle assumed a stance that would become commonplace for him in the years that followed, serving as an authoritative, after-the-fact manager of public debate.

Although this orientation was largely thrust upon him by historical circumstances, it ultimately served his personal ambitions and the institutional needs of the Academy extremely well. The secretary was first and foremost a citizen of the Republic of Letters, and for him the Academy was an institution beholden to the intellectual priorities of this wider international community. Open public contestation was anathema to the spirit of this learned community, and while learned journalism was starting to become more amenable to the staging of spirited intellectual debates, the editors of the *Journal de Trévoux* spoke from the heart and soul of the Republic of Letters when they described their mission in 1701 as that of the disinterested historian. These values were widely held among those self-identifying as *honnête savants* around 1700, and they were even more important to a royal official charged with presenting the Royal Academy in the best possible light. Accordingly, the voice that Fontenelle adopted, where he rendered judicious, after-the-fact assessments of royal science, was not only the one forced on him by circumstances, it was the voice that best fit with the obligations of his office.

In adopting a managerial stance, Fontenelle therefore imagined his work as a public service to the wider learned community, one that would focus on disseminating those dimensions of academic science suitable for fostering *honnête* intellectual community while eschewing the cantankerous to and fro of actual academic debate. Fontenelle embodied the Enlightenment idea that science itself was a public service, and the depth of his commitment to this mission can be gauged by the work he produced in this office.

Not only did he compose forty-one annual histories for the period 1699–1740, and see each of these narratives into print in a timely fashion, he also edited the roughly two dozen *mémoires* that appeared in each of these volumes, taking care of their transcription into the academic registers and then their transfer into print. He no doubt had countless invisible assistants helping him in this effort, yet even with help it was still a massive and monumental undertaking. Fontenelle also took on the project of writing and publishing retrospective histories of the Academy's work in each of the years preceding the 1699 reform, along with the publication of many of the *mémoires* that were written during these years when the Academy did not yet have a publication outlet. His retrospective *Histoire et mémoires de l'Académie royale des sciences depuis 1666 jusqu'en 1699* ultimately comprised thirteen thick folio volumes when it was published in the 1730s,[18] and it is a marker of the thankless service which this labor entailed that that only his sixty-nine *éloges* and his "Discourse on the Utility of the Mathematical Sciences," to be examined shortly, are contained in the modern critical edition of his complete works. Even if the thousands of pages that he wrote explaining and celebrating French academic science between 1666 and 1740 are not considered by his modern editors worthy of inclusion among his oeuvre, the writing was nevertheless a monumental accomplishment sustained by Fontenelle's belief in, and devotion to, the project of public scientific service.

In early 1702, the first of these efforts began to appear in print, and in his "Preface" to the inaugural volume, which he read to the audience assembled at the Louvre in April 1701, Fontenelle sketched the contours of the academic ideology that he would subsequently construct and defend persistently over the next four decades. His "Preface" began with an account of the new regulations, a brief explanation that articulated the precise function of the newly reformed Royal Academy within French society. "This academy was formed, in truth, by the orders of the king," Fontenelle explained,

> but without any act emanating from royal authority. The love of sciences itself produced almost all of these laws. . . . But to render this company durable and as useful as it possibly can be the most severe and precise regulations are necessary. . . . Realizing this, the king judged it appropriate to give the Academy these new regulations. . . . He charged M. de Pontchartrain, previously Minister and Secretary of State and now Chancellor of France, to give to the Academy of Sciences the

> most appropriate form for making it as useful as it promises to be. . . . M. l'abbé Bignon, longtime president of the Academy of Sciences, also communicated his light to M. Pontchartrain in this effort. . . . From this joint effort emerged an almost entirely new company similar in certain respects to those ideal Republics imagined by the sages where the laws give the greatest liberty imaginable and require only that one follow the dictates of reason.[19]

Having framed the Academy and its mission in terms of the republican, meritocratic values dear to both him and the royal patrons who supported him, Fontenelle continued by describing the nature of the science that an academy of this sort should pursue. As the *Mercure galant* reported, the text was "entirely concerned with the utility of mathematics and physics."[20] More precisely, Fontenelle constructed a detailed, if no less artful for its precision, argument justifying the place of abstract mathematics at the heart of a well constituted program for general academic science. Bourdelin called the work "perfectly well written and very well received," and the *Mercure galant* echoed his praise.[21] "It appeared that this discourse persuaded the public as much with the judiciousness of reasoning as it did with the pleasure of its style," its reporter enthused. "The applause after the reading was widespread."[22] When the "Preface" appeared in print in 1702, the *Journal de Trévoux* declared it "a masterpiece," printing large sections of it verbatim in their April 1702 issue.[23] Praise for the work in fact followed it wherever it was received, for the "Preface" was indeed a virtuoso performance that used only nineteen eloquent pages to articulate Fontenelle's entire conception of public science and the place of advanced mathematics within it. His public academic discourse would continue to navigate the pathways first mapped out in this text for many decades to come, so the "Preface" offers us a concise entry point into Fontenelle's broader public work.

He declared his precise focus at the opening of his discourse by asking two essential questions: "What is served by making the taste for mathematics and physics more widespread, and of what use are the occupations of the Academy in this regard?"[24] The subsequent paragraphs offered a set of elegantly argued answers, but in many respects the questions themselves outlined the main argument. Certainly, no one in this period doubted that the sciences, properly conducted, were useful. Support for the sciences was indeed universal. People, however, held widely different opinions about why the sciences were valuable and what their essential function should be. Fontenelle's "Preface" was not designed, therefore, to sell the public on the value of a scientific academy per se, but to move its audience toward the precise definition of science and utility that he and his fellow academicians prized.

One thread supported the administrative conception of utilitarian service dear to royal administrators. Here Fontenelle was following in the footsteps of Bignon, who charted his own course for the new public Academy in a speech delivered at the first public assembly, in April 1699. Announcing the goals of the recently reformed institution, Bignon reduced the mission of the Academy to serving as a utilitarian adjunct of the royal state. According to the *Mercure*, whose bias in this regard is important, the minister declared that "the Academy of Sciences aspires only to know the truth, and often seeks the most abstract and dry truths of all. It requires only that the truth be useful, and it does not care whether it is agreeable or not."[25] This description of academic science certainly satisfied many inside and outside the institution and the state. Its claim that utility could stand alone as the sole value of academic science also defined well the administrative rationale for the company's existence. Louvois had in fact defined academic science in similarly blunt terms in 1685. "I understand by useful research that which could relate to the service of the King and the State," he asserted.[26]

This understanding of academic science was not the ideal most likely to excite readers of the *Mercure galant*, however, and its stark focus on instrumental utility alone made it anything but the standard within the wider Republic of Letters. Since Bignon possessed all of the virtues admired by *gens des lettres*, his articulation of this position did not alienate the Academy from these constituencies, but Fontenelle saw the purpose of the Academy differently. In his own discourse, which was composed and delivered within a year of Bignon's, he implicitly responded to the minister's initial effort at public academic management by offering a more complex view of the public mission of the Academy.

Like the minister, Fontenelle described the Academy in terms of a marriage between the state and the wider learned world. Yet he conceived of the bond in very different terms. Fontenelle directly addressed in his discourse the questions, which he, like Bignon, knew were on the lips of die-hard utilitarians, but he did so in a way that balanced utility with the value of abstract theorizing. As he wrote: "We have a moon to bring light to our nights, but why, some ask, do we need to know that Jupiter has four? And what purpose is served by all the difficult calculations that allow us to chart their movements?"[27] Because such science contributes to the development of socially useful knowledge, the secretary retorted. Galileo had developed a way of determining longitude at sea by using the phases of the moons of Jupiter, and from this the secretary drew the lesson that much which appears useless at first ultimately turns out to be of deep practical value. He continued in this vein elsewhere in his oration, noting the connection between anatomy and improved surgical techniques and mechanics and the development of productive machinery. As he summed up in a clear appeal to the aesthetic sensibilities of worldly elites:

> It will always be easier for the public to enjoy the advantages acquired from such knowledge than the advantages that come from knowing itself. The determination of longitude by the moons, the discovery of the Torachique canal, a more accurate and effective balance—it is true that these are not achievements that are likely to create a stir in the same way that a pleasurable poem or a beautiful and eloquent oration does. But the utility of mathematics and physics, though somewhat obscure, is no less real as a result. Indeed, to consider man only in his natural state, nothing is more useful to him than the things that can conserve life and produce those arts that are such a great aid and mark of distinction for our society. Such is the destiny of the sciences diligently practiced by only a handful of people. The utility of their progress is invisible to the majority of people, . . . and the public enjoys the success that they offer with a sort of ingratitude.[28]

Here Fontenelle employs the classic connection between utilitarian science and the progress of material society as a trope for linking elites to the work of the Royal Academy. Eloquent and artful arguments such as these fused Fontenelle's public academic discourse with the wider political discourse about administrative monarchy offered by Bignon and the crown. Even more influentially, they also cut the standard suit of administrative utilitarian ideology in a way that made it fashionable to other constituencies as well, especially those aligned with the *mondain* public. Fontenelle's public academic discourse in fact came together most powerfully at the interface between his deeply held convictions about utilitarian scientific service, values shared by his academic colleagues and the royal administrative patrons that supported them, and his equally strong convictions about intellectual liberty, science, and truth, convictions that joined him with *gens du monde* and *gens des lettres* in France and abroad.

His "Preface" illustrates well this particular alchemy because the text is most often concerned with making the duties and burdens of administrative academic science appear appealing to elites not accustomed to finding such labors attractive. Readers of the *Mercure galant* were one audience that Fontenelle hoped to reach, and he knew that they were not immune to the right rhetorical appeals. In celebrating the public physics demonstrations of the Sorbonne's Professor Guillaume Dagoumer, for example, the *Mercure* called his demonstrations "rare and curious" and "full of pleasures." The journal also noted that, "the audiences were so large that spacious amphitheaters had to be constructed to hold everyone who wanted to have the pleasure of watching."[29] The journal also noted the educational value of these demonstrations, appealing to other university professors to emulate Dagoumer's "useful" initiatives.[30] In a similar vein, the public philosophy course of Professor Pierre Fleury at the Collège des Grassins was praised for offering clear explanations that were "very interesting and very useful for comprehending sound physics."[31]

Fontenelle's own public academic discourse was in large measure an extension of this kind of public scientific pedagogy. For given his understanding of *mondain* society, he knew that he could attract its members to the utilitarian work of the Academy as long as he also appealed to their worldly sensibilities. At times he spoke as a royal administrator in these efforts, such as when he declared that "by making the public more aware of the Academy" serious science would "circulate more widely and more easily." More often, however, he tempered his administrative orientation with a *mondain* conception of the same thing. There are "many amazing things before our eyes that we do not even see," he said in one representative passage. "The shops of the artisans sparkle at every corner with intellect and ingenuity, but this does not attract our attention. All that's lacking is an audience who recognizes the value of these instruments and these exceedingly useful and imaginative practices. Nothing is more marvelous than this spectacle to those who know how to be astonished by it."[32] Here the utilitarian labor of the technical artisan is connected with *honnête* notions of inventive genius, curiosity, wondrous spectacle, and the aristocratic pleasures of sumptuous materiality.

Other passages in his "Preface" were aimed in the same direction. Playing on the conception of *honnêteté* that was central to elite society, he offered a way of thinking about the academic *mémoires* that were published each year that appealed to these values. "The collection of papers that the Academy presents to the public is only composed of separate and independent works," he explained. "Each academician himself is responsible for guaranteeing the facts and the experiments reported, and the Academy only authorizes these reasonings with the caution of a sage Pyrrhonianism."[33] In a similar vein, the secretary assured his readers that "the Academy of Sciences only approaches nature in discrete blocks." "It embraces no general system because it fears the ease with which the impatient human mind can accommodate itself to precipitous systems that, once established, oppose themselves to truth."[34] Rigid, dogmatic thinking was the bane of independent *gens du monde*, and by distancing the Academy from preordained systems and authoritarian confessions of belief the secretary was connecting academicians with the ideals dear to *honnête savants* everywhere.

Like the ideal aristocratic sage of Old Regime France, Fontenelle's academicians were independent savants beholden only to an ethic of *honnête* integrity. Attaching themselves to the legacy of Montaigne and Stoic skeptics such as the moralist La Rochefoucauld, self-professing *honnête gens* in France often modeled their intellectual liberty on that of the classical skeptics. By making the Academy a collection of "prudent Pyrrhonians," therefore, Fontenelle was situating the academician in this milieu. The same theme was developed in Fontenelle's first *histoire*, which related the new practice, instituted in 1699, whereby each academician declared his intended research project before the Academy as a whole. The secretary likened the practice to a "species of oath" and noted the bonds of community that formed, especially between older and younger academicians, as a result of these solemn proclamations.[35] Here the Academy was presented as a community of specialized experts devoted to individualized, disciplinary research, an image consonant with the program of the 1699 reform. But it was also presented as a republican community of reason bound by a deep and virtuous commitment to personal honor and integrity. In each case, the meritorious service of the Academy is presented in ways that make it agree with the ideals and aspirations of independent elites while also serving the agendas of the administrative monarchy.

As he wrote in describing the Academy's program in the mechanical arts, a program that centered on the social elevation of certain practices still deemed "lowly" and "base" to many: "Because the Academy is more concerned with being useful to the public than being devoted to pleasure or attracting acclaim, it willingly undertakes the dry, arduous, and never sparkling work of describing the state of the arts as they are now found in France."[36] Earlier statements about the mechanical arts had emphasized their connection to genius, invention, and material splendor. In this passage, they are associated with a dutiful commitment to public service, one that reaps collective social rewards through the sublimation of selfish pleasure into communitarian labor that serves the greater good.

Fontenelle's efforts were often most successful, in fact, when he showed worldly elites how scientific service for the common good was compatible with liberty and pleasure. Indeed, what is noteworthy about his public academic discourse is how he wedded an elite desire for leisure and worldly diversion with a conception of academic science rooted in dutiful service to the public. Addressing the more hedonistic side of this equation, Fontenelle rebutted those who find the sciences "painful to study, barbaric, and difficult to penetrate." Is this sufficient ground for remaining ignorant of them? he asked.[37] He also provoked the pretentions of *mondain* elites by making an appreciation for the sciences an indicator of elite character and quality. Academic scientists, Fontenelle suggested in one smartly conceived passage, should be imagined as rich collectors amassing discoveries without regard for their usefulness. "Let us try to amass any and all of the truths in mathematics and physics that we can," he wrote. "There is very little risk in this, and it is certain that a deep foundation will be built from which a variety of useful truths are bound to emerge."[38] Here aristocratic sensibilities were provoked by making serious scientific inquiry just another form of luxurious collecting.

Fontenelle also celebrated the intellectual elitism of abstruse science as a means of making its practice admirable in the eyes of his intellectually pretentious public. "In the end, everything that elevates our reflections, and everything that, although purely speculative, is grand and noble, has a utility all its own," he declared. "The mind has its needs, and they often exceed those of the body. It wants to know, and everything that can be known is necessary to it. Nothing shows so well the destiny of the mind to know the truth, and possibly nothing is more glorious for it, than the charm one finds, sometimes in spite of oneself, in the most dry and arduous studies."[39]

The mathematical sciences in France were especially in need of discursive support of this sort, for no science struggled more in turn-of-the-seventeenth-century France to overcome its appearance as *épineuse* and *sauvage*. Especially challenged in this respect was the new cryptic and symbol laden species of it practiced by the analysts. Yet for Fontenelle, no science was more important to his overall vision of academic science. Given his general devotion to advanced mathematics as a whole, and his particular affection for the most advanced and *épineuse* specimen of them all, the infinitesimal calculus, the secretary was driven by both personal and professional passion to channel his formidable rhetorical talents into a general public defense of the overall value of such mathematics. And, more importantly for the precise arguments of this book, he was also led by the same motives toward a particular defense of analytical mathematics and its sciences as discipline worthy of inclusion at the heart of a modern scientific academy. This defense did important work in securing the value of analytical mechanics within the eighteenth-century Académie Royale des Sciences.

Fontenelle's use of his annual *histoires* to shape the debates about analytical mechanics after 1702, which we will discuss shortly, illustrates well the nature of his efforts, but he introduced many of the frameworks that he would later deploy in his introductory "Preface," making it a useful entry into this work. Much of his advocacy on behalf of advanced mathematics worked by merging it seamlessly with his overall presentation of the value of academic science as a whole. The utility of abstract mathematics was often celebrated in his general writings about the utility of science, for example, and he often made it the foundation of general scientific progress. In his "Preface," he illustrated his approach when he criticized those who "want to restrict mathematics to only those useful products" that have "an immediate and tangible" application while relegating all other work to "the realm of vain theory." Nothing could be more false than this understanding, he wrote, for "the art of navigation is inseparably linked to astronomy, . . . and astronomy in turn requires optics for its long-range lenses. In a similar fashion, all of the mathematical sciences are linked to one another, and all are founded on geometry which is itself founded on algebra."[40] Here, along with a plug for the progressive superiority of analytical mathematics over other traditional mathematical forms, the utility of mathematics is made manifest through its connection to socially useful results.

Fontenelle also placed the value of mathematical science above any immediate calculation of productivity since it is placed, he argued, at the very foundation of all the sciences. Mathematical work of any kind, he suggests, is always useful because it is foundational to every other science. This general justification of mathematical utility was a commonplace in Fontenelle's academic writing, but by placing algebra at the root of the tree of knowledge, he was also articulating his more precise scientific convictions as well. Throughout the "Preface," in fact, the secretary was ever eager to implicate his Malebranchian convictions about the value of analytical mathematicization into his statements about the overall value of mathematical work as a whole. "Geometry and especially algebra are the key to all the studies one can make of magnitude," Fontenelle wrote in a richly Malebranchian passage.

> These sciences, which occupy themselves only with simple ideas and abstract relations might seem unproductive since they never leave, so to speak, the world of the mind. But the mixed mathematical sciences, which descend to the material world and consider the movement of the stars, the increase of moving forces, the different paths taken by rays of light, etc.; in a word all the sciences that seek out the particular

> visible quantitative *rapports* between things advance further and more convincingly to the extent that the art of discovering these relations in general is perfected.[41]

This passage articulates well a Malebranchian justification for the analytical sciences as a thread tying together all the most advanced sciences. Elsewhere in the "Preface," the secretary reiterated the point while adding specifically *mondain* appeals to it. On one occasion he connects Malebranchian scientific thinking to the aesthetics of elite comportment, arguing that, "a work of moral philosophy, politics, criticism, or even eloquence will be more beautiful, all things being equal, if it is composed with a Geometrical hand."[42] The "Preface" also posited a general trend toward increased "order, regularity, precision, and exactitude" in contemporary literature that was attributable to Descartes, who used geometry to "set the tone for the century" by establishing overall "a new art of reasoning."[43] Here Fontenelle echoed Lamy's rhetorical theories by making numerical *rapports* foundational to the art of speaking, writing, and thinking well. These concerns were of cardinal importance to elites in Old Regime France, and elsewhere Fontenelle offers a specifically Malebranchian twist to his call for civil comportment through mathematics. "It is always useful to think correctly," he wrote, "even about non-useful subjects. Even when the numbers and lines lead to absolutely nothing, we know that this is the only certain understanding obtainable by natural reason. These truths thus give our mind more certain training in, and glimpses of, the truth. They teach us how to operate upon truths, and how to grasp their threads, so often delicate and imperceptible, and to follow them as far as they can lead us."[44]

This complex discourse about the value of even the most abstract, theoretical mathematics served a number of objectives simultaneously. It appealed to the state by justifying its commitment to the abstract, and not immediately utilitarian, work of analytical mathematicians like l'Hôpital and Varignon. It also appealed to the academicians themselves because it offered them an ideology that made even their most abstruse work valuable and worthy of esteem. Ideally as well, it also appealed to the wider public by convincing them that the most *sauvage* and *épineuse* mathematics was still of value to them. Ever attuned to the anxieties of this audience, however, (not to mention their patience), Fontenelle was also careful to spice his discourse about mathematics with more directly enticing arguments. One strategy was to play on the elitism of the *mondain* public by inviting individuals to share in the elitism of abstract mathematical work. Fontenelle asked the readers of his "Preface" to think of abstract mathematics not as a disagreeable and isolating practice, but as a practice worthy of their special intellectual freedom. So long as one conducted one's life with geometric order and rigor, he argued, one was leading the good life. Furthermore, so long as one's thoughts were disciplined with mathematical reason, even the most abstract flights of fancy were worthwhile and productive.

Those who approached the world mathematically, therefore, were the real *honnête gens*, or so Fontenelle artfully suggested. They correctly saw that the rigors of advanced mathematics were productive rather than destructive of the comportment that people of quality sought.

Fontenelle did not, of course, invent this equation of mathematics with elite comportment.[45] One finds similar arguments in conduct books and educational literature throughout this period, not to mention in Malebranche's influential and widely read *Recherche*. The Oratorians also made links such as these a central part of their public pedagogy. In short, belief in mathematical reasoning as a royal road to aristocratic right-mindedness was widespread in France in the years around 1700. But by activating these wider sentiments in conjunction with the specific work of the analytical mathematicians at the Royal Academy, Fontenelle was forging a new alliance between this community and the wider *mondain* public, one with important consequences.

The Cartesianism implicit in this appeal also served Fontenelle's official duties. According to the Cartesian assumptions that flowed throughout his academic writings, any individual could ascend to elite understanding by merely following the chains of evidence that reason provided. The academicians themselves, taking these links for granted, often left the details of their work hidden. But because all good science contained well-reasoned chains of argument, the links were always available to be exposed and utilized. Once recognized, the steps also created a ladder linking the larger public to serious science even if this access in no way erased the divide separating the Academy from its amateur audience. Few actually followed the ladder of clear and distinct evidence all the way to the top, but the genius of the appeal rested in the widespread belief that such an ascent could be made by anyone. Thus, as long as confidence in the Cartesian notion of evident knowledge remained widespread, Fontenelle's public academic discourse was a powerful tool for forming the unity about academic science that he hoped to achieve.[46]

Securing the *Pax Analytica*: Fontenelle's Defense of Analytical Mechanics

Fontenelle's desire for reasoned consensus, however, was tempered by his equally strong desire to persuade the public of his particular scientific views. Accordingly, within the contours of his larger public discursive work, Fontenelle began to advocate for the science he particularly supported, namely Varignon's analytical mechanics. In the spring of 1702, at roughly the same time as Rolle's first *Journal des savants* article, Fontenelle's inaugural Academy *histoire* appeared, the one that also included his opening "Discourse on the Utility of Mathematics and Physics." This volume was followed by the appearance of the *histoire* of 1700 within months of Saurin's rebuttal of Rolle in the summer of 1702, and the *histoire* for 1701 less than a year later. The *histoires* of 1702 and 1703 were published in 1704 and 1705, respectively. Thus, between the start of the Rolle-Saurin dispute in 1702 and its resolution in early 1706, Fontenelle effectively brought his public academic discourse into direct dialogue with the most recent mathematical debates occurring in the wider public sphere. Moreover, appearing as they did during the second phase of the debate about infinitesimal analysis, these works allowed Fontenelle to participate directly in the discussion, shaping its outcome in ways that were influential for the long-term legacy of French academic science and the particular place of analytical mechanics within it.

Bignon's discipline was certainly a crucial agent as well in bringing about this outcome, and the case of another related academic mathematical dispute at precisely this time illustrates clearly the crucial role that direct administrative discipline also played in the establishment of this particular outcome.

At the center of this controversy was Antoine Parent, a young mathematician who entered the Academy in 1699 as the *élève* of the newly appointed *pensionnaire méchanicien* des Billettes.[47] Parent's mentor was an important member of Bignon's early group of *méchaniciens*, and once admitted he also became a key figure in the mechanical arts initiatives crucial to Bignon's program for the Academy. These allegiances suggest a close attachment to Bignon, but Parent also pursued other alliances, and overall he appears to have been an ambitious young man seeking opportunities wherever they appeared.[48] He established ties with Sauveur at the Collège Royale and de la Hire at the Royal Observatory, and he was probably one of the most active academicians of his generation, continually presenting papers at the assemblies on a wide array of topics. This was the sort of diligence encouraged by the 1699 regulations, but the quality of Parent's work did not always match its quantity. When the abbé Louvois evaluated his performance for Pontchartrain in 1702, he simply noted that, "he is a very active member."[49] Bourdelin was more critical, using phrases like "neither useful nor well done" and "confusing and difficult to follow" to describe Parent's work in his diary.[50] Varignon bluntly called him "vain and worthy of humiliation."[51] Since he never moved above the rank of *élève* at the Academy, Parent also appears to have struggled for his professional and personal livelihood in exceptional ways as a result of his reputation.[52] His activities as a public savant are, therefore, intriguing when viewed from the perspective of the new professional imperatives created by the 1699 academy regulations. As an academician, he established a reputation early on as an antagonist. As Fontenelle described: "The breadth of his knowledge, together with his naturally impetuous character, gave him a tendency to criticize everything, sometimes in a rash way and very often without restraint."[53] Perhaps for this reason, or perhaps because his work generated so little praise, he did not have much success publishing in the vehicles provided by the Academy. He thus began to seek other outlets, and in 1703, these ambitions intersected with the vogue for mathematics then raging in France, not least because of the calculus wars. In March, he issued the first volume of his proposed monthly journal, *Recherches de mathématiques et de physique*. The only extant copy of the original volume contains an authorization from Bignon.[54]

The periodical itself offered a platform for Parent's mathematical work, and the range of the authors and topics treated reflects the breadth of his interests.[55] The polemical nature of his journal was also clear, and soon after its release the *Journal des savants* published a review critical of Parent's editorial voice.[56] Bignon was editing the official French periodical by this time, and consequently it offers a revealing window into his attitudes about public, intellectual disputation. It started by quoting at length from Parent's opening *Avertissement*, paying particular attention to his assertions that he would not serve the public by acting as a neutral journalist reviewing the latest scientific work but instead as "a critical analyst" of current and past scientific achievements. Here Bignon and his editorial staff cried foul. "Criticizing the work of famous authors is no easy task," the review declared. "Few in the world are disposed toward those who claim more confidence in this project than enlightenment and judiciousness of spirit."[57] Using this caution as a guide, the review then exposed the weaknesses of Parent's first volume, using its faults as evidence against Parent's critical authority. Concluding, the journal declared: "his project would be more agreeable to the public if [the editor] chose to publish *mémoires* that could teach us something new or could not be found anywhere else."[58]

Bourdelin noted that the elder de la Hire also raised objections against Parent's journal in the Academy a month after the review in the *Journal des savants* appeared. He did not record the precise response, nor is it recorded in the Academy registers, but since Bignon often employed academicians as part of his editorial team at the journal, a linkage between the two interventions is not unlikely.[59] Parent's journal also stopped appearing after its first issue, and all signs suggest, though no documentation confirms, that Bignon revoked Parent's journalistic privilege as part of a disciplinary action against him. Parent resumed his usual activity at the Academy, and whether through a reprieve for good behavior or some other reason, he was granted permission to resume his journal in 1705. The second volume appeared just as the Saurin-Rolle dispute was reaching its climax in the *Journal des savants*, and this time the first issue contained a censorial authorization signed by Fontenelle.

Once again, however, Parent ran afoul of Bignon and the Academy as a result of his excessive contentiousness. In his journal, Parent criticized the work of Amontons, another *élève méchanicien* of the Academy, and the accused academician demanded either a public retraction from Parent or permission to publish his own defense.[60] The two academicians debated the matter before the members, and in July an academic committee was appointed to resolve the dispute.[61] It ruled against Parent on July 18, recommending that either Parent's journal be shut down or a retraction be printed. They passed their judgment on to Bignon, recommending that he "proceed as he deemed appropriate."[62] No record of an actual response exists, but Parent's journal never appeared again, and Amontons died soon after, taking whatever lingering animosity he had toward Parent to the grave.

The outcome of Parent's confrontations with the French scientific establishment reveals Bignon's capacity to manage intellectual consensus through the exercise of his ministerial power. His handling of the calculus dispute was less authoritarian, but this says more about the esteemed status of its combatants when compared to Parent than it does about the nature of the disciplinary mechanisms themselves. As a royal officer charged with managing French culture within an absolutist system, Bignon's capacity to influence intellectual life in the kingdom was immense. Importantly, he did not rule over this world as an outsider and executive overlord. His reputation as a man of letters was unimpeachable, and when he exercised his discipline (as he often did), he rarely acted without wide support. Consent and unanimous agreement, however, are not the same thing, and his management of the intellectual controversies of the early eighteenth century toward consensus reveals how political power, and the agendas guiding it, contributed to the creation of intellectual cohesion within this differentiated and contentious social field. Analytical mechanics was another area where Bignon's management was especially decisive in shaping outcomes.

As the Academy secretary, Fontenelle was technically Bignon's assistant in these managerial efforts, and he sometimes found himself assuming a disciplinary role on occasion. The discipline he more commonly exercised, however, was "soft" and discursive in nature, for his judgments never carried the political punch of the minister, even if Fontenelle's esteemed reputation as a man of letters certainly added force to his assertions. The relationship between the minister and the Academy secretary worked best when they shared a similar point of view, for this allowed the Academy's public spokesman to channel his eloquent and persuasive writing toward the solidification of the minister's agendas. Since each identified deeply with the values of *honnête* meritocracy central to both administrative monarchy and the Republic of Letters, agreement between Fontenelle and Bignon also occurred far more often than it did not. In this respect, the consensus that they cultivated was a product of their shared intellectual understanding and collaboration.

Nevertheless, Bignon and Fontenelle had different intellectual orientations, and they especially occupied different positions within the political hierarchy of French learning. The Academy secretary also approached his role as official manager of the public discourse with a different set of priorities than the minister. Especially important for the history of analytical mechanics was the fact that Fontenelle was not a neutral participant in the scientific culture of his time. His management of the public discourse of the Academy in fact drew deeply from his particular intellectual allegiances in ways that the minister's did not. Bignon never publicly defended a philosophical or scientific position at any point in his life, while Fontenelle was, among other things, a close friend of Varignon, an avid member of the "Malebranche circle," and an ardent defender of the new infinitesimal analysis. He was also deeply committed to the empirical and nonmetaphysical strand of Cartesian philosophy then powerful in France. For Fontenelle, then, the job of the perpetual secretary was not just to build harmony between the Academy, the crown, and the Republic of Letters as a whole, as it was for Bignon, it was to build a consensus around his as opposed to other scientific ideas.

Fontenelle conceived his public academic discourse as a vehicle for realizing these agendas, and his public defense of mathematical analysis pursued a multitude of agendas simultaneously. Perhaps his most direct and potent strategy was his tendency to simply lavish praise whenever possible upon the work he liked and the mathematicians who pursued it. Fontenelle particularly rewarded Varignon in these efforts. The first of his papers in analytical mechanics were produced in 1699 and published in 1702, and Fontenelle devoted several pages of his first *histoire* to a lengthy summary of their importance. The work focused on the mathematics of the cycloid, and Varignon's efforts were directed at generalizing an approach developed first by Leibniz and then by Jakob Bernoulli. Fontenelle invoked these influences in order to raise Varignon to the level of these recognized mathematical titans. "It is in this way that geometry, whenever it is practiced by the greatest geniuses, always raises itself from the particular to the universal and then to the infinite," he exclaimed.[63] Varignon also claimed to move beyond Bernoulli and Leibniz in his work, and the secretary turned this progress into an argument for the infinitesimal analysis he used to make these leaps. "The advantage of general methods [such as those used by M. Varignon] is that they give all the truths of a given species all at once. Everything that before had only been discovered in particular is now discovered contained in a single whole."[64]

Celebrations such as these praising the power of mathematical analysis and the intellectual quality of those who practiced it were a staple of Fontenelle's public academic discourse. Yet Rolle's nonanalytical work was also published in the *mémoires* of 1699, and the secretary's treatment of it reveals the other side of his discursive management. Neither eager nor permitted to engage in open contestation in the pages of the Academy's official *histoire*, Fontenelle instead let his preferences be expressed through other, more implicit means. Varignon was given a lengthy and glowing account of his achievements, while Rolle's work during the same year was reduced to a perfunctory two-sentence synopsis. Yet even here, the astute reader would have felt the sting of Fontenelle's judgment. "M. Rolle," he wrote, "offers here the first piece of a new work that he claims will put algebra on a solid foundation. His work begins from the premise that the current methods of the algebrists are false and defective."[65] No other commentary is offered, but given Fontenelle's praise for the new analysis elsewhere, and the wider awareness of the Rolle-Varignon dispute present in the public sphere when the published volume appeared, no more commentary was needed. Fontenelle would continue to use indirect criticism like this in his public academic discourse.

In subsequent *histoires*, Fontenelle also employed other strategies for shaping the public discourse about analytical mechanics. The *histoire* of 1700 gave Fontenelle his first opportunity to deploy his characteristic rhetoric of explanation and clarification with respect to infinitesimal analysis, and he used the technique to good effect. This had been a key year in the development of analytical mechanics, and the published *mémoires* included an important mathematical paper by l'Hôpital and Varignon's first three papers on central-force mechanics. Fontenelle grouped all four papers together under one title, "On Centrifugal Forces," and he built upon each *mémoire* in writing a detailed and clear explanation of the new analytical approach to mechanics as a whole.[66]

He began with the mathematics, explaining how curves are treated as "an infinite assemblage of infinitely small lines." He made no reference to the problems that mathematicians such as Rolle raised against this approach. Rather, he presented the work in the simplest empirical terms, striving to persuade the reader of its validity by making its assumptions clear and evident. To fully explain the reduction of curves to infinitesimals, Fontenelle invoked a physical model. He first explained Galileo's principle of natural, inertial motion along a straight line, and then asked his readers to consider how such a law could be used to account for curvilinear motion as well. The answer, he suggested, arose from the recognition that curvilinear motion is nothing other than a composite of infinitely small rectilinear motions subjected to constant change of direction. Fontenelle further explained that one could ignore the cause of this change of direction, because the behavior would be the same regardless of how it was produced. Drawing the essential point, he then argued that this mechanical picture of curvilinear motion is exactly what the calculus assumes in considering curves as composed of infinitely many infinitesimally small line segments.[67]

Having invoked this mechanical picture to explain the mathematical approach to curves used in analytical mechanics, Fontenelle then introduced the more overtly physical aspects of the new science. The work of Varignon in particular, he explained, was directed toward understanding the motion of bodies acted upon by centrifugal forces, or those that tend to pull bodies away from their centers during rotation. Fontenelle described the essential difficulties in some detail. He noted that Johann Bernoulli was the first to frame the problem correctly. He also singled out Christiaan Huygens's *Horologium oscillatorium* as an important influence, and noted that "the illustrious M. Newton had explored one dimension of the problem before leaving the rest to others."[68] No one had fully accounted for all the phenomena yet, Fontenelle claimed, and the reason for the failure was mathematical. "The problem is impossible to tackle with ordinary geometry," he explained. "M. le Marquis de l'Hôpital attempted to conquer the difficulties with the help of his method of infinitesimal analysis, and it appears that his method has defied all the others by getting to the bottom of it."[69] Varignon's work pushed the marquis's efforts even further. Whereas l'Hôpital had assumed uniform motion in his work, Varignon removed this constraint, treating any motion whatsoever. By using the same infinitesimal approach, he was able to maintain l'Hôpital's generality while still resolving the problem. As Fontenelle summed up the achievement: "It was by the geometry of the infinitely small that Varignon reduced variable motion to the same rules as constant motion. It does not seem that he could have succeeded by any other method."[70]

One is reminded here of Varignon's challenge to Rolle that he produce an equally powerful central-force mechanics without using the differential calculus. Fontenelle makes a similar point, introducing his readers to the intricacies of this complex scientific achievement while at the same building a powerful argument on behalf of it. Having clarified the nature of l'Hôpital's and Varignon's work, Fontenelle concluded by drawing the essential conclusion. "All methods should support infinitesimal geometry," he explained,

> since all geometry consists only in the art of discovering the *rapports* between magnitudes and of deducing one from the other. This art becomes more perfect the more we can use a small number of known *rapports* to deduce a large number of unknown relationships. The previous examples, however, offer ample evidence to conclude that there are *rapports* that we are only beginning to see and capture. These arise when we follow magnitudes all the way to their most essential and infinitely small parts. Indeed, sometimes we must follow them to the infinitely small parts of the infinitely small parts and onward as far as is necessary.[71]

This deeply Malebranchian synopsis makes infinitesimal analysis the very anchor of the mathematical physical sciences as a whole. Analysis is foundational because it captures the universal *rapports* implicit in all quantitative relations, be they mathematical or physical. All other mathematics, therefore, follows from it, and all other sciences, in turn, follow from this kind of quantitative mathematical analysis. Furthermore, used here at the climax of a systematic clarification of analytical mechanics as a whole, Fontenelle's precise conceptualization effectively transforms the practitioners of the analytical sciences into the defenders of universal science tout court while implying that their opponents are stubborn defenders of a narrow-minded literalism. This was precisely Fontenelle's point. Moreover, appearing as it did within the official history of the Royal Academy, the discourse solidified a cultural consensus that publicly linked the practitioners of analytical mathematics inside the Academy with the many constituencies in early eighteenth-century France that were sympathetic to the Academy and its scientific authority. It also marginalized those voices that had recently risen in opposition to this constellation. Through public discursive practice such as this, deployed via the authoritative public organs of the Royal Academy, a potent cultural consensus supporting analytical mechanics was established and reinforced.

Solidifying support for the new mathematical analysis in this way was one of Fontenelle's major goals throughout his lifetime. On the one hand, he was motivated in these efforts by a deeply held belief, inspired by Malebranche and others, that this form of mathematical reasoning was supremely valuable. His equally powerful commitment to a unified field of public of science joined through this kind of rationalist thinking also pushed him in this direction. As the *histoire* of 1700 reveals, the secretary was convinced that the public could be won over to the new analytical science if its inherent rationality could be presented with sufficient clarity. In the face of critics who charged that the calculus was irrational, therefore, Fontenelle set out to demonstrate that it was in fact reasonable as long as one understood its justification properly. This same urge would reach a climax in 1727 when Fontenelle published his one and only narrowly scientific book, a massive treatise on the philosophical foundations of infinitesimal analysis that was published as an official work of the Royal Academy.[72]

A good illustration of Fontenelle's approach early on is his description of Varignon's work on falling bodies in the 1704 *histoire*. The secretary began with a general and deeply Cartesian epistemological pronouncement. "It is not enough to discover a truth," he declared, "one must also know what produced it or from where it came. For if one is mistaken about the nature of the cause, one can believe that it is acting in a place when in fact it is not. Likewise, one can also extend the validity of the truth far beyond its domain. . . . Even a geometric demonstration is capable of being thrown into errors by the application one makes of it unless one has first climbed to the source of the truth and exposed its first principles."[73] This remark recalls Descartes's famous critique of Galilean mechanics as providing only mathematical description, not the deeper rational explanation that is required, and the *Port-Royal Logic* articulated a similar viewpoint in its treatment of correct scientific reasoning.[74] In this passage, Fontenelle invoked this widely French epistemic virtue to justify Varignon's revision of Galileo's "true and incontestable" principles of free fall.

As Varignon had shown in 1693, Galileo's principles were not wholly accurate: A set of errors not readily apparent in Galileo's work could be exposed. According to Fontenelle, this demonstrated that "the first principles had not yet been seized."[75] By contrast, Varignon's mathematical work achieved the true, evident understanding of free fall because his mathematics captured the foundational principles at work. "Varignon's general equation captures the necessary relationships in their entirety," the secretary explained. "The essential reason for this is metaphysical. The curve is only a composite that results from the interaction of two forces that have between them a certain *rapport* of magnitude or quantity. The parabola, for example, is the composite that results from a uniform velocity and a constant velocity interacting according to the squares of the heights. This composite is determined necessarily by the *rapport* between the two velocities that formed it." Thus, in Fontenelle's presentation, Varignon's method succeeded where those of Galileo and other "able geometers" failed because his mathematics captured the essential physics underlying projectile motion far more precisely and powerfully.

While Fontenelle employs here a "metaphysical" explanation to justify Varignon's mathematical work, he also uses it to emphasize the mathematical character of Varignon's achievement. "Famous geometers," he writes, "have already tried to conceive of curves in terms of composite motions. But those who have neither known nor admitted the geometry of infinitesimals were soon frustrated, or at least severely hindered in their research. It is only by considering curves as infinite polygons that one discovers that each infinitely small side of the curve is the diagonal that produces a composite motion, and this idea resolves everything."[76] The key point is Fontenelle's explanation of Varignon's achievement in terms of the superior power of his analytical mathematics. As he wrote elsewhere: "It is good when a general metaphysics precedes the calculation which directs and clarifies it; however it is the calculation in turn that gives the precision and the details. . . . Geometry agrees here, as it does everywhere else, with pure metaphysical theory."[77] This Cartesian/Malebranchian harmony between analytical mathematics and metaphysics was central to Fontenelle's own scientific worldview. It was central to Varignon's as well. In making it central to the public academic discourse of the Royal Academy, the secretary was not only articulating his own intellectual convictions and those of his allies, he was also trying to persuade others of the veracity of this scientific work, while offering it as a scientific center around which academic science could converge.

Systematic, rational explanation of this sort was one method that Fontenelle employed in his public academic discourse, but as his introductory

"Preface" indicated, he possessed a variety of rhetorical strategies in his toolbox. He employed each in his defense of the new analysis. One interesting approach was to use the widely discussed opposition between the Ancients and Moderns as a weapon in his efforts. Leibniz, it will be recalled, sought to erase the difference between Ancient and Modern mathematics in his defense of the calculus in the *Journal de Trévoux*. Fontenelle, not surprisingly perhaps given his partisan participation in the wider Modern campaigns, took the opposite tack. He emphasized the distinction between Ancient and Modern mathematics so as to identify the practitioners of the new analysis with the triumphant Moderns who were continuing to wage their culture war against the backward-looking Ancients in France.

A good example of this recurring rhetorical strategy is Fontenelle's account of Varignon's work on the mathematics of spirals. "If one wanted to make a parallel between Ancient and Modern geometers so as to compare their relative merits, the spirals of which we are going to speak here offer perhaps the best case that can be imagined."[78] Starting with the Ancients, Fontenelle assigned the necessary judgments. Writing wryly with respect to Archimedes's work on spirals, he noted:

> We have his demonstrations, but they are so long and so difficult to grasp that M. Bouillard, as noted in the Preface to [l'Hôpital's] *l'Analyse des Infiniment petits*, vowed that he could never understand them. Viète unfairly suspected them of logical inadequacies because he could never follow them all the way to the end. But all the testimonies that can be offered about the difficulty and obscurity of these demonstrations actually highlight the glory of Archimedes: for what vigor, what single-mindedness was required to produce a set of demonstrations that several of our greatest geometers cannot follow despite the devotion and attention that they possess?[79]

One recalls Whewell's similar description of Newton's antiquated mathematics as astonishing not because of its brilliance, but because of its skill with such a cumbersome and awkward method of reasoning.

This presentation stresses above all else the complexity and awkwardness of Ancient mathematics. As such, it resonates with the Modern criticism of Ancient art and literature at the time as similarly crude, unpolished, and barbaric. In equally Modern fashion, Fontenelle also celebrated the elegant clarity and economy of the new and modern mathematical style. "The spirit of modern geometry consists in raising all truths, be they ancient or modern, to the greatest universality possible," the secretary enthused. By this standard, Fermat was the first fully Modern student of the spiral, because he generalized and simplified the study of this curve in ways unmatched by any Ancient mathematician. "M. Varignon," however, "has found a general equation that encompasses every possible spiral to infinity." For this reason, the secretary argued, he was the most Modern of all. The superiority of this Modern approach is further confirmed when it is noted that Varignon's equation is so universal that it includes Archimedes's spiral as one special case.[80]

Given the importance of the Moderns' campaign to the urban elites of interest to Fontenelle, the inscription of the mathematical debates of the period into the terms of this struggle was a powerful rhetorical move. It allowed the perpetual secretary to fuse two roles into one—that of a defender of Modern culture and that of a defender of the new analytical mathematics. It also worked to channel the appeal of Modern culture into support for infinitesimal analysis by making the latter an important component of the former. Fontenelle also drew upon other discourses forged in this culture war to appeal to his audience. One was the normative discourse of civilization that also circulated widely among Old Regime elites. The secretary understood that elite self-conception was grounded in convictions about the cultural superiority of modern European civilization over that of "primitive" peoples. In his public academic discourse, he often provoked these prejudices to serve his own agendas. "Here is the greatest advantage that the modern geometers have over the ancients," the secretary wrote in one illustrative passage.

> It is possible for us to discover an infinitely large number of truths at infinitely less cost. This is not because we have superior genius, but because we have better methods. The glory of the ancients resides in their discovery of what little they did discover without the help of our art. The glory of the moderns, by contrast, resides in discovering the marvelous art [of discovery] itself. The ancients, therefore, can be likened to the residents of Mexico or Peru, who, despite the absence of construction machines and similar instruments, and lacking any knowledge of scaffolding, nevertheless raised buildings with their own hands. The moderns are like the Europeans who build incomparably better buildings because they have improved machines.[81]

Here Fontenelle makes analytical mathematics the mark of advanced civilization while those who resist it are made into "backward" and "barbaric" people blocking the advance of progress. Elsewhere Fontenelle developed a similar defense when presenting the elder de la Hire's work on "magic squares" in the *histoire* of 1705. He began by offering his usual celebration of universality and utility: "M. de la Hire offers here a general method for treating the squares of odd numbers. It has relevance also for the theory of composite movements which is such a useful and fecund part of mechanics."[82] Connecting de la Hire's discoveries with the progress of modern civilization, he added that "if one wanted to contrast the cultured and uncultured human mind, one need only imagine the distance that separates his approach to these problems from those of the savages who can only count to ten because this is all the fingers that they have."[83] Arguments such as these channeled the cultural prejudices of his intended audience—learned elites—toward sympathy for his primary object of interest—abstract mathematics. As such, they served Fontenelle's wider agendas for public science masterfully.

Overall, the secretary's program was overwhelmingly positive in the manner just described, striving through eloquence and brilliant rhetorical acumen to lead the public pleasurably toward the positions he held dear. Fontenelle was not averse, however, to using critical means as well in the service of his agendas. Both the external demands of his office and the inner pulls of his own self-conception made overt contestation something to be avoided. Yet for this very reason, the presence of so much criticism in Fontenelle's early *histoires* reveals how important the issues that he treated in this way were to him. Not surprisingly, Rolle was the recipient of much of Fontenelle's critical attention. He published mathematical *mémoires* in each of the Academy volumes for the years 1700–1704, but nowhere was his work given anything other than a brief, perfunctory description. Furthermore, when Rolle's papers overtly critical of the calculus were published in 1705, as part of the *mémoires* for the year 1703, Fontenelle did not utter a word about any of them in his *histoire*, leaving it to Saurin to refute them in the *Journal des savants*.

The Academy's volume for 1704, however, opened with the following *Avertissement*: "The reflections that a variety of persons have offered regarding [M. Rolle's paper], the principles that are advanced in it, and the consequences that one can pull from them oblige us to state that although the work was deemed appropriate to be among the other works published by the Academy, it has never been the intention of the institution to adopt any of the ideas that are to be found there."[84] This was an unprecedented disclaimer that was not repeated again in any other academy volume of which I am aware. Whether Fontenelle authored the text himself, or whether, as is more likely, Bignon ordered it to be printed as part of an institutional arrangement is not clear. But it appeared in print after the calculus commission had issued its report and Rolle had declared his conversion to infinitesimal analysis. Maybe, therefore, the text was produced as part of the final resolution of the debate. Whatever its origins, the mere publication of the declaration within the context of Fontenelle's overall neglect of Rolle's work in the Academy's *histoires* accomplished his agendas.

The secretary used similar critical tactics against other opponents of the new analysis. In late 1701, as the calculus dispute was beginning to erupt, the foreign associate of the Academy Walter Tschirnhaus visited Paris and exercised his prerogative to present work at the academic assemblies held during his visit. Bourdelin described Tschirnhaus's paper of December 10 as offering "a new method for squaring curves without the use of infinitesimals." On December 17, he read a second paper that pursued a similar agenda.[85] The papers were published in the *mémoires* for that year, and in them Tschirnhaus, like Rolle, attempted to demonstrate the redundancy of infinitesimal analysis by showing how its results could be achieved through other, more rigorous means.[86] In June 1701, the *Journal de Trévoux* also cited Tschirnhaus's "more rigorous geometry" in its indictment of the "mysterious method of infinites" used by Bernoulli. It was this criticism, it will be remembered, that had led Leibniz to publish his defense of infinitesimal analysis in the same journal later that year. In a similar fashion, Tschirnhaus's arguments in support of traditional geometry before the Academy triggered a response from Fontenelle published in his *histoire*.

Summarizing the German's work, Fontenelle wrote sarcastically: "M. Tschirnhaus alleges, in support of his method, that he avoids altogether the infinitesimal calculus. Apparently this calculus has become so general and so fashionable these days that it is now a sort of honor to be able to avoid using it in important research. Nevertheless, M. Tschirnhaus conflates an infinitely small arc with its cord, and he does not treat the two magnitudes so confused as real magnitudes. This is entirely in the spirit of infinitesimal geometry. I guess it is not so easy to escape very far from the really great discoveries."[87] This was the full extent of Fontenelle's report on the first paper, but he continued in the same vein in his report on the second. Overall, the secretary argued, Tschirnhaus kept his own method of reasoning hidden from view while unwittingly employing infinitesimal analysis in his work. Varignon had accused Rolle of doing the same thing, and here Fontenelle charges another opponent of the new mathematics with the same subterfuge. He also criticized the German for asserting that the "method of infinitesimals is nothing more than a useful and convenient abridgement of his own method." "In returning to first principles," the secretary wrote, "he claims to have found only a tributary flowing from the source of his own method."[88]

Framed this way, Tschirnhaus's work did not challenge the value of infinitesimal analysis, it confirmed it. His critical stance also worked unintentionally to praise the work of Varignon and other practitioners of the calculus. One of Bernoulli's tactics, displayed among other places in the brachistochrone contest, was to derive the same solution twice, once with ordinary geometry and a second time with infinitesimal analysis. This allowed for a demonstration of the validity of the more innovative method while simultaneously revealing the greater power and simplicity of the analytical approach.[89] Here Fontenelle used his *histoire* to send precisely the same message. According to the secretary, Tschirnhaus had used a different method to produce results more easily found by the infinitesimal calculus. This merely confirmed that the new analysis was grounded in sound mathematical truth irrespective of Tschirnhaus's intimations to the contrary. No real challenge to the calculus was accordingly offered by his work, the secretary argued. The rest of Fontenelle's account further supported this position. While Rolle's work of 1702 was reduced to a one-line statement announcing his "study of geometric lines," Bernoulli's new work in integral calculus was given an extensive and glowing review. His collaboration with Leibniz was especially celebrated, and the narrative ended by calling both savants "the practitioners of the most sublime geometry."[90]

In these and similar statements combining subtle criticism with vigorous advocacy, Fontenelle used his position as Academy secretary to advance a particular mathematical agenda and to challenge those of its opponents. To succeed most fully in his ambitions, however, he needed to be more than just a successful intellectual combatant in the newly public debates about science. His ultimate goal was *honnête* harmony, not sectarian strife, and while he was aggressive in placing his own views at the center of the unity he hoped to achieve, he remained devoted to the cultivation of consensus in the public sphere. This conciliatory side of Fontenelle's mission was threatened most openly by the calculus debate since it placed him into a deeply partisan position within a conflict that was anathema to his vision of a harmonious republic of science. Yet even here, Fontenelle was able to solidify a crucial compromise between partisan advocacy, *honnête* decorum, and neutral arbitration.

Most important in this respect was the public presentation of the Varignon/Rolle academy debate in the *histoire* of 1701.[91] By the time that this volume appeared in print in early 1704, the public had become accustomed to partisan bickering about the validity of infinitesimal analysis.[92] Thus, rather than use the presentation as an opportunity to further fuel the fires of contestation, Fontenelle instead opted for conciliatory approach, using his narrative as a vehicle for consensus building. No doubt, Bignon and his royal supervisors would not have tolerated anything less, but the secretary's work was nevertheless his own. As the bias that runs throughout the account reveals, he wanted to neutralize the conflict while at the same time ensuring that the antagonists he supported were favorably perceived in the public mind.

Adopting from the outset the authoritative, historical voice appropriate to the narration of a distant event, Fontenelle began by setting the Academy debate within a wider historical context. He opened by once again invoking the categories of the Ancients-versus-Moderns battle. Noting briefly the ancient method of dealing with the infinite, he then quickly moved on to the "more modern" methods developed by Barrow, Newton, the Bernoullis, and "especially Leibniz," framing these developments in terms of a narrative of progress. Connecting this story directly to the Academy, the secretary concluded by calling l'Hôpital's 1696 *Analyse des infiniment petits* the "great book" that brought these developments to a climax.[93] With this background in place, Fontenelle then introduced the academic debate itself. The new work on the calculus, he claimed, "created for the first time a regular body of geometry where an infinity of different solutions all depended on the same principle." "Several solutions that ancient geometry could never even have attempted were now derived with ease," he continued, "and even those truths common to both the old and new systems are now derived with far greater ease."[94]

This presentation framed the new analysis as a triumphant step beyond the past, setting up the anti-calculus camp, who were introduced next, as the defenders of a venerable but decaying mathematical antiquity. "M. Rolle and M. l'abbé Gallois rose in opposition to this new method that promised so many advantages," the secretary wrote.[95] He then summarized the basis of their attacks. They challenged the system "because [the method] supposes that one can move perpetually toward the infinite" and because it "admits infinitely small magnitudes that can be resolved into still other infinitely small magnitudes that are themselves composed of still more infinitely small magnitudes and so on to infinity." From this, they argued that this regress "produces contradictions."[96] At this juncture, Fontenelle employed his discourse of clarification to offer a brief explanation of why the method was in no way contradictory. Yet rather than allow his narrative to flow into a detailed account of the intellectual debate itself, he adopted a different strategy, turning the battle into a theatrical drama that made it a parable about scientific innovation within the context of public science.

Framing the debate this way accomplished a number of things simultaneously. First, it allowed the secretary to sidestep the difficult and largely irreconcilable intellectual difficulties that had made the calculus such a source of controversy. Second, it allowed him to focus on the meaning of the debate for the overall culture of public science, a strategy that served his larger ambitions. Consonant with his narrative strategy, Fontenelle offered no further discussion of the criticism which Rolle and Gallois levied against the calculus. Instead, he focused on the responses of the defenders of the calculus beginning with l'Hôpital. He was dramatized as the very personification of the Stoic sage. The marquis remained perfectly silent in the face of the Rolle–Gallois attack, Fontenelle explained, and his reasons for doing so were complex. "Either he was content with the evidence offered by all the great geometers of Europe, or he believed that geometric truths, once revealed, have no need of human support. Or perhaps he readily left to one side those principles that produced murky metaphysical questions preferring instead the easier path of demonstrable geometrical paralogisms."[97] Whatever his motives, he did not fight, and this made l'Hôpital a hero in Fontenelle's story, one who's conduct personified the Stoic values of ***honnête*** independence that were appropriate to the marquis's aristocratic rank.

Varignon, however, was anything but restrained in his response to Rolle, and Fontenelle accomplished his agenda by making his partisanship the result of a deep and passionate commitment to the new mathematics. As the narrator explained, Varignon had been practicing the new geometry "almost from its birth." He defended it, therefore, out of "a zeal to correct the errors offered in its name."[98] In this way, zealotry became a sincere and heartfelt reflection of his authentic quest to know the truth. The narration of the Academy debate itself continued in this same moralizing vein. "[The calculus debate] consumed almost all the time devoted to mathematics during this year's academic sessions," the secretary lamented, "time that would have gone to discussing the new research capable of perfecting or improving geometry." Trying to justify such distractions, the secretary offered that "it is the destiny of new developments, whatever they may be, to produce contradictions, and debating these contradictions is a key vehicle for establishing firm truths." Recognizing the importance of this principle, "M. l'abbé Bignon allowed the debate to go on in a free way. He did so because he understood that the academic spirit demands that all voices be heard and that no viewpoint, no matter how marginal, be suppressed."[99]

Here Fontenelle countered those who believed that partisan disputes reflect only destructive and vain pedantry by defending open debate as a valid and necessary source of insight. Leibniz did the same in his correspondence, ultimately praising critics of the new calculus like Nieuwentijt and Rolle since their criticism had helped to perfect the fledgling mathematics.[100] In his eulogy of Rolle, Fontenelle made a similar point, criticizing the deceased academician only for his zeal in wanting to roll back the new mathematics altogether, not for his urge to expose its failings for critical scrutiny.[101] Public criticism of this sort was in fact essential to scientific progress, on this all parties agreed. Since it could also descend into agonistic bickering that ran against the grain of *honnête* commerce, however, Fontenelle was quick to instruct his readers about the appropriate limits to which debates must adhere. *Honnêteté* itself was the central constraint, and to his credit, Varignon never violated these cherished maxims of gentlemanly decorum, at least not in Fontenelle's telling of the story.

Equally crucial was the account of the debate's resolution since here Fontenelle used the figure of Bignon to represent the appropriate role for honorable authority in a meritocratic republic of savants. At first, Fontenelle explained, Bignon let the debate run freely because unfettered discussion can aid the pursuit of truth. But once discussion had become counterproductive, the minister invoked a needed discipline to protect the community. No narration of the bitter ad hominem attacks between Varignon and Rolle was offered, nor any account of Saurin's intervention and the acrimony with Rolle that this created. Instead a moralizing summary sealed the presentation. "In the end," the secretary wrote, "since the argument was going on too long and was becoming charged, as is ordinary, with personal and useless issues that cannot be determined by exact demonstrations—in short, because the passions were overwhelming the geometry—M. abbé Bignon named Father Gouye and Mrs. Cassini and de la Hire to decide the question."[102] This presentation justified Bignon's intervention as a necessary dose of authority at a moment when destructive passions were threatening to subvert the common good. Fontenelle made a similar point in his eulogy of the extremely contentious Parent, writing that "the pursuit of truth in the Academy demands [that academicians have] the freedom to contradict. But every society must place certain limits on contradiction, and sometimes it is too quickly forgotten that the Academy is a society as well."[103]

The secretary also made it clear from where Bignon's authority derived. "He was perhaps hoping to calm the tensions a bit by deferring judgment," Fontenelle explained, "for in the end the only entity worthy of deciding in such matters is the public. If the new geometry proves to be ungrounded, it will know when to retract the great popularity that the public has begun to offer it."[104] In this way, the public became the real hero of Fontenelle's parable.

The lesson of the calculus debate in his telling was that science was ultimately beholden to the values and authority of those it addressed: the public at large. The actual public was riven with competing interests and constituencies, and conflicts in the field of public science were therefore inevitable. As perpetual secretary of the Royal Academy of Sciences, however, Fontenelle came to his job with a belief that reason combined with *honnête* decorum could be a basis for harmony within this space. In narrating the history of the calculus wars, a battle that had challenged these goals deeply and early in the history of the new public academy, Fontenelle set out to show how the creation of *honnête* consensus was consonant with the union of reason and public order that he believed public science could and would foster

When successful, discourse such as this created the appearance of a naturally formed consensus. This same discourse also disseminated Fontenelle's particular philosophical convictions as well. But these two faces were never really opposed. So long as the universal Cartesian values of clear and distinct reasoning were shared by the wider French public, and so long as marrying this reason with *honnêteté* remained a widespread cultural goal, Fontenelle's public discourse supported his claim to embody the public voice directly. No doubt, this carefully orchestrated rhetoric of clarity, republican virtue, and *mondainité* frustrated those opposed to his scientific views, but their frustration only attests to the effectiveness of the strategy. Accordingly, as the new Academy secretary's public academic discourse became established in France, the precise unity embedded in these public presentations became increasingly widespread with it. Fontenelle was certainly not solely responsible for creating the consensus that secured analytical mechanics in France, but his singular role in fostering its establishment cannot be overstated. The widespread support for analytical mechanics in France after 1706 was one of many tangible outcomes of his decisive influence.

The *Pax Analytica* Established

By the summer of 1707, Fontenelle had succeed in getting the Academy *mémoires* delivered between 1699 and 1706 into print, along with a historical narrative contextualizing all of them in terms of the overall year in academic science that they had contributed to. By this date as well, the Academy's ruling ending *la querelle des infiniment petits* was a distant memory, and Rolle's conversion to the new mathematics was more than six months old. Accordingly, the increasing uniformity in the public discourse about the mathematical sciences in France after this date, a uniformity that bears the distinctive mark of Fontenelle's influence, is not surprising.

The sudden death of the Marquis de l'Hôpital in 1704 and Jakob Bernoulli in 1705, both at relatively young ages (l'Hôpital was forty-three and Bernoulli fifty) gave Fontenelle an opportunity, however bittersweet, to begin building a deeper foundation for the newly emerging consensus. He employed his most widely read vehicle—the funeral *éloge*—to exploit this opportunity.

The eulogy for l'Hôpital was especially masterful. In it, Fontenelle brilliantly integrated a poignant account of the marquis's move from the king's army into a career as a mathematician with an incisive analysis of how his brilliant and sophisticated mind took him to the highest ranks of the European mathematical community. Playing on all of the cultural biases of his elite audience, he in effect used l'Hôpital to personify the new French mathematical elite that he was attempting to fashion.[105] Bernoulli's *éloge* achieved a similar effect, and he reinforced the same themes in his *éloge* for the former "Malebranche circle" member Carré in 1711, and in his particular remembrance of the "anti-infinitesimalist" Gallois in 1707.[106]

Fontenelle's *histoires* during these years echoed the themes found in these eulogies, and the result was an increasing stabilization of the public discussion of analytical mechanics in ways sympathetic to the goals of the secretary and his academic supporters. In the decade after 1710, no overt challenges to the new analysis appeared in France while the presence of Fontenelle's discourse in the public sphere (combined with many declarations of praise for it) increased. The power and presence of the various discourses opposed to Fontenelle's also subsided. The Jesuits at the *Journal de Trévoux*, for example, ceased after 1705 to use their journal as an explicit organ of opposition to the calculus and analytical mechanics. Furthermore, between 1702 and 1709 the journal employed the non-Jesuit Pierre-Jacques Blondel to serve as their eyes and ears at the Academy's twice-yearly public assemblies.[107] Blondel's highly neutral reports on academic mathematics were a departure from the open polemicism of the journal's first two years, and the Jesuits adopted a similar editorial line in its own reviews of the Academy's published *histoires et mémoires*.[108]

The *Mercure galant* adopted an even more sympathetic stance, and by 1707 Bignon's reforms at the *Journal des savants* had made this journal an exceedingly neutral organ of intellectual commentary. The polemics that had raged in its pages throughout the previous two decades completely disappeared, replaced by a steady stream of book reviews noteworthy for the anonymity of the reviewers and the judicious nonpartisanship of the editorial tone. The solidification of the Royal Academy's own publishing mechanisms also reduced the need for the journal to serve as an outlet for academic work. After 1710, the periodical also became a monthly rather than a weekly publication, and it established in this new format its familiar eighteenth-century form by serving as a deeply serious and neutral reviewer of the wider discourse of the Republic of Letters.

Within this new setting, Fontenelle often found his own academic discourse echoed rather than challenged in the wider public sphere, and this reinforced the larger consensus that he and the Academy were striving to create. Other institutional maneuvers also solidified the new peace. Bignon's academy commission had been caught in 1706 trying to discipline an individual, Saurin, who was not beholden to the institution. Changes soon after, however, made these inconveniences disappear. Saurin had clearly established his scientific credentials by 1707, for not only had he demonstrated his mathematical acumen in his battles with Rolle, he had also produced other work that demonstrated the range of his talents. An article he published in the *Journal des savants* in January 1703, for example, offered a solution to one of the principal objections made by Christiaan Huygens against the Cartesian vortical system of celestial mechanics. This situated Saurin at the heart of the important vortical mechanics discussions that were beginning at the time, and would become more vigorous in the next decade.[109] His other published work only reinforced his increasing prominence as a savant, including his mathematical pieces published in the *Journal de Trévoux* and elsewhere. Not surprisingly, therefore, when academic seats became vacant, Saurin's name began to appear as a candidate for admission.[110] In August 1706, he was nominated for the first time to fill Carré's *associé géomètre* seat when the latter was promoted to *pensionnaire*.[111] He was also nominated in January 1707 to fill the *associé géomètre* seat left vacant by the death of Régis.[112] In the latter election, Bourdelin described Varignon as Saurin's "evangelist."[113] In each case, however, Saurin was passed over for an *élève* of the Academy who was in line to be promoted.

Institutional obstacles such as these were in fact the major impediments to Saurin's entrance into the Academy. At forty-seven years of age, and with life experiences that made him seem much older than that, Saurin was a grossly inappropriate candidate to become an academic *élève* under the wing of a senior academician. Nevertheless, this method of appointment had become standardized in the reform of 1699, and it could not be bypassed. In February 1707, with the promotion of Varignon's *élève* Guisnée to fill his former *élève* Carré's *associé* seat, Varignon appointed Saurin as his new "student."[114]

Bourdelin noted the appointment in his diary without comment, and while the arrangement likely frustrated (or perhaps it amused?) the two colleagues, it turned out to be short-lived.[115] Ironically, it was the death of one of Saurin's adversaries in the calculus wars, the abbé Gallois, which provided the opportunity for his promotion. On May 7, the Academy debated about the replacement for Gallois's *pensionnaire* seat, and in the end they nominated Saurin, passing over Varignon's former *élève* Guisnée in the process. The king approved the appointment on May 18, and Saurin moved up the Academy hierarchy to its highest rank only three months after his initial admission.[116]

The new *pensionnaire* delivered his first academic *mémoire* in July, a discourse on the barometer, and while the paper did not allude to it, a shift of importance had occurred within the Academy. One of the leading opponents of the new infinitesimal analysis, and a stalwart of the old mathematical culture who traced his lineage back to the initial founding of the Academy, had died only to be replaced by an ardent proponent of the calculus and an aggressive mathematical modernist.[117] The position of the analytical mathematicians in the company was further reinforced in the summer of 1707 when a young astronomer named Bomie also became an *élève* of the Academy.[118] He was appointed to replace the deceased *méchanicien* Amontons, and he was thus chosen by those attached to the mechanical arts programs and the Observatory. In his first academic paper of August, however, he delivered a work on "centrifugal and centripetal forces" that revealed a tremendous debt to Varignon's new science of motion.[119] The paper opened with a historical summary of the work on this question that traced developments back to the work of Huygens, Newton, and Leibniz before culminating with Varignon's achievements of 1700–1706. Bomie then offered some ideas of his own, employing throughout the analytical method pioneered by Varignon. In its support, he declared: "The new system, or the new explanation of the movement of the planets, is entirely founded upon his ideas."[120]

Varignon's precise relationship with Bomie is not clear, for he was neither Varignon's *élève* nor a known member of the Malebranche circle. Indeed, Bomie was an astronomer, which made his commitments to an analytical, mathematical approach to the discipline all the more intriguing.[121] He and Varignon nevertheless pursued very similar scientific agendas in the years after 1707. Varignon delivered his own *mémoire* to the Academy in 1706 focused on "the *rapport* between central forces and the gravity [*pesanteur*] of bodies," and this work featured significantly in Fontenelle's *histoire* for that year.[122] The secretary gave an equally extensive treatment of central-force astronomy in 1707, and in this *histoire* the work of Varignon and Bomie was presented side by side.[123] Bomie also entered the public sphere with this work, witnessing the publication of his first academic paper in the Academy's volume for 1707 and appearing at the Academy's public assembly of April 1708, delivering a paper titled "The Physical and Geometrical System of the Movement of Planets."[124] Varignon's former *élève* Guisnée also continued to do work sympathetic to the analytical program during these years, and they could count on the sympathy of other Malebranche circle members in these efforts, most notably Carré and his *élève* François Nicole, who was appointed in 1707 after publishing works using the differential and integral calculus in among other places the *Journal des savants*.[125] Among the *honoraires*, only Father Gouye remained from the Old Style mathematicians, and his position was checked admirably by Malebranche, who remained an active participant in the Academy until his death in 1714.

Meanwhile, outside the Academy a similar consolidation was occurring. Saurin's entrance into the Academy coupled with his wider reputation in the Republic of Letters helped to solidify the perception that the new mathematics had been accepted despite its difficulties. The publication in 1708 of Father Reyneau's *Analysis Demonstrated, or the Method for Resolving Mathematical Problems* reinforced this consensus.[126] Reyneau's treatise (which was also conceived as an Oratorian textbook) offered a systematic explanation of the new analysis accessible to a wide audience. It neither attempted nor succeeded in actually demonstrating the validity of the new analysis according to the canons of geometrical rigor, but it at least showed that the new method was sound if used correctly. The book was widely praised by savant and amateur audiences alike, and it became the standard reference work in the education of the next generation of analytically minded mathematicians and *méchaniciens*. As late as the 1750s, d'Alembert called it "indispensable" and claimed it was "the most complete work that we have on analysis."[127] In 1708, its importance was even more pronounced, for its brilliant clarity and accessibility helped to secure the practice of differential analysis in France at a time when its status remained in question.

In May 1708, Varignon wrote to Johann Bernoulli that "since the loss of his ally Gallois, [Rolle] has not dared to say anything against the infinitesimal calculus." But, he added, "since I hear that he has not ceased to decry it with even greater vehemence outside the Academy, I remain nervous."[128] Indeed, even inside the Academy, the debates were not entirely over. As late as the fall of 1709, Rolle was delivering papers with titles such as "Remarks and difficulties concerning the disappearance of unknown quantities in analytical geometry."[129] Saurin was likewise continuing to rise in opposition to such challenges, declaring in rebuttal to one that, "there is nothing either real or solid in the difficulties that M. Rolle finds with the received methods."[130] These flare-ups notwithstanding, the peace regarding analytical mechanics was solidifying, with both the frequency and the intensity of the disputes rapidly diminishing.

More representative of the intellectual climate both inside and outside the Academy was a debate between Parent and Saurin in the spring of 1708. Between February and June of this year, the two academicians sparred regarding the mathematics of falling bodies. After Saurin's third rebuttal, however, Fontenelle was able to write in the Academy registers that "M. Parent was convinced by this reading and with good faith announced that M. Saurin had been correct."[131] This courteous resolution of a debate between two battle-scarred mathematical pugilists is indicative of the harmony that took hold within French academic mathematics after 1710.

Varignon's fears about the resumption of conflicts regarding analytical mechanics were unwarranted, therefore. By the end of the first decade of the eighteenth century, the practice of infinitesimal analysis had been established in France. Its metaphysical difficulties remained an object of fascination for many and a source of anxiety for some, but explicit indictments of the new analysis on these grounds ceased to be an important part of French scientific discussion. Furthermore, as established analysts such as Varignon, Saurin, Malebranche, Bomie, Guisnée, Carré, Bernoulli, and new arrivals such as Nicole continually demonstrated, the value of the new mathematics in the solution of difficult problems in mathematics, astronomy and mechanics was unquestionable. Fontenelle also remained vigilant in placing his powerful mixture of *mondain* eloquence and Cartesian clarity behind the new mathematics. Through these means, its status as an established French academic science was secured. The austere abstraction and complexity of the new mathematics assured that it would remain a butt of jokes in *le beau monde*, but its recognized potency guaranteed that it would never again be threatened as a legitimate scientific practice. Accordingly, by the time that King Louis XIV breathed his last breath in September 1715, analytical mechanics had become an unassailable centerpiece of French academic science, and it would remain in that position for the remainder of the century.

CHAPTER 11 *Coda* Newton and Mathematical Physics in France in the Twilight of the Sun King

How did the emergence of analytical mechanics amid the changes at the Académie Royale des Sciences in the 1690s, the rancorous public debate about infinitesimal analysis after 1698, and its managed consolidation by 1715 ultimately shape the practice of mathematical physical science in eighteenth-century France? One obvious, yet fundamental, outcome was the simple establishment in France of calculus-based mathematical physics as a legitimate and state-supported scientific pursuit. Bignon jokingly called Varignon a "mathematical conjuror" after his first public presentation of his calculus-based mechanics, and the reality sustaining the humor was the absence of large numbers of people in France or anywhere else in 1699 capable of explaining how Varignon's work was not a cryptic and symbol laden sleight of hand. The battles over the calculus were provoked by worries over precisely this slipperiness and opacity, and the acceptance of calculus-based science was not achieved through a widespread clarification of the rationality of the new mathematics, but through a narrow political settlement that gave expert practitioners the authority they needed to do their work despite their inability to explain its rigor more fully. It took another hundred years before the calculus acquired a set of rigorous, demonstrative foundations, and yet in that same century the calculus was used to lay the foundations of what we today call mathematical physics.

A broad public understanding supported this work in France, but it was not an understanding rooted in widespread appreciation for scientific rationality and technical advantages of the science. Instead, it was a commonsense understanding, forged largely by Fontenelle, that made the French academician acceptable, in a wholly innovative way it must be stressed, as a mathematical expert to be trusted despite the arcane and esoteric nature of his scientific work. The French academician Alexis-Claude Clairaut would reward the public for its trust in 1759 when he predicted the return of Halley's comet to within a few weeks of its actual appearance, a result derived from a stunning deployment of the differential and integral calculus to extract the comet's orbit from reams of empirical astronomical data.[1] This prediction, which made the former child prodigy—Clairaut gave his first mathematical paper to the Royal Academy in 1729 at the age of thirteen—a celebrity, perhaps fatally so,[2] was a great confirmation of the power of analysis to produce the results that its practitioners always claimed for it. But Clairaut's position as a trustworthy state-supported savant was not supported by earlier public demonstrations of the triumphant accuracy and efficacy of calculus-based celestial mechanics. They were based instead on the settlement of the *pax analytica*, which had secured the practice of this kind of mathematical physics in the French Academy despite the absence of any clear, rational, and publicly accessible justification for it.

What the particular historical development of analytical mechanics in France produced, therefore, was an ironic, yet productively so, relation between the advanced mathematical scientist and the public. Forged out of the drive to make academic science in France more publicly visible and accessible, it created a new public identity for the mathematician: that of a specialized expert, supported by the state to practice a narrow and arcane specialization free of distraction from the broader public. In short, the identity of the expert mathematical scientist of today.

To see the historical novelty of this outcome, consider the contrast between the developments in France described in this book and the development of calculus-based mechanics and astronomy in Britain during the same years. Eighteenth-century British mathematicians came to share with Newton a vehement distaste for what they came to call "Continental analysis," which is to say both the Leibnizian calculus and its application to questions of mechanics and physics. Newton's precise views on this matter were made clear during the calculus priority dispute that erupted in 1709,[3] and like the critics of the calculus in France, British mathematicians defended Newton's geometrical approach to mechanics in the *Principia* in terms of its superior adherence to ancient standards of rigor. They also came to view the analytical treatment of physical questions in the manner of the French in terms of a double violation: wrong in its mathematics, and wrong in the physical claims derived from them. But since they also began to openly use Newton's fluxional calculus, which they claimed was more geometrical, and thus more rigorous, than Leibniz's, to pursue similar questions in celestial mechanics in the way that Varignon had made common in Paris, the two communities also pursued roughly parallel mathematical physical projects in the eighteenth century despite these differences.

Yet while the British mathematical community pursued their work in universities and in relation to the Royal Society of London, which was never comfortable with abstract theoretical mathematics and became even less so as the eighteenth century progressed, it was the work centered at the French Academy rather than in Britain that produced the great advances toward what we now call "classical Newtonian mechanics," including celestial mechanics.[4] By 1800, the French advantage in this respect was recognized by Charles Babbage and his colleagues at Cambridge University, leading to the formation of the Analytical Society in 1812 with the express purpose of converting British mathematicians to the Leibnizian calculus and the broader analytical sciences pursued with it.[5] Leaving aside the irony of their importation from France of what would later come to be called "Newtonian mechanics" into Newton's very own home of Cambridge University, the point to stress is the role played by the peculiar institutional arrangements of French mathematical science in the eighteenth century in bringing about this circuitous historical development.

The failure of British mathematicians to embrace the apparent advantages of Continental analysis in the eighteenth century is often attributed to their slavish devotion to the legacy of their hero Isaac Newton. This is certainly part of the story, but also important is the way that an ironic sympathy between calculus-based mathematical physics and public support for it in France worked to create an environment supportive of this work at a moment when no clear consensus existed that this program was as potentially successful as it would ultimately turn out to be. Or, to spin the nationalist argument in a different direction that is perhaps more palatable to John Bull, it was not the slavish devotion of Britons to the Englishman Newton that pointed them in the wrong direction regarding the future of mathematical physics, it was rather the failure of the British state to construct a culture of absolutist public science in the manner of Louis XIV that explains this divergence.

To see the point another way, consider the following counterfactual: What if the opponents of the infinitesimal calculus had won in France? The British opponents of Continental analysis often made very similar arguments against it, and Bishop Berkeley's *The Analyst*, published in 1734, offers in many respects a more theologically tinged version of the epistemological attack that Rolle, Gallois, Gouye, and their allies sustained in their battles against the calculus in France.[6] Had the Old Style mathematicians carried the day, French mathematical science may have developed more closely in tandem with developments in Britain, making the whole history of modern mathematical physics completely different. Of course, this is not what happened, but by viewing the reality in terms of this possible counterfactual alternative, one sees the role played by the historical contingencies in France in the decades around 1700 in creating the eighteenth-century French mathematical physics that is today seen as foundational in the development of modern science.

The Peculiar Understanding of Newton's *Principia* in Early Enlightenment France

The contingent creation of a peculiar French cultural and institutional climate supportive of calculus-based mechanics and physics in the eighteenth century was, therefore, an important historical outcome of the French development of analytical mechanics before 1715. A second outcome was the way that the same history shaped in equally contingent and idiosyncratic ways the place of Newton's *Principia* within this scientific environment. Theoretically at least, Varignon's science posed questions about the nature of forces, their action, and the role of mathematics in scientifically capturing terrestrial and celestial motions that were akin to those posed by the *Principia* when read in terms of its full argument for universal gravitation. This was in addition to the way that his analytical mechanics challenged the existing canons of mathematical rigor. Varignon's science also pushed at the disciplinary divide separating mathematics from physics, while calling into question the traditional boundaries distinguishing geometry, arithmetic, mechanics, and astronomy. Yet because of the peculiar nature of the scientific contestation in France, which reduced the controversial aspects of Varignon's science to its mathematical provocations only, analytical mechanics was established in France without triggering any discussion of the potentially attendant physical, metaphysical, and epistemo-scientific issues, in effect postponing such discussions for later.

Whatever Varignon may have originally intended with his science, Rolle and its other critics viewed it in terms of its mathematics alone, and they attacked it on mathematical grounds only while ignoring the other claims bound up in it. Varignon likewise forced Newton's work into the same interpretive sieve since he and others continually invoked the *Principia* as a treatise in infinitesimal mathematics while ignoring all the other claims that the book made. Early in the process, other outcomes seemed possible. When summarizing Varignon's first public-assembly paper in November 1700, before *la querelle des infiniment petits* had taken over all discussion in its vicinity, the *Mercure galant* reduced the argument to a mathematical demonstration of central forces and their physical effects in planetary motion.[7] Bourdelin likewise noted in his diary of 1700 that Varignon was "continuing his demonstrations of the centripetal or central forces of M. Newton."[8] Each of these descriptions suggested work on the quantitative physics of forces, yet by 1702 this more explicitly Newtonian understanding of Varignon's work as a mathematical analysis of physical forces ceased to appear. Instead, his "new science of motion" was conceived and criticized mathematically, while the physical and mathematico-physical questions that analytical mechanics, along with Newton's *Principia* in its entirety, raised were ignored altogether. Nowhere in France during this period does one find a detailed discussion of the physical or metaphysical problems posed by Newton's, or Varignon's, cardinal category of "central forces" or a discussion of the disciplinary validity of using pure mathematical analysis to make claims about mechanics and astronomy. This particular outcome, which was anything but inevitable, shaped in important ways the larger history of analytical mechanics in France, and Newton's legacy within it.

After 1709, a new and more deeply mathematized approach to the study of moving bodies in terrestrial and celestial space was established in France, one comparable with at least one understanding of the mathematical approach to natural philosophy offered in Newton's *Principia*. Newton's work in this area was also a visible and acknowledged source for these French developments. Yet at the same time, and without in any way challenging the ascent of this first program, the Cartesian vortical-mechanical approach to celestial mechanics that Newton had explicitly refuted in Book II of the *Principia* continued to grow and prosper as well.[9] In this second scientific discussion, the one focused on causal, physical explanations of celestial mechanics, Newton's name and legacy, even as a critic, was rarely invoked.

In sum, two approaches to celestial mechanics were established in France in the decades after 1700, both connected in certain ways to the *Principia*. Yet rarely did the two generate any friction between them. In fact, the practitioners of each program were very often the same, and they mostly moved between the two endeavors without articulating any fundamental tension between them. Newton's *Principia*, which explicitly used mathematical mechanics to undermine vortical physics, was read and discussed by those savants capable of doing so without generating any tensions at all with the French vortical program. The treatise was likewise referenced authoritatively as a work of mathematical mechanics without any acknowledgment of, or anxiety about, the way that the treatise was also a critique of Cartesian vortical physics.

Fontenelle was especially representative of this peculiar French outcome.

He was fully aware of all the arguments in the *Principia*, yet he combined unwavering support for the analytical program in mathematical mechanics derived from the mathematical reading of Newton's work with equally vigorous support for the vortical system of celestial mechanics that Newton's treatise had explicitly set out to refute. Others shared his outlook, and the complexities of this stance are vividly illustrated by the reception in Paris of a vortical system of natural philosophy published in 1707 by a Lyon savant named Philippe Villemot.[10]

Wanting to ensure a wide distribution for his work, Villemot sent a copy of his "little treatise on astronomy" to the Paris Academy hoping to earn at least a review in the *Journal des savants*.[11] Noting in his letter of introduction that "the public considers your illustrious company as the sovereign court of the Republic of Letters," he also sought academic approval as a way to launch his treatise into the wider public sphere.[12] Villemot also sent a copy to the *Journal de Trévoux* for exactly the same reasons. Since he had the favor of a Lyon cleric named Father Vial, he asked Vial to write to Paris on his behalf. Villemot's treatise and letter of introduction, together with Vial's letter of support, ended up in the hands Father Sebastien Truchet, a leading member of Bignon's coterie of mechanical arts advisors and an academic *honoraire* after 1699 with connections to the *méchaniciens*. He may have been asked by the Academy to do an evaluation of the work, or Bignon may have asked him to produce the review in the *Journal des savants*, which appeared in late 1707.[13] Whatever the explanation, these letters, and a draft of his own review of the book (which is very close to the one published in the *Journal des savants*), are found in Truchet's papers. The ensemble illustrates well the working practices of official French science after 1699, and the peculiar French approach to mathematics and physics that became dominant in this context in the final years of Louis XIV's reign.

Vial called Villemot's work a "masterpiece," and he indulged in similar hyperbole in describing what he saw as the book's many achievements. "There may be no other mathematician in the academies of Paris or London that is his equal," he gushed. His reference to London also led him to discuss the critical engagement with Newton's *Principia* found in Villemot's text.[14] The author was primarily a vortical physicist, eager to reform the Cartesian system in light of the recent critiques offered against it. He therefore read the *Principia* with special attention to its refutation of vortical cosmology in Book II. Villemot's Cartesianism also led him to cast a critical eye upon the immaterial attractionist theory of universal gravitation that Newton offered as a replacement for Descartes's evident vortical mechanisms. Villemot was also a student of Varignon's analytical mechanics, however, and Vial wrote of the latter's system saying that it "makes as many millions of things visible in a small action as the microscope makes one see in the eye of an insect."[15] What Vial's description of Villemot's relationship with the *Principia* reveals is how Newton's reputation had been transformed in France by the widespread influence of Fontenellian statements such as these about Varignon's analytical mechanics.

"He paid too much respect to Newton in citing him," Vial wrote of Villemot. "This English mathematician, who only appears sublime because he is so convoluted and looks profound because of his obscurity, wrote on [the] subject [of celestial mechanics]. But what gibberish in comparison to the clarity, the ease, not to mention the elegance and the *je ne sais quoi* of M. Villemot's work."[16] Vial's assessment does two things at once. First, it reinforces the reduction of Newton's overall argument in the *Principia* to nothing more than a display of obscure and idiosyncratic mathematics. This reduction had first appeared in the early 1690s during the formative years of analytical mechanics, and here it is deployed in a newly triumphant way to make Villemot's affection for French analysis a key ingredient in his superior approach to the science of celestial mechanics. Unlike Varignon, however, Villemot offered a fully physical and mechanical account of celestial mechanics, one that employed differential analysis, but only as a tool of calculation. Vial was not, therefore, evaluating the relative merits of these two styles of mathematical mechanics as Bernoulli, Varignon, and Fontenelle had done. He was using praise for Villemot's analytical mathematical style to impugn both Newton's mathematical *and* his natural philosophical work. He further avoided any discussion of the physical and philosophical differences that separated Villemot's system from Newton's, subsuming these contrasts within the now standard aesthetic critique of Newton's mathematical style. This allowed him to effect, like his predecessors, a displacement of Newton's actual physical and philosophical arguments in the *Principia* while adding a new critical and polemical judgment against Newton's mathematical work in the *Principia*.

In particular, Vial argued strongly that Villemot's science was superior to Newton's as a result of his superior mathematical methods. Such polemics against Newton were extremely rare before 1715, even though the terms that Vial used were commonplace by this date. More representative was an approach to Newton's *Principia* that mimicked Vial in questioning the overall quality of the mathematical arguments in the text while avoiding any overt critique of Newton or his treatise. Varignon, for example, approached Villemot's refutation of Newton in a similar way, calling Villemot's treatise a work of "physics" but then criticizing the author for having "ample intellect but not enough mathematics." He further exposed some of the book's errors, triggering the author to suppress it until corrections could be made.[17] The errors were all mathematical, however, and Varignon engaged in no way with Villemot's mechanistic physics or critique of Newton, which constituted the heart of his treatise. Truchet approached the text in a similar fashion in his evaluation. "This is a physical more than a geometrical demonstration," he noted, but he claimed to follow the author in ultimately assessing the book on mathematical grounds. "He preferred to offer [a physical demonstration] instead of one rooted in the method of infinitesimals," Truchet wrote, "because this approach has the advantage of rendering sensible to the imagination that which the other can only make sensible to the mind." Here infinitesimal mathematics and Cartesian physics are offered as parallel and interchangeable modes of scientific argumentation, approaches that achieve the same end while targeting different dimensions of human understanding.[18]

To conflate analytical mathematics with mechanistic physics in this way was to harmonize out of existence a host of complex scientific and epistemological differences. Yet many in France around 1710 were comfortable doing just that, including Fontenelle who effected a similar conflation in his treatment of Villemot's work. The author of the *Nouveau système* was not an academician, yet the secretary was so pleased with his treatise that he used the Academy's *histoire* of 1707 to praise it. Varignon and Bomie had also presented papers developing their analytical celestial mechanics to the Academy in 1707, so the secretary saw an opportunity to integrate the explicitly mechanical and anti-Newtonian physics of Villemot with the explicitly mathematical and Newton-inspired celestial mechanics of Varignon and Bomie. In his *histoire* for 1707, Fontenelle wove all three works together as if they constituted a seamless whole. This effectively erased the explicit critique of Newton's theory of universal gravitation offered by Villemot and highlighted by Vial. While Villemot had offered his treatise as a rival physical system that challenged the one offered in the *Principia*, Varignon and Bomie conceived of their analytical central-force astronomy as a generalization of and a movement beyond Newton's work. Fontenelle offered a third understanding by collapsing these distinctions, and then making all three exemplary of a new and unified mathematico-physical approach to celestial mechanics.[19] Truchet erased the polemics with Newton in his evaluation as well, suggesting, like Fontenelle, that Newton's *Principia* was less a work to be rebutted than one to be surpassed. For his part, Villemot found these characterizations anything but uncomfortable, expressing gratitude for Fontenelle's "kind treatment of his work" in a letter of thanks he wrote to the Academy in August 1708. In it, he also sent his best wishes to "Malebranche, Varignon, Saurin, and the rest," revealing where his scientific sympathies lay.[20]

The understanding of Newton's *Principia* in France was profoundly shaped by ironic outcomes such as these, results conditioned by the particular history of analytical mechanics after 1700. Villemot's explicitly anti-Newtonian treatise could have triggered a debate about the physics of central-force mechanics, one comparable to the debates that did erupt around precisely these topics two decades later. But in early eighteenth-century France, Villemot's anti-Newtonian treatise was instead absorbed quietly into the perceived unity of analytical and vortical mechanics that characterized French science around 1710.

Other work was absorbed in a similar fashion, and French savants likewise pursued research that revealed the perceived unity of these two seemingly opposed understandings of celestial mechanics. Saurin, for example, began developing what would become a systematic alternative to Villemot's vortical cosmology in a *Journal des savants* article of 1703 even as he was becoming the champion of Varignon's calculus-based central-force mechanics in the French public sphere.[21] In physical terms, Saurin adopted a more complex understanding of the fluid mechanism that vortical cosmologists made operative upon bodies. Whereas Villemot assumed that fluid vortices produce movement through impact in the manner of a wave pushing a cork onto a beach, Saurin conceived of their motion in terms of a hydrostatic displacement in the manner of a cork receding in a vessel along with the water in which it swims.[22] This different physical model also supported Saurin's introduction of the differential calculus into his explanation. Since causal change for Saurin was rooted in the infinitesimal changes in fluid density, the new analysis provided a ready means for mapping these *rapports* mathematically. In April 1709, Saurin presented his work to the public in his first academic public assembly.[23] He earned a glowing review in the *Mercure galant*,[24] and he published the more complete paper in the *mémoires* of that year.[25] As late as 1730 Johann Bernoulli praised the work as the inspiration for his own vortical yet analytical explanation of gravity.[26]

Saurin's work demonstrated how analytical mathematics and vortical physics could be wedded into a coherent science, and his was not the only such solution offered in these years. In the sixth and final edition of *De la recherche de la verité* published in 1712, Malebranche offered a different vortical explanation of planetary motion and terrestrial gravity, one that implicitly challenged Saurin's model.[27] Malebranche began by declaring some of

Saurin's assumptions "manifest contradictions." Most important was Saurin's claim that fluid mechanisms are so infinitely subtle that they exert no resistance on moving bodies but are nevertheless responsible for causing their motion. How could both be true at the same time? Saurin's assumption was crucial since it allowed him to maintain Kepler's laws while still making celestial motion the result of vortical mechanisms. Saurin's understanding was not as patently impossible as Malebranche believed since Newton had also theorized the existence of a similarly all-pervasive but nonresisting medium as a possible cause for celestial motion and terrestrial gravity. Later thinkers, including Euler and Bernoulli, also did the same.[28] Malebranche found such an idea contrary to reason, however, and he replaced Saurin's nonresisting fluid with a different model that returned Villemot's (and Descartes's) "dense vortex" to the center of action.

Like Saurin, however, he did so in a way that allowed for the introduction of differential analysis into his physics. Malebranche's innovation rested in his theory of the *petit tourbillon*, a "mini vortex" nested within the larger streams of the main vortical fluid that allowed him to conceive of the macroscopic motions of bodies in terms of infinitesimal microscopic changes, and then to map these changes mathematically using the calculus. Or, to state the same point another way, the *petit tourbillon* allowed Malebranche to save Saurin's mathematics while avoiding the perceived problems of his physics. It also allowed the two mathematical colleagues to turn rival physical positions into the same project of mathematical analysis. Where Saurin saw infinitesimal changes in fluid density, Malebranche saw infinitesimal vortical impacts, yet each used the same mathematics to model these changes. In the end, the similarities that united Saurin and Malebranche, like those that united these two academicians with Villemot, were much more important in France than the different physical theories that separated them.

The same attitude also shaped the understanding of Newton's *Principia* during these years. Actors such as these saw no apparent contradiction in accepting Newton as a distinguished contributor to the advanced mathematical physics that they practiced while also dismissing out of hand his particular (and particularly absurd by early eighteenth-century epistemological standards) physics of universal gravitation. Two decades later, these rival vortical systems, along with their successors, would compete with one another in and around the Paris Academy in a struggle that also included Newtonian gravitational theory as a viable physical alternative. In the decades around 1710, however, no such battle had yet erupted. And while vortical mechanics did become an increasingly active zone of research within the Academy during these years, a dialogue with Newton's *Principia* was not the drive train of this development. Vortical mechanics also prospered during these years without in any way challenging the equally active program in analytical mechanics present in the same space. Indeed, the really significant outcome was the peculiar fusion of these two ostensibly different physical explanations, and their seemingly different relationship to Newton's *Principia*, into one largely harmonious understanding of French celestial mechanics.

This fusion allowed Malebranche, whose writings played such an important role in creating the philosophical context for French science after 1690, to remain a central philosophical touchstone for French mathematical science well into the eighteenth century. When first introduced to the idea of the *petit tourbillon* in a paper that Malebranche read to the Academy in April 1699, Bourdelin wrote that "this system was found to be very elegant. I really liked it for it contained many good things."[29] Many shared Bourdelin's affection, and this outcome allowed the discourse of rational beauty and simplicity that had been developed around Malebranche's mathematical philosophy to pass seamlessly over to his vortical cosmology as well. This conflation occurred even when the precise vortical system in question was not Malebranche's theory of the *petit tourbillon*.

Consequently, in a manner similar to Vial's conception of Villemot as a superior cosmologist to Newton because his mathematical style was more simple and elegant, French savants increasingly conceived of and defended vortical cosmology through appeals to its mathematical order and clarity. This despite the way that such a defense could be construed as using advanced mathematical apples to evaluate and judge physical and metaphysical oranges. On the one hand, this happened by reading the mathematico-physics involved more in terms of its mathematics than its physics, a tendency that Malebranche encouraged in his defense of physical skepticism and mathematical phenomenalism. On the other hand, it also occurred because the physics itself was judged according to the cherished canons of Cartesian *évidence* that made simple and elegant deductions from clear and distinct physical principles (in this case the idea of point-contact fluid mechanism) the benchmark of rigorous science. Newton's theory of universal gravitational attraction had no credibility when viewed from this epistemological vantage point, as his first French reviewer had noted in 1688, and here the veritable absence of any sustained discussion of Newton's *Principia*, or any use of the term "Newtonian" to describe science or scientists in this period, finds its explanation.

So too does the parallel veneration of Malebranche and the increasing use of the term "Malebranchianism" after 1690 as a marker of the most modern and sublime French scientific thought.

In sum, what occurred in France in the decades around 1700 was the widespread acceptance of one understanding of Newton's achievement in the *Principia* through a filtration of his overall argument through Malebranchianism. This in turn created a peculiar French approach to the science of terrestrial and celestial mechanics that enjoyed great institutional and intellectual support while moving Newton and his actual work ever more to the margins. Overall, the complex mix of institutional and intellectual changes described in this book was responsible for this outcome. And once established, Fontenelle's public academic discourse, which began in earnest in 1702 and continued without interruption until 1740, played a powerful role in sustaining this particular historical outcome. Not only did the secretary's public voice help to create a public consensus supporting this understanding, it did so while reinforcing the distance that separated Newton's *Principia* from it. This divorce was neither premeditated nor irresponsible since Fontenelle, like his colleagues, saw no necessary link between Newton's *Principia* and the larger French scientific developments of the period. He cannot, therefore, be characterized as acting in an anti-Newtonian manner in doing what he did. A link with Newton was important and widely acknowledged in the development of analytical mechanics, but after that the connections were displaced through the celebration of the analytical turn in France, a move that Fontenelle and virtually everyone else in France before 1715 believed to have been authentically original and in no way derived directly from the *Principia*.

Fontenelle's public work as Academy secretary during these years both illustrated and sustained this particular result. He first articulated the relationship between Newton's *Principia* and French analytical mechanics in his second *histoire* of 1700. Varignon's papers in central-force mechanics were published in this volume, and since he frequently invoked Newton's *Principia* in this work, Fontenelle was led to address the connection. His subsequent accounts in later *histoires* differed little from the conceptualization offered here, so a discursive tradition was established with this volume that lasted for many decades.

The presentation began by celebrating the mathematical sophistication of Varignon's work. Overall, the calculus dispute, as was discussed already, provided the most important context for Fontenelle's description, but he nevertheless addressed the physical dimension of Varignon's science and its Newtonian ancestry directly. "By this path," Fontenelle explained, "M. Varignon fell upon the principal propositions of M. Newton's learned work. What ennobles his science the most, however, are the consequences for astronomy and for the different systems of the heavens that he pulls from it."[30] Explaining these "more noble consequences," Fontenelle followed Varignon in making the essence of his achievement mathematical. "M. Newton and M. Leibniz were the first, and still the only others, to research the different relations of gravity [*pesanteurs*] of a planet toward the sun at different points in its orbit. [Their theory, however] is restricted only to the conic sections, but M. Varignon extends it to all possible curves. He also derives the relationships [*rapports*] between the times necessary to describe the different arcs in the orbit."[31] Varignon's essentially mathematical approach to these questions was also invoked in the secretary's final conclusion. Here, he wrote, "all that remains to be done is the collection of exact observations about the actual movements of the planets. . . . [Geometry] is now prepared to offer whatever curves one likes, and it is up to astronomy to choose."

This presentation situates Varignon's work very closely to Newton's, at least according to one interpretation of the central argument of the *Principia*. But it also ignores entirely another, namely the quantitative empirical argument in Book III that planets do not move "according to whatever curves one likes," but in ellipses governed by an inverse square law of universal gravitation. In other words, Fontenelle makes Newton a mathematical phenomenalist pure and simple, and Varignon his better when viewed from this perspective. Fontenelle also situates Varignon alongside Leibniz's alternative vortical account of celestial mechanics found in the *Tentamen* in ways similarly favorable to his French colleague. Leibniz also treated celestial phenomena in terms of the mathematical laws governing them, and since Varignon's work did so as well, Fontenelle's unification imagined a veritable conformity between them. Leibniz and Newton, however, conceived of their systems as physical and metaphysical rivals, and they made nonmathematical claims for their conception of force that Varignon never made. In seamlessly connecting the Frenchman's work with each of these alternatives, Fontenelle was effectively erasing these differences in the name of Varignon's claim to a superior universality of perspective. Fontenelle appears to have closely followed Varignon's own understanding of his work in framing his presentation, for drawing his own connection between Newton's mechanics and his own, he stated in one of his papers from 1700 that, "a very simple formula for central forces, be they centripetal or centrifugal, follows, one that is the primary foundation of the excellent work of M. Newton called *De Phil. natur. Princ. Mathem.*"[32]

Many could and would disagree with this particular claim about the "primary foundations" of Newtonian physics. But before 1715, Varignon's view dominated scientific thinking in France. In echoing it in his public presentation of analytical mechanics, Fontenelle was revealing both his own sympathy with Varignon's viewpoint and his interest in using it to build a public consensus about the science of mechanics in France.

"Cartesianizing" Analytical Mechanics: Fontenelle's Explanatory Pictures

Other forces also pushed Fontenelle away from making any strong, genealogical connection between Newton's *Principia* and Varignon's analytical mechanics. Most important were the obligations of his academic office and his wider ambitions for public science. These responsibilities often pushed his discourse in directions that the science itself did not compel. In particular, to achieve his overall aspirations, Fontenelle needed to provide clear and evident explanations of academic science to his public audience, explanations which analytical mathematics rarely offered. In the face of this dilemma, the secretary often responded by constructing his own explanations, sometimes breaking with the actual content of the science itself in doing so.

Most important in this context was his use of the Cartesian vortices to represent abstract mathematical mechanics even when this was not in fact part of the science he was publically presenting. Varignon never addressed the question of the vortices explicitly after 1690, and his work in analytical mechanics was completely devoid of any references to *tourbillons* of any sort.[33] Furthermore, when he spoke of "centripetal" and "centrifugal" forces, he meant these to be understood in mathematical terms only since he wanted his mechanics to be applicable to any physical system whatsoever. He was, in fact, most proud of the way that his mechanics could account for any motion whatsoever, no matter what physical system was assumed, and Fontenelle often emphasized the neutrality of his mathematical mechanics with respect to physics and metaphysics in his explanations of it. In this way, he connected Varignon's science to the Malebranchian mathematical philosophy that both he and his colleague admired.

The requirements of his office, however, often created different agendas, and these often led Fontenelle to make different choices in his public academic discourse. Indeed, when confronted with the problem of making analytical mechanics clear and evident to the public at large, Fontenelle often abandoned the language of Malebranchian mathematical phenomenalism even though it most accurately described the actual intent of the science that he was describing. In these moments, he often adopted instead the evident pictures of Cartesian vortical mechanics even when they were not precisely present in the science he was describing. His inaugural 1700 *histoire* again illustrates the point, for here Fontenelle introduced another set of formulations destined for a long life in Fontenelle's public scientific discourse even if his attribution of them to Varignon was incorrect.

Describing Varignon's central-force mechanics, the secretary offered the following picture: "Everything that turns around a center tends to move away from it, and M. Descartes founded the hypothesis of the vortices upon this principle. All the planets contained within the solar vortex turn around this star and they thus tend to move away from it. The ethereal matter in which they spin, however, is more subtle and active than the planets themselves. It is consequently more disposed to move away from this center. In this way it pushes upon the planets continuously and keeps them always upon the circumference of the same curve that they describe around the sun."[34] This explanation physically accounts for the centripetal forces at the center of Varignon's mechanics. In the *mémoire* itself, however, not only is no such physical account offered, the precise vortical model offered in Fontenelle's account is explicitly eschewed since the paper claims to calculate the nature of central forces irrespective of the physical causes that produce them. Varignon wanted his mechanics to be neutral with respect to any precise physical system, yet in his public histories Fontenelle created a direct link between his analytical mechanics and Cartesian vortical cosmology.

An even more direct link is offered in Fontenelle's 1703 *histoire*:

> According to the innovative system of M. Descartes and the most powerful appearances that physics can offer, what we call the gravity of earthly bodies is only a particular effect of a more general principle that acts in all curvilinear motion. If this principle gives all the planets in our vortex a tendency toward the sun comparable to the tendency of earthly bodies toward the center of the earth, then the theory of M. Varignon explained in the *histoire* of 1700 gives a solution to all the problems one can imagine regarding the gravity of the planets toward the sun and the inequalities of its action at different points in its orbit.[35]

To present Varignon's analytical mechanics through these vortical explanations was to add a physical dimension to Varignon's work that was not explicitly present in the science itself. In short, it was to make Varignon into a vortical mechanist in spite of his analytical mathematical self.

Fontenelle's ambitions as perpetual secretary were instrumental in shaping this specific discursive outcome. Describing another set of Varignon's analytical papers treating central-force mechanics in the 1705 *histoire*, Fontenelle adopted a different narrative strategy. After explaining the mathematics in the opening paragraphs, he stated that "if one wants to form an idea of all this in terms of physics, and according to some system of the heavens, one can conceive of the shape of the vortex dominated by our sun as determined by the differing forces of the neighboring vortices which surround it."[36] The rest of the presentation was couched in vortical explanations such as those found in his earlier histories. In his narrative for 1706, by contrast, Fontenelle invoked a different illustrative picture. Here he asked readers to imagine that the sun contained an inherent power capable of pulling the planets toward it and that this gravitational attraction was responsible for the centripetal force that Varignon explained mathematically. Quickly qualifying, however, Fontenelle added that, "this idea conforms less well to the rules of physics even if we prefer it because our imagination finds it readily accessible."[37]

In both of these examples, Fontenelle invokes recognizable physical pictures as a way of explaining to the public Varignon's pictureless mathematical work. Varignon himself neither needed nor developed pictures of this sort in his own work, and in fact his science explicitly eschewed them, a feature that is in fact, as was discussed earlier, one of its major innovations. Varignon's analytical mechanics was innovative precisely because of its move away from mimetic representations of nature and toward an abstract mathematical, or logical to use Peter Galison's terms, mechanics. For him, pictureless symbols and direct mathematical reasoning were enough. Point-contact understandings of causal change were not necessary for his analytical mathematical account of motion in nature. Fontenelle, however, pushed by his public ambitions as Academy secretary, returned the old pictures that Varignon had eliminated when explaining his science to the public. This thwarted the actual reception of Varignon's mathematical phenomenalist understanding of celestial mechanics in the wider public sphere. In presenting Varignon's science in this of all ways, Fontenelle also softened some of the more radical claims of the new science of motion while also helping to create a public understanding of it that was often out of step with how its practitioners conceived of the science itself.

Toward the Enlightenment "Newton Wars," 1710–30

Each of these developments had a profound effect on the place of Newton in eighteenth-century French science. After 1715, the mathematical approach to mechanics that we today associate with Newton's name and call "classical Newtonian mechanics" was thriving in France, but it bore little if any direct attachment to Newton's legacy, or to that of his *Principia*, in the minds of those who practiced it. This was a direct result of the peculiar public reception of Newton's work within the debates about infinitesimal analysis and the reception of these disputes within the new public academy of the eighteenth century. By entering France through a more purely mathematical approach to mechanics than the one offered in the *Principia*, Newton's category of "centripetal force" was shorn of its more physical and metaphysical connotations and not taken seriously at first as a physical category. Similarly, because Malebranchianism allowed French central-force mechanics to be wedded harmoniously with Cartesian vortical astronomy, and against Newton's rival understanding of universal gravitation, the French were able to quietly absorb crucial features of Newton's work in the *Principia* while otherwise remaining detached from his natural philosophy as a whole. Fontenelle reinforced precisely this harmony in the mind of the public by channeling his own deeply held Malebranchian commitments into a public discourse that solidified this particular arrangement.

In his public-assembly presentation of 1709, Saurin offered a very different account of the relationship between Newton's *Principia* and French mathematical physics as it had developed since 1687. He publicly, and rather exceptionally, indicted Newton's physical theory of universal gravitation, calling it "unreasonable" and an abandonment of "clear mechanical principles." He also added that Newton's science threatened to return physics to the "occult qualities" and "ancient darkness of Peripatism from which heaven wants to save us."[38] This direct indictment of the physics of the *Principia* was a harbinger of things to come. But couching his critique within the usual praise of Newton's mathematical "exactitude" and "precision," Saurin's paper also echoed the first French review of 1688 in praising Newton's mathematical mechanics. Accordingly, Saurin's critique was absorbed quietly, like Villemot's anti-Newtonianism before him, into the overall consensus. Virtually no one else before 1715 offered similar attacks, and overall Saurin's *mémoire* was exceptional among academic works of the period in being one of the very few that actually devoted space (albeit little more than a paragraph in a twenty-page *mémoire*) to an explicit refutation of Newton's physical theories. Far more common was the application of the most powerful rhetorical attack of all against Newton's idea of universal gravitation: unperturbed silence.

The fact that nothing called "Newtonianism" took hold in France in this period, or that no group of "Newtonians" emerged, is to be explained by the historical contingencies of this particular French outcome. Newton's own views, of course, were much more complex and many other interpretations of his work were possible. Indeed, as the eighteenth century progressed Newton's *Principia* became a site of increasing contestation as savants, including Newton himself, began to make different claims about the work and its significance.[39] In 1700, many on both sides of the English Channel held a similar understanding of the significance of Newton's treatise,[40] but by 1715 new divisions had started to emerge.

The appearance of Newton's *Opticks* in 1705 was one important step toward eighteenth-century Newtonianism and its attendant battles, for here the physics of gravitational attraction was argued for explicitly, giving a new frame for reading the mathematico-physical arguments in the *Principia* in different and more explicitly physical ways. Newton also added a series of queries to the new editions of the *Opticks* published after 1705, texts that further framed his science in more physicalist and attractionist terms. Also crucial was the rise of John Toland's self-consciously Newtonian materialist philosophy, and the vigorous refutation of it that self-proclaimed Newtonians such as Samuel Clarke began to issue through the Boyle Lectures and other forums after 1705.[41]

The calculus priority dispute, which erupted in 1709, played an especially important role in making the new "Newtonian" creed as well. This bitter, public battle gave Newton and his followers a cause to fight for, while also triggering the invention of many of the key rhetorical tropes that would define the discourse of eighteenth-century Newtonianism. Newton's continuing work as a savant also played a role, since he brought out a new edition of the *Principia* in 1713, complete with a new "General Scholium" that explicitly posed the question of gravitational attraction for discussion. A polemical preface by Roger Cotes, which framed natural philosophy in terms of a battle between Scholastics, Cartesians, and Newtonian "experimental philosophers," also added a charge to the physical discussions of the treatise.[42] However, since Newton in no way revised or updated the mathematical character of his treatise in 1713, although he did correct many errors, French mathematicians found no reason to read the new edition differently, or to change the mathematical mechanics that they had built from it twenty years earlier.

By 1720, attentive savants were starting to perceive that a great philosophical storm was brewing, one that pitted self-conscious Cartesian mechanist-vorticists against newly self-conscious and aggressive "Newtonian attractionists." This outcome, however, was in no way obvious in 1710, and if it was not yet apparent in 1710, it was even less determined by the events that greeted the initial appearance of the *Principia* after 1687. Indeed, in the two decades around 1700, few would have predicted that a great struggle over a scientific entity called Newtonianism would ensue in the coming decades. Accordingly, the fact that before 1710 few talked about Newton in the polemical terms used by Vial and Saurin, and that no self-conscious Newtonians or anti-Newtonians existed, in France or anywhere else, is to be explained by the very different understanding of Newton and his achievement that was prevalent before this date.

In short, no Newtonianism yet existed because the conceptual referent for such a term had not yet been developed historically. Meanwhile, the foundations for what later commentators would call classical Newtonian mechanics were in place even if they had been developed in France by mathematicians with no self-conscious devotion to Newton and no self-conception of themselves as Newtonians. All the more familiar Enlightenment understandings of Newton and Newtonianism were yet to be assembled, and when they arrived they would be merged with the very different legacies left by the analytical mechanics debates of the period 1690–1710.

Conclusion: Enlightenment Legacies of the French Development of Analytical Mechanics

Two episodes from this later history can serve to conclude this book by illustrating what is gained by viewing the eighteenth-century history of Newton's French reception in terms of this complex and multifaceted series of twists and turns. Both involve the science of celestial mechanics as practiced through the deployment of the Leibnizian calculus, which is the most important continuity flowing through the whole history of the eighteenth-century French engagement with Newton's *Principia*. Each also centers on a particular understanding of the relationship between analytical mathematics and the science of physics that was also, I argue, an important legacy of the peculiar early history of Newton's French reception.

A figure who has appeared frequently in this book is at the center of the first episode: Johann Bernoulli. He remained scientifically active well into his seventies, and this allowed him, along with Fontenelle, who remained intellectually active into his nineties, to participate significantly in both the early analytical mechanics debates around 1700 and the later-Enlightenment "Newton Wars" of the 1730s and '40s. A constant in all of his scientific activity was Bernoulli's intense need to receive scientific recognition and acclaim for his work, along with his willingness to wage intellectual battle whenever respect for his work was not extended in the manner he deemed appropriate. In 1720, the Paris Academy initiated a new practice of awarding cash prizes for the best papers written in response to particular scientific questions. These prize contests became for Bernoulli a recurring occasion to satisfy his need for visible markers of his recognized scientific acclaim. He was therefore a vigorous participant in all of these contests and very often a sore loser when they did not turn out as he would have liked.

A case in point occurred in 1730. Bernoulli sent a paper to Paris in pursuit of the prize for the best explanation of why the planets move in elliptical orbits. This was a timely question because it forced the participants to consider a hotly debated issue of the moment: whether Cartesian vortical impacts or Newtonian gravitational attraction offered the best account of the celestial motions. By 1730, the Cartesian vortical approach to celestial mechanics pursued by Villemot, Saurin, Malebranche, and others had developed into a powerful scientific paradigm in France, and many defenders of this approach existed in the Academy and in France more generally. The Newtonian alternative was also being defended in the public sphere, and Bernoulli was not alone in thinking that this prize contest was a kind of test of the two explanations. He also shared the view, increasingly held by many at the time, that the French academy had become something of a Cartesian enclave, and that the vortical explanation was serving as the official paradigm for celestial mechanics within the Royal Academy.

There were, of course, other points of view in the Academy at this time, and Bernoulli knew this since one of the leading anti-vorticists was his own student and protégé (in his mind at least): Pierre Louis Moreau de Maupertuis. Maupertuis, like many others soon to be called "French Newtonians," had cut his mathematical teeth by learning analytical mathematics and mechanics from works such as Reyneau's *Analyse démontrée* and Varignon's lectures on mechanics delivered at the Collège Royale, which were published posthumously in 1725 as his *Traité de méchanique*. In 1732, Maupertuis would become the first royal academician to openly defend the Newtonian theory of universal gravitation, arguing not that this is in fact the actual physical law governing the universe, but rather, in good Malebranchian fashion, that nothing rules out its possibility so long as the explanation agrees with the empirical phenomena and the laws of quantitative mathematical analysis.[43]

Maupertuis was, therefore, a Newtonian attractionist of a very phenomenalist and Malebranchian sort, and Bernoulli was no die-hard attractionist or vorticist either, even if he manifested this tendency differently in 1730. Winning the Academy's prize was his primary goal, so he developed a cunning approach based on this reading of the intellectual winds in Paris. Convinced, as he wrote in a letter to Gabriel Cramer, that "the attractions and void of M. Newton would only generate horror among Messieurs the French," he developed a vortical account of the elliptical orbits of the planets even though this was not his preferred theory. "I adroitly reproduced the vortices of Descartes in a slightly modified form and added the appearance of a new luster to them," he told Cramer. Whatever the accuracy of his assessment of the prejudices that would guide the academic-prize committee, his ploy appeared to work. He won the contest.[44] Yet Bernoulli's move was not as cynical as it may appear, for his ability to change physical explanations in ways that satisfied this or that preferred physical hypothesis depended crucially on the consistency of his use of analytical mathematics to solve the problem in either case. In short, it was because the actual foundations of his science were in fact mathematical that he could change the physical explanation offered as needed.

Whether Bernoulli was presenting himself as a Cartesian vorticist or a Newtonian attractionist, he never wavered in being a practitioner of the differential and integral calculus in developing his science. What this episode reveals, therefore, was Bernoulli's comfort with treating mathematical analysis as something separate from physical explanation, and as a foundation for the science of physics that precedes and supersedes it. This general tendency was in fact essential to the French Enlightenment Newtonianism that developed self-consciously after 1730, and the mathematical phenomenalism that this way of practicing mathematicized physical science assumed was, I argue, a direct consequence of the particular way that Newton's work in the *Principia* was received and translated via Malebranche and the analytical mechanics debates around 1700 into eighteenth-century French mathematical physics.

The second episode adds another illustration of the same conclusion. By the mid-1740s, the debates in France about the validity of Newtonian gravitation as a physical theory had begun to subside, and mathematicians in the Paris Academy, trained like Maupertuis in the French tradition of analytical mechanics, were beginning to work on the great project, completed in Laplace's *Treatise on Celestial Mechanics* published around 1800, of accounting for the known phenomena of the heavens using the force laws and calculus-based mathematical analysis of what was then starting to be called "*la physique Newtonienne*." Clairaut's prediction of the precise date of the return of Halley's comet in 1759 was one of many achievements of this "Newtonian" project, yet what these academicians were really practicing was the amalgam of Newtonian, Leibnizian, and Malebranchian science that had congealed into French mathematical physics in the ferment of the French public sphere in the decades around 1700. It was also the science that was developed and then established with the aid of Fontenelle's management at the Royal Academy after 1700.

Along with Clairaut and Maupertuis, a third leading practitioner of these analytical sciences was Jean le Rond d'Alembert. Leonhard Euler, who hailed from Bernoulli's Basel, also matched their eminence and was their colleague outside Paris in all of these efforts. Each of these mathematicians had been educated by studying the foundational work of the first generation of analytical mathematicians, and like Bernoulli, who was a teacher to each of them, they also conceived of the relationship between mathematical analysis and physical explanation in the same way that he did.

This fact was revealed in late 1747, when all three men were trying to use the common analytical formulation of Newton's law of universal gravitation to precisely establish the mathematical physics of the moon's orbit.[45] Through an accident that was no doubt rooted in the deep similarity of their basic working methods, all three came to the same erroneous conclusion at the same time, believing that Newton's force law was inaccurate by a precise factor of two. They realized later that they had all unwittingly made the exact same mathematical error, but for a few months they were convinced that Newton's law of force was in contradiction with the empirical phenomenon of the lunar orbit. The interesting aspect of this episode is what they did in the face of this apparent anomaly.

Clairaut asserted his conclusion decisively at a public assembly of the Academy in November 1747. He declared that Newton's force law was wrong and had to be modified. Euler asserted the same, and d'Alembert was led in the same direction, although he was more hesitant than his colleagues about drawing this radical conclusion. Others in the Academy called foul, however; notable among them was the Comte de Buffon, who read a paper, "Reflections on the Law of Attraction,"[46] at the Academy the following January. Buffon was becoming an outspoken critic of the mathematicized science favored by so many of his academic colleagues, and he argued that they could not simply change the force law when it did not agree with the mathematical analysis being performed. If the mathematics is not matching the physics, he contended, it's the mathematics that needs to be changed. This was especially true, he argued, given all the empirical, experimental, and physical evidence supporting Newton's force law.[47]

The willingness of Clairaut, Euler, and d'Alembert to diminish such physicalist considerations when matching their mathematics with empirical phenomena is the point to take from this example. Like Bernoulli, and many other French Enlightenment mathematical scientists, the practice of "Newtonian physics" as it had developed by 1750 was about matching systematic mathematics with empirical data in a way that rendered the physical world rational. The rationality that mattered most in this project was mathematical rationality, and true to the Malebranchian and Varignonian sources of their orientation, they were flexible when thinking about the physics that explained this mathematical order.

Newton's contribution to their science, moreover, was not based on his understanding of the universal laws of force as derived from rigorous empirical and experimental data analysis, even if Buffon was right to find exactly this in the *Principia*. It was rather his new mathematical approach to natural philosophy, which his *Principia* modeled even if later mathematicians were led to liberate Newton's mathematical insights from the archaic mathematics he used to develop them. Accordingly, and despite Buffon's protestations, academic French Newtonianism became emblematic in the Enlightenment of a new mathematicization of the physical sciences, and a new comfort with explicitly phenomenalist mathematical renderings of the natural order. Elsewhere, especially in Britain, a very different situation prevailed. The source of this characteristically French feature of Enlightenment mathematical physics was not the *Principia* directly and absolutely, but the reception and mediation of the idiosyncratic innovations found in this treatise by French mathematical science and culture in the decades around 1700.

I. B. Cohen had this precise Enlightenment understanding of mathematical physics in mind when he located a new "mathematical style" in Newton's *Principia*, one that laid the foundations, he argued, for the highly mathematicized approach to physics characteristic of modern science today. Cohen was right about the innovative nature of eighteenth-century mathematical physics in charting this path toward modern science. But it has been the project of this book to argue that these innovations were not born directly, and fully formed, out of the genius of Newton's *Principia*. Nor were they the result of any Newtonian Revolution said to have been triggered by the publication of this book. These outcomes were instead produced historically after 1690, through a crooked and contingent process of change, one that involved interactions among actors and ideas propelled by mathematical, philosophical, cultural, social, and political logics.

A gnarled and messy historical path, one strewn with many contingent forks taken and a few unexpected detours experienced, actually connects Newton's work in the *Principia* with the eighteenth-century science that we today call by this author's name. This book has tried to follow this crooked historical path in all of its adventitious detail by reconstituting the actual historical twists and turns that made "Newtonian physics" what it had become in France by the middle of the eighteenth century.

After this journey, Newton should no longer appear as the demigod, or oracle, that charted the providential path toward Enlightenment, and even less as the singular, superhuman genius who in one solitary and prophetic act revealed the foundations of modern mathematical physics for all to see. However, his presence in this history as a key participant should also be evident as well. The history of science should, I believe, offer complex accounts of the multidimensional and contingent processes by which science as we know it is made by fully human beings struggling within fully embodied intellectual, political, and cultural predicaments to make progress in projects of knowing and understanding. Histories that seek to canonize singular visionary founders are contrary to this mission, so this book has taken an opposite tack with respect to one often-celebrated scientific saint. Rejecting the legend of the revolutionary Newtonian foundation of modern mathematical physics, this book has tried to understand its beginnings in France by connecting Newton's contributions to it with those of a host of other men and women who were just as crucial in bringing this outcome about. It has also done so by stressing the complex intellectual motivations driving these changes, and the equally powerful social, cultural, and institutional dynamics that were essential in bringing them about.

Ultimately, the contingencies of history produced what we now call classical Newtonian mechanics. The contributions of Newton's *Principia* to this historical outcome are many, and they are best appreciated when they are given a fully human and messily historical accounting. Newton's status as a brilliant individual who produced profoundly important science is in no way challenged by this approach. But if this book has accomplished its work, it should be enough to say that while Newton was certainly a key player in the eighteenth-century foundation of modern mathematical physics, he was only one such player, and not the singular agent of the collective and multidimensional changes that created what we now call "classical Newtonian mechanics." There was no such thing as the Newtonian Revolution, in other words, and the French, caught in the changing political cultural of Louis XIV's France after 1690, triggered it.

Notes

CHAPTER ONE

1. Owen Gingrich, *The Book Nobody Read: Chasing the Revolutions of Nicolaus Copernicus* (New York: Penguin, 2004); Arthur Koestler, *The Sleepwalkers: A History of Man's Changing Vision of the Universe* (London: Hutchison, 1959), 191–92.
2. The Latin title was "In Viri Praestantissimi Isaaci Newtoni opus hocce mathematico-physicum seculi gentisque nostrae decus egregium." My citations for this and all other parts of Newton's *Principia* are drawn from Isaac Newton, *The Principia: Mathematical Principles of Natural Philosophy; With a Guide to Newton's Principia by I. Bernard Cohen*, I. Bernard Cohen and Anne Whitman trans., with the assistance of Julia Budenz (Berkeley: University of California Press, 1999). Hereafter cited as Newton, *Principia*.
3. Newton, *Principia*, 380. The original Latin reads as follows:

   Talia monstrantem mecum celebrate camoenis,
   Vos O coelicolum gaudentes nectare vesci,
   Newtonum clausi reserantem scrinia veri,
   Newtonum Musis charum, cui pectore puro
   Phoebus adest, totoque incessit numine mentem:
   Nec fas est propius mortali attingere divos.

4. Clifford Truesdell, "A Program toward Rediscovering the Rational Mechanics of the Age of Reason," *Archive for History of Exact Sciences* 1 (1960–62): 5.
5. For Saverien's biography, see *Biographie universelle, ancienne et moderne* (Paris: L. G. Michaud, 1825), 40:513–15; Hoefer, ed., *Nouvelle biographie générale* (Paris: Firmin Didot Frères, 1864), 43:394–95.
6. Richard Westfall, *Never at Rest: A Biography of Isaac Newton* (Cambridge: Cambridge University Press, 1980), 473.
7. Alexandre Saverien, *Histoire des philosophes modernes avec leur portrait gravé dans le goût du crayon*, 4 vols. (Paris: de Brunet, 1762), "Discours préliminaire," 3:l–li.
8. See, for example, James Gleick, *Isaac Newton* (New York: Pantheon, 2003), 147.
9. Cited in Simon Schaffer, "Somewhat Divine," *London Review of Books*, November 16, 2000.
10. *HARS* (1754), *Hist.*, 177. "Le hasard le conduisit chez Mylord Devonshire dans le moment où M. Newton venoit de laisser chez ce Seigneur un exemplaire de

ses Principes. Le jeune Mathématicien ouvrit le livre, &, séduit par la Simplicité apparente de l'ouvrage, se persuada qu'il alloit l'entendre sans difficulté; mais il fut bien furpris de se trouver hors de la portée de ses connaissances, & de se voir obligé de convenir que ce qu'il avoit pris pour le faîte des Mathématiques n'étoit que l'entrée d'une longue & pénible carrière qui lui restoit à parcourir. Il se procura cependant le livre, & comme les leçons qu'il étoit obligé dedonner l'engageoient à des courtes presque continuelles, il en déchira les feuillets pour les porter dans sa poche & les étudier dans les intervalles de ses travaux. De quelque façon qu'il s'y fût pris, il n'auroit jamais pu offrir à ce grand Mathématicien un hommage plus digne ni plus flatteur que celui qu'il lui rendoit en déchirant ainsi fes ouvrages."

11. Westfall, *Never at Rest*, 470–71.
12. John Locke, *Some Thoughts concerning Education*, § 194, in *The Works of John Locke: A New Edition, Corrected*, 10 vols. (London: Thomas Tegg, 1823), 9:186–87.
13. See Christaan Huygens to Leibniz, November 18, 1690; and "Observations de 1689 sur quelques passages des 'Principia' de Newton," *O.H.* 9:538 and 21:413–26.
14. J. B. Shank, *The Newton Wars and the Beginning of the French Enlightenment* (Chicago: University of Chicago Press, 2008).
15. H. Floris Cohen, *The Scientific Revolution: A Historiographical Inquiry* (Chicago: University of Chicago Press, 1994), 21–150.
16. I. Bernard Cohen, *The Birth of a New Physics* (New York: Anchor Books, 1960; revised and updated ed. New York: W. W. Norton, 1991); I. Bernard Cohen, *The Newtonian Revolution, with Illustrations of the Transformation of Scientific Ideas* (Cambridge: Cambridge University Press, 1980); A. Rupert Hall, *The Scientific Revolution, 1500–1800: The Formation of the Modern Scientific Attitude* (London: Longman, 1954), reissued in a revised 3rd edition as *The Revolution in Science, 1500–1800* (Boston: Addison Wesley, 1983); A. Rupert Hall, *From Galileo to Newton, 1630–1720: The Rise of Modern Science* (London: Collins, 1963); Alexandre Koyré, *Newtonian Studies* (London: Chapman & Hall, 1965; rpt. Chicago: University of Chicago Press, 1968); Alexandre Koyré, *From the Closed World to the Infinite Universe* (Baltimore: Johns Hopkins University Press, 1957).
17. Ernst Mach, *The Science of Mechanics: A Critical and Historical Account of Its Development*, T. J. Macormack trans. (New York: Open Court, 1988), 226.
18. A. R. Hall, *The Scientific Revolution, 1500–1800: The Formation of the Modern Scientific Attitude*, 2nd ed. (Boston: Beacon Press, 1966), 244–45.
19. I. Bernard Cohen, *The Newtonian Revolution*, 16.
20. George Smith, "Newton's *Philosophiae Naturalis Principia Mathematica*," *Stanford Encyclopedia of Philosophy* (Winter 2008), http://plato.stanford.edu/archives/win2008/entries/newton-principia/.

21. Distinguished books exemplifying this recent philosophical literature include Phillip Bricker and R. I. G. Hughes eds., *Philosophical Perspectives on Newtonian Science* (Cambridge, MA: MIT Press, 1990); Steffen Ducheyne, *The Main Business of Natural Philosophy: Isaac Newton's Natural-Philosophical Methodology* (Dordrecht: Springer, 2012); Andrew Janiak, *Newton as Philosopher* (Cambridge: Cambridge University Press, 2008); Andrew Janiak and Eric Schliesser eds., *Interpreting Newton: Critical Essays* (Cambridge: Cambridge University Press, 2012); and William Harper, *Isaac Newton's Scientific Method: Turning Data into Evidence about Gravity and Cosmology* (Oxford: Oxford University Press, 2011). George Smith's essay "How Newton's Principia Changed Physics" in the Janiak and Schliesser volume *Interpreting Newton* is especially emblematic of the general assumptions of this literature.
22. Useful review essays that survey the recent philosophical literature include Eric Schliesser, "The Methodological Dimension of the Newtonian Revolution," *Metascience* 22 (July 2013): 329–33; Steffen Ducheyne, "Mathematical and Philosophical Newton," *Metascience* 20 (2011): 467–76; and Mary Domski, "Putting the Pieces Back Together Again: Reading Newton's *Principia* through Newton's Method," *HOPOS: Journal of the International Society for the History of Philosophy of Science* 3 (2013): 318–33.
23. For a representative example of the influence of Cohen's *Newtonian Revolution* across the disciplines, see John Bender, "Enlightenment Fiction and the Scientific Hypothesis," *Representations* 61 (1998): 6–28.
24. Ibid., 15–16.
25. Robert E. Schofield, *Mechanism and Materialism: British Natural Philosophy in the Age of Reason* (Princeton, NJ: Princeton University Press, 1970), 4.
26. On "genealogical history" as the term is used here, see Friedrich Nietzsche, *The Use and Abuse of History for Life*, Adrian Collins trans. (New York: Macmillan, 1957); Michel Foucault, "Nietzsche, Genealogy, History," in *The Foucault Reader*, Paul Rabinow ed. (New York: Pantheon, 1984), 76–100. The methodology to be deployed here is also identical in spirit to the one offered in Bruno Latour, *Science in Action: How to Follow Scientists and Engineers through Society* (Cambridge, MA: Harvard University Press, 1987).
27. Truesdell, "Program toward Recovering the Rational Mechanics of the Age of Reason," 5.
28. Ibid., 4.
29. Ibid.
30. Ibid., 17.
31. This has been the approach taken by Craig Fraser in his nevertheless insightful corpus of historical work on eighteenth-century mathematics and mechanics after

Euler. See, as a representative example, "Mechanics in the Eighteenth Century," in Jed Z. Buchwald and Robert Fox eds., *The Oxford Companion to the History of Physics* (Oxford: Oxford University Press, 2014).

32. Ivor Grattan-Guinness ed., *From the Calculus to Set Theory, 1630–1910: An Introductory History* (London: Duckworth, 1980), 75–76.
33. Jens Martin Knudsen and Poul Georg Hjorth, *The Elements of Newtonian Mechanics* (New York: Springer, 1995).
34. For an interesting and ironic discussion of Euler's claim to have discovered the F = ma law of acceleration himself in 1750, see Thomas Hankins, "The Reception of Newton's Second Law of Motion in the Eighteenth Century," *Archives internationales d'histoire des sciences* 20 (1967): 43–65.
35. The central works of this large and still growing literature are Jed Z. Buchwald, "Discrepant measurements and experimental knowledge in the early modern era," *Archive for History of Exact Sciences* 61 (2006): 565–649, which is further developed with explicit reference to Newton's work in the period when he produced the *Principia* in Jed Z. Buchwald and Mordechai Feingold, *Newton and the Origin of Civilization* (Princeton, NJ: Princeton University Press, 2013), esp. 8–105; Niccolò Guicciardini, *Isaac Newton on Mathematical Certainty and Method* (Cambridge, MA: MIT Press, 2011); and George Smith, "The Newtonian style in Book II of the *Principia*," in Jed Z. Buchwald and I. Bernard Cohen eds., *Isaac Newton's Natural Philosophy* (Cambridge, MA: MIT Press, 2000), 249–313.
36. Henk J. M. Bos, "Philosophical Challenges from History of Mathematics," in Tinne Hoff Kjeldsen, Stig Andur Pedersen, and Lise Mariane Sonne-Hansen eds., *New Trends in the History and Philosophy of Mathematics* (Odense: University Press of Southern Denmark, 2004), 51–66.
37. Ibid., 65.
38. Bos, "Philosophical Challenges from History of Mathematics," 65; emphasis in original.
39. Ibid.
40. Ian Hacking, *Why Is There Philosophy of Mathematics at All?* (Cambridge: Cambridge University Press, 2014).
41. Bruno Latour, *Pandora's Hope: Essays on the Reality of Science Studies* (Cambridge, MA: Harvard University Press, 1999), 311.
42. Hans-Jörg Rheinberger, *On Historicizing Epistemology: An Essay*, David Fernbach trans. (Stanford, CA: Stanford University Press, 2010), 59–60. Also influential upon the methods pursued in this book is Hans-Jörg Rheinberger, *Toward a History of Epistemic Things: Synthesizing Proteins in the Test Tube* (Stanford, CA: Stanford University Press, 1997).
43. Guicciardini's most important scholarship is now conveniently contained in three

exceptionally insightful books: *The Development of Newtonian Calculus in Britain, 1700–1800* (Cambridge: Cambridge University Press, 1989); *Reading the Principia: The Debate on Newton's Mathematical Methods for Natural Philosophy from 1687 to 1736* (Cambridge: Cambridge University Press, 1999); and *Newton on Mathematical Certainty and Method*. Dozens of articles further complement the core themes of these three books. Niccolò has also been a very kind and generous teacher, colleague, and friend, and I am grateful to him for the many ways, both personally and professionally, he has supported this book.

44. Two review essays that make the "contextualist" versus "technical internalist" divide central to contemporary Newton studies are Margaret C. Jacob, "Constructing, Deconstructing, and Reconstructing the History of Science," *Journal of British Studies* 36 (1997): 459–67; and Mary Domski, "Putting the Pieces Back Together Again: Reading Newton's *Principia* through Newton's Method," *HOPOS: Journal of the International Society for the History of Philosophy of Science* 3, no. 2 (Fall 2013): 318–33.
45. Niccolò Guicciardini, "Author's Page," Amazon.com, http://www.amazon.com/Niccolò-Guicciardini/e/B001HPBJ98/ref=ntt_dp_epwbk_0.
46. Larry Stewart, *The Rise of Public Science: Rhetoric, Technology, and Natural Philosophy in Newtonian Britain, 1660–1750* (Cambridge: Cambridge University Press, 1993); Mary Terrall, *The Man Who Flattened the Earth: Maupertuis and the Sciences in the Enlightenment* (Chicago: University of Chicago Press, 2002); Andrew Warwick, *Masters of Theory: Cambridge and the Rise of Mathematical Physics* (Chicago: University of Chicago Press, 2003).
47. *Journal des savants* (August 1688): 153–54.
48. Aiton, *The Vortex Theory of Planetary Motions* (New York: American Elsevier, 1972), 200.
49. Michel Blay, *La Naissance de la mécanique analytique: La science du mouvement au tournant des XVIIe et XVIIIe siècles* (Paris: Presses Universitaires de France, 1992), 221.
50. Michel Blay, *Les Raisons de l'infini: Du monde clos à l'univers mathématique* (Paris: Gallimard, 1993), 173; emphasis in original. This work is translated as *Reasoning with the Infinite: From the Closed World to the Mathematical Universe*, M. B. DeBevoise trans. (Chicago: University of Chicago Press, 1998).
51. Helpful to me in thinking about Varignon beyond the sources already cited are Michel Blay, "L'Introduction du calcul différentiel en dynamique: L'Example des forces centrales dans les mémoires de Varignon en 1700," *Sciences et techniques en perspective* 10 (1985–86): 157–90; Blay, "Quatre mémoires inédites de Pierre Varignon consacré à la science du mouvement," *Archives internationales d'histoire des sciences* (1989): 218–48; Pierre Costabel, *Varignon et la diffusion du calcul en*

France (Paris: Palais de la Découverte, 1965); and Michael S. Mahoney, "Pierre Varignon and the Calculus of Motion" (Unpublished paper). I am grateful to the late Professor Mahoney for sharing this unpublished work with me, and to John Detloff for facilitating the exchange.

52. Indicative of this situation is the treatment of Varignon in Koffi Maglo, "The Reception of Newton's Gravitational Theory by Huygens, Varignon, and Maupertuis: How Normal Science May Be Revolutionary," *Perspectives on Science* 11, no. 2 (2003): 135–69.
53. Guicciardini, *Reading the Principia*, 201. Admittedly, to expect Guicciardini to have done otherwise in a book titled "Reading the *Principia*" is unfair, but I make the point simply to note the new departure offered by this book.
54. This claim needs to be qualified by referencing the essential work of André Robinet, the French editor of Malebranche's collected works, and the author of several important historical works about his thought, including his science. Robinet's scholarship lays the groundwork for everything that I will argue, and it is accordingly cited extensively in the chapters that follow. Robinet's work also informed the work of the late Michael Mahoney during his brief period of work in the 1990s on Varignon and his new calculus-based mechanics. Mahoney left much of his own work unpublished, however, and while he cited Robinet's influence within it, neither his work nor Robinet's overall has been incorporated into the history of eighteenth-century mechanics in the way that I think it should be. This book attempts to fill that gap, and in this way it is innovative.

INTRODUCTION TO PART ONE

1. Pierre Varignon, "Règle générale pour toutes sortes de mouvements de vitesses quelconques variées à discretion," PVARS, July 5, 1698. This paper and three others by Varignon that were never published by the Academy are reproduced in Michel Blay, "Quatre mémoires inédites de Pierre Varignon consacrés à la science du mouvement," *Archives internationales d'histoire des sciences* (1989): 218–48.
2. Pierre Varignon, PVARS, September 6, 1698; January 24, 1699; March 7, 1699. All these papers are published in Blay, "Quatre mémoires inédites de Pierre Varignon."

CHAPTER TWO

1. In 1635, Cardinal Richelieu directed the new Académie française to produce a comprehensive and authoritative dictionary of the modern French language, but by 1690 the volume was still not complete. The appearance of Antoine Furetière's

nonacademic *Dictionnaire universel, contenant généralement tous les mots français, tant vieux que modernes, et les termes des toutes les sciences et des arts* (The Hague: A. and R. Leers, 1690) provoked a change in this situation, triggering an explosion of dictionary publication over the next two decades. In 1694, the Academy released its own *Dictionnaire de l'Académie française dedié au roy* (Paris: Jean Baptiste Coignard, 1694) along with a separate *Dictionnaire des sciences et des arts par M.D.C de l'Académie française* (Paris: Jean-Baptiste Coignard, 1694). The "M.D.C." referred to M. Thomas de Corneille, who edited this appended volume of scientific and technical terms. Subsequent new editions of the *Dictionnaire de l'Académie française*, which began to appear in 1718, would combine these two dictionaries into one volume, but the isolation of scientific terminology from ordinary French language in 1694 is a significant indicator of the knowledge hierarchies prevalent in Old Regime France at this time. In 1704, the Jesuits joined the dictionary game, issuing *Dictionnaire universel françois et latin* (Trévoux: Chez Étienne Ganeau, 1704). This dictionary, which appeared in many new editions throughout the eighteenth century, came to be known, like the journal published at the same Jesuit printing house, *Journal de Trévoux*, as the *Dictionnaire de Trévoux*. My discussion of the semantics of French scientific terminology in the late seventeenth century will draw upon all of these sources.

2. "*méchanique*," in *Dictionnaire de l'Académie française*.
3. Ibid.
4. "*méchanique*," in Corneille, *Dictionnaire des sciences et des arts*. Corneille was the uncle of Fontenelle, and when he was coming to the end of his service as perpetual secretary of the Académie Royale des Sciences, having been a major figure in the mathematical developments in France since his appointment in 1697, he was asked to draft a new set of entries for the mathematical terms for the 1731 edition. These are found in *O.F.* 9:449–561.
5. "*méchanique*," in Furetière, *Dictionnaire universel*.
6. Ibid.
7. Ibid.
8. Ibid.
9. Ibid.
10. Important sources for this discussion include Mario Biagioli, "The Social Status of Italian Mathematicians, 1450–1600," *History of Science* 27 (1989): 41–95; Nicholas Jardine, *The Birth of History and Philosophy of Science: Kepler's A Defence of Tycho against Ursus, with Essays on Its Provenance and Significance* (Cambridge: Cambridge University Press, 1984); Pamela O. Long, *Artisan/Practitioners and the Rise of New Sciences, 1400–1600* (Corvallis, OR: Oregon State University Press, 2011); Paul Lawrence Rose, *The Italian Renaissance of Mathematics: Studies on*

Humanists and Mathematicians from Petrarch to Galileo (Geneva: Droz, 1975); Robert Westman, "The Astronomer's Role in the Sixteenth Century: A Preliminary Survey," *History of Science* 18 (1980): 105–47.

11. David C. Lindberg, *The Beginnings of Western Science, 600 B.C. to A.D. 1450* (Chicago: University of Chicago Press, 1992), esp. ch. 9.
12. Roger Hahn, *Anatomy of a Scientific Institution: The Paris Academy of Sciences, 1666–1803* (Berkeley: University of California Press, 1971).
13. The tension between the aristocratic-dynastic and meritocratic-administrative conceptions of monarchy under Louis XIV is a major theme of François Bluche, *Louis XIV*, Mark Greengrass trans. (Oxford: Basil Blackwell, 1990).
14. The latter impulse is stressed in Robin Briggs, "The Académie Royale des Sciences and the Pursuit of Utility," *Past and Present* 131 (1991): 38–88.
15. In addition to Hahn, the literature on the founding of the Académie Royale des Sciences includes Harcourt Brown, *Scientific Organizations in Seventeenth Century France, 1620–1680* (New York: Russell and Russell, 1967); John Milton Hirshfield, *The Académie Royale des Sciences (1666–1683): Inauguration and Initial Problems of Method* (New York: Arno, 1981); David S. Lux, "Colbert's Plan for the *Grande Académie*: Royal Policy toward Science, 1663–67," *Seventeenth-Century French Studies* 12 (1990): 177–88; David J. Sturdy, *Science and Social Status: The Members of the "Académie des Sciences," 1666–1750* (Woodbridge: Boydell & Brewer, 1995); and René Taton, *Les origines de l'Académie royale des sciences* (Paris: Palais de la Découverte, 1966).
16. The House of Salomon is imagined in Francis Bacon, *New Atlantis* (London: Ewald, 1627).
17. Edouard Meaume, *Sébastien Le Clerc et son oeuvre, 1637–1714* (Paris: Baur, 1877).
18. Ibid., 55.
19. Insight into Colbert's wider program of aligning scientific learning with royal statecraft is found in Jacob Soll, *The Information Master: Jean-Baptiste Colbert's Secret State Intelligence System* (Ann Arbor: University of Michigan Press, 2011).
20. Fontenelle, *Histoire de l'Académie royale des sciences: Depuis son établissement en 1666 jusqu'à 1699*, 2 vols. (Paris: La Compagnie des Libraires, 1733), 1:4.
21. The fullest account of these struggles is found in Hirshfield, *The Académie Royale des Sciences (1666–1683)*, 18–21.
22. Kathleen Wellman, *Making Science Social: The Conferences of Theophraste Renaudot, 1633–1642* (Norman: University of Oklahoma Press, 2003).
23. On the broader history of academicism in early modern France, see Brown, *Scientific Organizations*, and Frances A. Yates, *The French Academies of the Sixteenth Century* (London: Warburg Institute, 1947).
24. C. D. Andriesse, *Huygens: The Man behind the Principle*, Sally Miedema trans.

(Cambridge: Cambridge University Press, 2005); Yoella G. Yoder, *Unrolling Time: Christiaan Huygens and the Mathematization of Nature* (Cambridge: Cambridge University Press, 2004).

25. Sturdy's *Science and Social Status* offers convenient summaries of the life and work of all of the early academicians, so I will draw heavily upon his work in my account here.
26. Rose, *The Italian Renaissance of Mathematics.*
27. The details of the early modern mathematical epistemology are most fully worked out in Henk J. M. Bos, *Redefining Geometrical Exactness: Descartes' Transformation of the Early Modern Concept of Construction* (Dordecht: Springer, 2001).
28. On Roberval, see Léon Auger, *Un savant méconnu: Gilles Personne de Roberval (1602–1675); Son activité intellectuelle dans les domaines mathématique, physique, mécanique et philosophique* (Paris: A. Blanchard, Paris, 1962), and Kokita Hara, "Gilles Personne de Roberval," *DSB*. On Frénicle de Bessy, see Ernst Coumet, *Mersenne, Frenicle et l'élaboration de l'analyse combinatoire dans la première moitié du XVIIe siècle* (Paris: s.l., 1968) and H. L. L. Busard, "Bernard Frénicle de Bessy," *DSB*. Pierre de Fermat was another like-minded luminary in this circle, and he was the correspondent and mathematical interlocutor of both Roberval and Frénicle de Bessy. He died in 1665, however, making him unavailable for appointment to the new Royal Academy.
29. See H. L. L. Busard, "Pierre de Carcavi," *DSB*. Carcavi's close associations with Colbert are featured in Soll, *The Information Master*.
30. Sturdy, *Science and Social Status*, 84–86.
31. See Anne Goldgar, *Impolite Learning: Conduct and Community in the Republic of Letters, 1680–1750* (New Haven, CT: Yale University Press, 1998).
32. Sturdy, *Science and Social Status*, 87–89.
33. The early working practices of the Academy are most intensively studied in Alice Stroup, *A Company of Scientists: Botany, Patronage, and Community at the Seventeenth-Century Parisian Royal Academy of Sciences* (Berkeley: University of California Press, 1990). See also Hirshfield, *The Académie Royale des Sciences (1666–1683)*, 111–98.
34. This is especially Hahn's argument, but it is echoed in Stroup and in Sturdy.
35. See Stroup, 70, where she notes among other things that it was Huygens who suggested to the other academicians the initial focus on natural history.
36. Sturdy, *Science and Social Status*, 111.
37. Jacques Buot, *L'Usage de la roue de proportion, sur laquelle on Pratique promptement et facilement toutes les règles de l'arithmétique* (Paris: Mondière, 1647).
38. Cited in Sturdy, *Science and Social Status*, 182.
39. Ibid., 111–13. See also Michael S. Mahoney, "Edme Mariotte," *DSB*; and Guy

Picolet et al., *Mariotte, savant et philosophe: Analyse d'une renommée* (Paris: J. Vrin, 1986).

40. On Mariotte's place in the general history of mechanics, see René Dugas, *A History of Mechanics*, J. R. Maddox trans. (New York: Dover, 1988), ch. 5; and Ernst Mach, *The Science of Mechanics*, Thomas J. McCormack trans. (La Salle, IL: Open Court Publishing, 1989), 148–50, 403.
41. Cited in Sturdy, *Science and Social Status*, 182.
42. What little is known about the *élèves* is summarized in ibid., ch. 8. Most of what follows about the *élèves* is drawn from Sturdy's account.
43. Cited in ibid., 127.
44. Ibid., 127–28.
45. PVARS, December 22, 1667, and April 11, 1668.
46. Sturdy, *Science and Social Status*, 130.
47. Ibid., 133–37.
48. As an example, in 1667 Carcavi received a pension of 2,000 livres, and Auzout, du Hamel, Perrault, and Roberval each received 1,500 livres. Pivert, by contrast, received 600 livres, and Niquet and La Voye-Mignot 800 livres. Couplet received no pension at all until 1671, and then only 800 livres. Ibid., 154–55.
49. Sturdy, *Science and Social Status*, 130–33.
50. Chandra Mukerji, *Impossible Engineering: Technology and Territoriality on the Canal du Midi* (Princeton, NJ: Princeton University Press, 2009).
51. Ian Thompson, *The Sun King's Garden: Louis XIV, Andre Le Notre and the Creation of the Gardens of Versailles* (New York: Bloomsbury, 2006), 229–60.
52. Hélène Vérin, *La Gloire des ingénieurs: L'Intelligence technique du XVIe au XVIIIe siècle* (Paris: Albin Michel, 1993).
53. These efforts are summarized in James Edward King, *Science and Rationalism in the Government of Louis XIV* (Baltimore: Johns Hopkins University Press, 1949).
54. Briggs, "Académie des Sciences and Utility," 48.
55. Sturdy, *Science and Social Status*, 73–75.
56. Ibid., 149–53.
57. Charles Wolfe, *Histoire de l'observatoire de Paris de sa fondation à 1793* (Paris: Gauthier-Villars, 1902).
58. Stroup, *A Company of Scientists*, 47–51. Stroup offers a more detailed account of royal funding for academic science in "Royal Funding of the Parisian Académie Royale des Sciences during the 1690s," *Transactions of the American Philosophical Society* 77, pt. 4 (1987).
59. The extensive literature on astrology as a premodern scientific practice at princely courts is summarized in Darrel Rutkin, "Astrology," in Lorraine Daston and Katha-

rine Park eds., *The Cambridge History of Science*, vol. 3, *Early Modern Science* (Cambridge: Cambridge University Press, 2006), 541–61. See also Mario Biagioli, *Galileo, Courtier: The Practice of Science in the Culture of Absolutism* (Chicago: University of Chicago Press, 1992); Robert S. Westman, *The Copernican Question: Prognostication, Skepticism, and Celestial Order* (Berkeley: University of California Press, 2011).

60. On astronomical clocks, see Dava Sobel, *Longitude* (New York: Walker, 1995), 25–29.
61. John North, *Cosmos* (Chicago: University of Chicago Press, 2008), 380–81.
62. Michael S. Mahoney, "Christian Huygens: The Measurement of Time and of Longitude at Sea," in Henk J. M. Bos ed., *Studies on Christian Huygens* (Lisse: Swets & Zeitlinger, 1980), 234–70.
63. On the history of the French navy, see Charles de la Roncière, *Histoire de la marine française*, 6 vols. (Paris: Plon, 1899–1932); Etienne Taillemite, *L'Histoire ignorée de la marine française* (Paris: Perrin, 1988); Geoffrey Symcox, *The Crisis of French Sea Power 1688–1697: From the guerre d'escadre to the guerre de course* (The Hague: Springer, 1974); Geoffrey Symcox, "The Navy of Louis XIV," in P. Sonnino ed., *The Reign of Louis XIV: Essays in Celebration of Andrew Losky* (Atlantic Highlands: Humanities Press, 1990), 111–26; and Jean Meyer, "Louis XIV et les puissances maritimes," *XVIIe siècle* 31 (1979): 155–72.
64. On the history of these three ports as well as Cherbourg and Lorient, see Georges Cabanier ed., *Les Ports militaires de la France* (Paris, 1867). A wonderfully detailed history of Rochefort is offered in Martine Acerra, *Rochefort et la construction navale française, 1661–1815*, 4 vols. (Paris: Librairie de l'Inde, 1992). See also Bluche, *Louis XIV*, 225–230, 299–300, 496, 621–28; and Emmanuel Le Roy Ladurie, *The Ancien Régime: A History of France, 1610–1774*, Mark Greengrass trans. (Oxford: Basil Blackwell, 1998), 152.
65. Meyer, "Louis XIV et les puissances maritimes," 155. See also Bluche, 225–30. I am grateful to Brett Steele and Jeremy Whiteman for confirming these assessments and for sharing with me their expertise regarding the technical capabilities of the French navy in this period. For a dissenting view, which stresses the fragility of French ships despite their technical sophistication, see N. A. M. Rodger, "Form and Function in European Navies, 1660–1815," in L. Aktveld et al. eds., *In Het Kielzog* (Amsterdam: Amsterdam University Press, 2003), 85–97. I am grateful to Robin Briggs for this reference.
66. See Jean Meyer, *Colbert* (Paris: Hachette, 1981), 217–60 and passim. Older but still useful are two works on mercantilism by C. W. Cole, *Colbert and the Century of French Mercantilism*, 2 vols. (New York: Columbia University Press, 1939) and *French Mercantilism* (New York: Columbia University Press, 1943).

67. On the early history of efforts to document and publicize the mechanical arts, see Jacques Proust, *Diderot et l'Encyclopédie* (Paris: Colin, 1962), 177–88.
68. Robert S. Westman, "The Astronomer's Role in the Sixteenth Century: A Preliminary Study," *History of Science* 18 (1980): 105–47.
69. For a terrific account of the complexities of instrumental astronomical measurement in the seventeenth century, see Nicholas Dew, "Scientific Travel in the Atlantic World: The French Expedition to Gorée and the Antilles, 1681–1683," *British Journal for the History of Science* 43 (2010): 1–17.
70. For a representative example of Newton's use of the data produced by the astronomers at the French Royal Observatory, see *Principia*, Book III, Propositions XIX–XX.
71. The astronomer Delisle struggled in this period to both earn a living and to establish his reputation as an astronomer. His correspondence, preserved at A.N., Archive de la Marine, 2JJ 60–69 and Obs., B1–8 therefore offers a rich window into the financial character of astronomical practices circa 1720. These numbers are drawn from this correspondence. A good discussion of eighteenth-century astronomical instrumentation is also offered in Maurice Daumas, *Scientific Instruments of the Seventeenth and Eighteenth Centuries*, Mary Holbrook trans. (London: B. T. Batsford, 1972), 173–87.
72. See Florence Hsia, *Sojourners in a Strange Land: Jesuits and Their Scientific Missions in Late Imperial China* (Chicago: University of Chicago Press, 2009).
73. For some representative examples of the public acclaim given the royal astronomers, see *Merc. gal.* (April 1699): 245–52; (September 1699): 173–80; (July 1700): 167–69; (November 1700): 134–35; (June 1701): 181–85.
74. Cassini's work did indeed trigger a flurry of attempts to try to explain comet behavior according to the theory of the vortices. On this, see Pierre Brunet, *L'Introduction des théories de Newton en France au XVIIIe siècle* (Paris: Blanchard, 1931), 29–39.
75. PVARS, March 21, 1716.
76. Le Roy Ladurie, *The Ancien Régime*, 206.

CHAPTER THREE

1. The Phélypeaux were one of the great ministerial families of the Old Regime. A Phélypeaux served as secretary of state under every monarch from the reign of Henry IV until the revolution. Other members of the family served the royal state in other capacities. The minister Maurepas, for example, an important figure in the history of eighteenth-century political and cultural life, was also a member of the Phélypeaux family. Scholarship on the Pontchartrains, however, is not

abundant. Sara E. Chapman's work is the most recent and extensive. See *Private Ambition and Political Alliances in Louis XIV's Government: The Phélypeaux de Pontchartrain Family 1650–1715* (Rochester, NY: University of Rochester Press, 2004). Before her, Charles Frostin worked to resuscitate interest in the Pontchartrain ministry. His research is summarized in *Les Pontchartrains ministres de Louis XIV: Alliances et reseau de l'influence sous l'Ancien Regime* (Rennes: Presses Universitaires de Rennes, 2006). Other, less focused accounts are found in Bluche, *Louis XIV*; Le Roy Ladurie, *The Ancien Régime*; Collins, *The State in Early Modern France*; and Thomas Schaeper, *The French Council of Commerce, 1700–1715: A Study of Mercantilism after Colbert* (Columbus: Ohio State University Press, 1983). Extremely useful for the explicit political history of the Académie Royale des Sciences during these years is Elmo Stewart Saunders, "The Decline and Reform of the *Académie des sciences à Paris*, 1676–1699" (PhD diss., Ohio State University, 1980).

2. Colbert's unified program of administrative reform that placed the Academy of Sciences at the heart of the administrative monarchy had only been possible because Colbert held in his hands the three ministerial portfolios that most mattered: the navy, which was responsible for overseas trade, the colonies, and the royal fleet; the *surintendance des batiments du roi*, which controlled the royal manufactories and the academies; and the *contrôle générale*, which controlled the entire fiscal and financial apparatus of the monarchy. When Colbert died in 1684, these posts were fragmented, but the Pontchartrains reunified them in 1691. On this see Saunders, 19, 28, 132–33; Schaeper, 11–12.
3. On Maurepas, see John C. Rule, "Jean-Frederic Phélypeaux, Comte de Pontchartrain et Maurepas: Reflections on His Life and His Papers," *Journal of the Louisiana Historical Association* 6 (1965): 365–77. See also Rule's posthumous book, coauthored by Ben S. Trotter, *A World of Paper: Louis XIV, Colbert de Torcy, and the Rise of the Information State* (Montreal: McGill-Queens University Press, 2014).
4. Pierre Clément, "Les Successeurs de Colbert: Pontchartrain," in *Revue des deux mondes*, 2e ser., 46 (1863): 918–19.
5. Bluche, 497.
6. On the *capitation*, see François Bluche and Jean-François Solnon, *La Veritable Hierarchie sociale de l'ancienne France* (Geneva, 1983); Bluche, *Louis XIV*, 502–4; and A. Guéry, "État, classification sociale, et compromis sous Louis XIV: La Capitation de 1695," *Annales ESC*, no. 5 (1986): 1041–60. The best account of the thought of Vauban and Boisguilbert in relation to the French state is found in Keohane, 327–31, 350–57. Boisguilbert was a cousin of Fontenelle, and Thomas Legendre, a third figure involved in the introduction of the *capitation* in France,

was also related to Fontenelle. Legendre was a Norman merchant who had known the Fontenelle family for many years. On this, see Niderst, *Fontenelle*, 182.

7. Solnon, 394.
8. Ibid., 338.
9. Ibid., 340.
10. Ibid.
11. Marie-Madeleine Pioche de la Vergne, comtesse de la Fayette, *Histoire de Madame Henriette d'Angleterre, suivie de mémoires de la cour de France pour les années 1688 et 1689*, G. Sigaux ed. (Paris, 1982), 141; cited in Solnon, 340.
12. La Bruyère, *Oeuvres complètes*, J. Benda ed. (Paris: Gallimard, 1951), 397; cited in Solnon, 339.
13. Solnon, 340.
14. On these shifts, see J.-M. Perouse de Montclos, *Histoire de l'architecture française: De la renaissance a la revolution* (Paris: Éditions Mengès, 1991); Michel Gallet, *Les Architectes parisiens du XVIIIe siècle* (Paris: Éditions Mengès, 1995); Solnon, 347–48; Crow, 39–44. Crow's work overall also supports the general argument being developed here. Especially interesting is the revival of the official public art salon in 1699 under J. H. Mansart. This initiative grew out of the same developments, both political and social, being explored here. Elias also documents the rise of Parisian sociability in this period, but he draws from it very different conclusions.
15. A good, recent re-assessment of the rise of the eighteenth-century Parisian salons is Antoine Lilti, *Le Monde des salons: Sociabilité et mondanité à Paris au XVIIIe siècle* (Paris: Fayard, 2005).
16. On this development, see De Jean, esp. ch. 2. On the *Mercure galant* more generally see Vincent; and François Moureau, *Le Mercure galant de Dufresny (1710–1714), ou journalisme à la mode* (Oxford: Voltaire Foundation, 1982).
17. On Mme. de Lambert, see Roger Marchal, *Madame de Lambert et son milieu* (Oxford: Voltaire Foundation, 1991). On the cultural importance of the Parisian salons more generally in the eighteenth century, see Lilti, *Le Monde des salons*.
18. A recent study that considers this question of rupture is Laurent Lemarchand, *Paris ou Versailles? La monarchie absolue entre deux capitales (1715–1723)* (Paris: Éditions du Comité des travaux historiques et scientifiques, 2014).
19. Michel Antoine, *Le Conseil du roi sous le règne de Louis XV* (Geneva: Droz, 1970). Solnon also explores many of these themes in the final section of his book, titled "La cour déclinante," 471–563.
20. Solnon suggests that Louis XV even allowed the pre-Versailles court system to reemerge through his habit of locating courtly gatherings in a variety of royal venues rather than at the royal palace of Versailles exclusively. On administrative governance under Louis XV, see Antoine, *Le Conseil du roi sous le règne de Louis XV*.

21. Jean-Jacques Dortous de Mairan, "Éloge de M. Bignon," in *HARS* (1743), *Hist.*, 189.
22. Françoise Bléchet, "Un précurseur de l'Encyclopédie au service de l'État: l'abbé Bignon," in Annie Beck ed., *Encyclopédisme: Actes du Colloque de Caen, 12–16 janvier 1987* (Caen, 1991). 395. On Bignon more generally see also Françoise Bléchet, "Le Role de l'abbé Bignon dans l'activité des sociétés savantes au XVIIIe siècle," in *Actes du 100e congrès national des sociétés savantes, Section d'histoire moderne et contemporaine et commission d'histoire des sciences et des techniques* (Paris: Bibliothèque nationale, 1979), 31–41; Françoise Bléchet, "Fontenelle et l'abbé Bignon, Du President de l'Académie Royale des Sciences au Secrétaire Perpétuel: quelques lettres de l'abbé Bignon à Fontenelle," *Corpus, revue de philosophie* 13 (1990): 51–61; and Jack A. Clarke, "Abbé Jean-Paul Bignon: 'Moderator of the Academies' and Royal Librarian," *French Historical Studies* 8 (Fall 1973): 213–35. David Sturdy was at work on a biography of Bignon before his unexpected death in 2009 and it is not clear what will happen now with this proposed work. I am grateful to Sturdy for generously sharing with me the results of his research and for his interest and support of my work overall. I am also grateful to Françoise Bléchet for personally sharing with me her expertise about Bignon.
23. "Jean-Pierre Crousaz to abbé Bignon, 1723," in Fonds Jean-Pierre Crousaz, Bibliothèque Cantonale et Universitaire de Lausanne, 2:288.
24. Most of Bignon's correspondence is in B.N. mss. f. fr. 22234–36; some is also found in Ac. Sci., "Dossier Bignon."
25. Sturdy, 421–23. The Academy had four officers. A president and vice president were chosen from among the honorary members and a director and sous-director from the regular membership. Once chosen, the officers appear to have ruled together as a sort of tetrarchy, and it is very hard to discern what, if any, distinctions existed in their power.
26. Bléchet describes Bignon's power this way: "[He] was solely responsible for the selection, recruiting, and editing of the papers read at the public assemblies. He also managed the proceedings. He further controlled the awarding of monopolies and certificates of approval for inventions. One must credit him, therefore, with both the successes and the challenges experienced by the institution in this period, and it is worth noting that the time of his most centralized control over the institution corresponds with its period of greatest expansion." "Un précurseur de l'Encyclopédie," 400.
27. Tournefort's appointment is noted, with reference to Bignon and Pontchartrain as the agents for it, in PVARS, November 21, 1691. The same is noted for Homberg in PVARS, November 24, 1691. On Tournefort and Homberg and these appointments, see Sturdy, 226–38.

28. PVARS, December 15, 1691.
29. On Gouye more generally, see Florence Hsia, *Sojourners in a Strange Land: Jesuits and Their Scientific Mission in Late Imperial China* (Chicago: University of Chicago Press, 2009), 88–93, 99, 117–20.
30. *Mémoires pour servir à l'histoire naturelle des animaux* (Paris: Imprimerie Royale, 1671); *Mémoires pour servir à l'histoire des plantes, dressez par M. Dodart de l'Académie des sciences* (Paris: Imprimerie Royale, 1676; 2nd ed. 1686). Edme Mariotte, *Essays de phisique ou mémoires pour servir à la science des choses naturelles* (Paris: Estienne Michallet, 1679; 2nd ed. 1681).
31. PVARS, December 19, 1691.
32. *Mémoires de mathématique et de physique tirés des registres de l'Académie royale des sciences*, 2 vols. (Paris: Jean Anisson, 1692–93), 1:1–17.
33. Ibid., 1:iii.
34. Describing the problems of monthly reporting to Martin Lister, the Marquis de l'Hôpital singled out the Academy's dearth of correspondence as the decisive factor. This fact is particularly interesting in light of the mandate requiring correspondence initiated in regulation XXVII of the 1699 reform. See Martin Lister, *A Journey to Paris in the Year 1698* (London, 1699; facsimile edition Urbana: University of Illinois Press, 1967), 97.
35. The appointment of the equally obscure La Coudraye to replace the obscure Sedileau after his death in 1693.
36. Alice Stroup, "Royal Funding of the Parisian *Académie Royale des Sciences* during the 1690s," *Transactions of the American Philosophical Society* (December 2007), appendix C.
37. As an illustration of the change, the Archives de l'Académie des Sciences have bound the registers into roughly equally sized volumes organized chronologically and numbered sequentially. Vol. 13 contains the records of the proceedings from December 17, 1689 to August 31, 1693; vol. 14 contains the records from November 14, 1693 to July 27, 1695; and vol. 15, which is divided in two, contains the material from July 1695 to July 1696 in the first half and the material from July 1696 to January 1697 in the second. Vol. 16 contains only the material from February 6 to June 1, 1697.
38. PVARS, December 9, 1694.
39. PVARS, January 8, 1695.
40. PVARS, March 12, 1695. The precise title likely selected by the Academy was Giovanni Alfonso Borelli, *De motionibus naturalibus a gravitate pendentibus* (Bologna: Regio Iulio, in officina Dominici Ferri, 1670).
41. Pierre Varignon, *Projet d'une nouvelle méchanique, avec un examen de l'opinion de M. Borelli, sur les propriétez des poids suspendus par des cordes* (Paris: Martin, Boudot, et Martin, 1687).

42. PVARS, January 12, 1695.
43. PVARS, January 15, 1695.
44. See for example PVARS, Tome 15, where Varignon's papers entered in his own hand on different-sized paper, are found after the sessions of April 20 and June 22, 1695, and then du Hamel's description of a mathematical paper by Lagny being given to Varignon to make a copy on May 12, 1696, and then the actual inscription of the paper, which appears to be in another hand.
45. Claire Salomon-Bayet, "Un préambule théorique à une Académie des arts," *Revue d'histoire des sciences* 23 (July/September 1970): 229–50. See also Saunders, "Decline and Reform," 171–73, 195–201.
46. PVARS, after the entry for the session of September 5, 1693. This was clearly a later edition and bears no connection to the records of the academic sessions it interrupts.
47. This document is reproduced in Salomon-Bayet, 244.
48. I owe this insight to Dave Smith, who noticed these practices in his work on the Bureau du Commerce. For Smith's work more generally, see David K. Smith, "Structuring Politics in Early Eighteenth-Century France: The Innovations of the French Council of Commerce," *Journal of Modern History* 74, no. 3 (2002): 490–537.
49. Proust, 184–85.
50. PVARS, November 14, 1699. For a brief review of this assembly, see *Merc. gal.* (November 1699): 179–84.
51. [Jean-Paul Bignon], *Les Aventures d'Abdalla, fils d'Hanif... Traduites en français sur le manuscrit arabe, trouvé à Batavia par le M. de Sandisson* (The Hague: G. de Voys, 1713; Paris: Pierre Witte, 1717).
52. Hirshfield, *The Académie Royale des Sciences (1666–1683)*, 79–96.
53. On Blondel, see especially Anthony Gerbino, *François Blondel: Architecture, Erudition, and the Scientific Revolution* (New York: Routledge, 2010). See also Sturdy, *Science and Social Status*, 171–81; "Nicolas-François Blondel," *DSB*; and Placide Mauclaire and Charles Vigoureux, *Nicolas-François Blondel, ingénieur et architecte du roi (1618–1686)* (Laon: Imprimerie de l'Aisne, 1938).
54. François Blondel, *L'Art de jeter les bombes* (Paris: L'Auteur et Langlois, 1683).
55. François Blondel, *Cours d'architecture enseigne dans l'Académie royale d'architecture* (Paris: L. Roulland, 1675).
56. Antoine Picon, *L'Invention de l'ingénieur moderne: L'École des ponts et chaussées, 1747–1851* (Paris: Presses de l'École nationale des ponts et chaussées, 1992).
57. "Jean Dominique Cassini," *DSB*; Sturdy, *Science and Social Status*, 182–84.
58. "Mémoir des gens des lettres qui sont dans les pays étranger et qui y sont d'un plus grand réputation." B.N. mss. f. fr. 23045.

59. "Cassini," *DSB*.
60. "Philippe de la Hire," *DSB*; Sturdy, *Science and Social Status*, 195–205.
61. Carl Goldstein, *Print Culture in Early Modern France: Abraham Bosse and the Purposes of Print* (Cambridge: Cambridge University Press, 2012); Jean-Claude Vuillemin, "Abraham Bosse," in Luc Foisneau ed., *Dictionary of Seventeenth-Century French Philosophers*, 2 vols. (London: Thoemmes Continuum, 2008), 1:176–79; Maxime Préaud and Sophie Join-Lambert eds., *Abraham Bosse: Savant Graveur, 1604–1676* (Paris: Bibliothèque Nationale de France, 2004).
62. Jean Dhombres and Jacques Sakarovitch eds., *Desargues en son temps* (Paris: A. Blanchard, 1994); Judith V. Field and Jeremy J. Gray, *The Geometrical Work of Girard Desargues* (Dordrecht: Springer, 1987).
63. *La Manière universelle de M. des Argues Lyonnois pour poser l'essieu & placer les heures & autres choses aux cadrans au Soleil* (Paris: P. de Hayes, 1643); *La Pratique du trait à preuve de M. des Argues Lyonnois pour la coupe des pierres en architecture pratiquer la perspective* (Paris: P. de Hayes, 1643); *Manière universelle de M. des Argues pour pratiquer la perspective par petit-pied comme le géométral . . .* (Paris: P. de Hayes, 1648); *Moyen universel de pratiquer la perspective sur les tableaux ou surfaces irrégulières . . .* (Paris: A. Bosse, 1653).
64. *De la manière de graver à l'eaux-forte et au burin* (Paris: Charles-Antoine-Joubert, 1645); *Traité des manières de graver en taille douce sur l'airin par le moyen des eaux-fortes . . .* (Paris: A. Bosse, 1645).
65. Michel Gareau, *Charles Le Brun: First Painter to King Louis XIV* (New York: Harry N. Abrams, 1992); Linda Walsh, "Charles le Brun, 'Art Dictator of France,'" in Gil Perry and Colin Cunningham eds., *Academies, Museums and Canons of Art* (New Haven, CT: Yale University Press, 1999), 86–124.
66. Norman Bryson, *Word and Image: French Painting of the Ancien Regime* (Cambridge: Cambridge University Press, 1983).
67. The dispute is discussed in Goldstein, *Print Culture in Early Modern France*, chs. 7, 8; and in Lyle Massey, *Picturing Space, Displacing Bodies: Anamorphosis in Early Modern Theories of Perspective* (College Park: Pennsylvania State University Press, 2007).
68. *Observations sur les points d'attouchement de trois lignes droites qui touchent la section d'un cone* (Paris: A. Bosse, 1672); *Nouvelle méthode en géométrie pour les sections des superficies coniques et cylindriques* (Paris: Thomas Moette, 1673); *Nouveaux élémens des sections coniques, les lieux géométriques, la construction ou effection des équations* (Paris: André Pralard, 1679).
69. René Taton, "Philippe de la Hire," *DSB*. See also René Taton, "La première oeuvre géométrique de Philippe de la Hire," *Revue d'histoire des sciences* 6 (1953): 93–111.

70. Fontenelle, "Éloge de M. de la Hire," *O.F.* 6:429–30.
71. The dynasty at the Royal Observatory that he founded after 1679 is summarized in Sturdy, *Science and Social Status*, 197–205.
72. David E. Cartwright, *Tides: A Scientific History* (Cambridge: Cambridge University Press, 1999), 56–57.
73. "Éloge de M. de la Hire," *O.F.* 6:432–33.
74. Sturdy offers a detailed social analysis of de la Hire family, including their assets, marriages, and domestic life. *Science and Social Status*, 197–205.
75. *La Gnomonique, ou l'Art de faire des cadrans au soleil* (Paris: Thomas Moette, 1682); *L'École des arpenteurs, où l'on enseigne toutes les pratiques de géométrie, qui sont nécessaires à un arpenteur* (Paris: Thomas Moette, 1689); *Traité de mécanique, ou l'on explique tout ce qui est nécessaire dans la pratique des arts, et les propriétés des corps pesants lesquelles ont un plus grand usage dans la physique* (Paris: Jean Anisson, 1695).
76. Jean Picard, *Traité du nivellement . . . Ouvrage posthume primitivement publié par M. de la Hire* (Paris: Estienne Michallet, 1684); Edme Mariotte, *Traité du mouvement des eaux et autres corps fluides, . . . mis en lumière par les soins de M. de la Hire* (Paris: Estienne Michallet, 1686); *Veterum mathematicorum Athenaei, Bitonis, Apollodori, Heronis, Philonis et Aliorum Opera* (Paris: Typographia Regia, 1693).
77. On Rolle, see Fontenelle, "Éloge de M. Rolle," *O.F.* 6:479–86; Jean Itard, "Michel Rolle," *DSB*; J. J. O'Connor and E. F. Robertson, "Michel Rolle," in MacTutor History of Mathematics, http://www-history.mcs.st-andrews.ac.uk/Printonly/Rolle.html; Sturdy, *Science and Social Status*, 217–20.
78. On the details of Diophantus's work, see Kline, *Mathematical Thought from Ancient to Modern Times* (Oxford: Oxford University Press, 1990), 138–44.
79. Ali Abdullah al-Daffa, *The Muslim Contribution to Mathematics* (London: Croom Helm, 1977); "Al-Khwaārizmī, Abu Jaʿfar Muḥammad ibn Mūsā," *DSB*.
80. Richard W. Hadden, *On the Shoulders of Merchants: Exchange and the Mathematical Conception of Nature in Early Modern Europe* (Albany: State University of New York Press, 1994).
81. Cited in Simon Singh, *Fermat's Last Theorem* (London: Fourth Estate, 1997), 66.
82. My discussion of Viète and his influence is especially indebted to Bos, *Redefining Geometrical Exactness*, and Mahoney, *Pierre de Fermat*. See also Paolo Freguglia, *Ars analytica: Mathematica e methodis nella seconda metà del cinquecento* (Busto Arsizio: Bramante, 1988).
83. René Descartes, *Discourse on Method, Optics, Geometry and Meteorology*, Paul J. Olscamp trans. (Indianapolis: Hackett Publishing, 2011), 177.
84. By comparison, Cassini in the same year earned 9,000 livres, and du Hamel,

Gallois, and de la Hire 1,500. These numbers are from 1688. Stroup, *A Company of Scientists*, table 1.

85. *Traité d'algèbre, ou principes généraux pour résoudre les questions de mathématique* (Paris: E. Michallet, 1690); *Démonstration d'une méthode pour résoudre les égalitez de tous les degrez; suivie de deux autres méthodes dont la première donne les moyens de résoudre ces mémes égalitez par la géométrie et la seconde pour résoudre plusieurs questions de Diophante qui n'ont pas encore esté resoluës* (Paris: E. Michallet, 1691).
86. *Méthode pour résoudre les équations indéterminées de l'algèbre* (Paris: Jean Cusson, 1699).
87. Michael S. Mahoney, "The Beginnings of Algebraic Thought in the Seventeenth Century," in Stephen Gaukroger ed., *Descartes: Philosophy, Mathematics, and Physics* (Sussex: Harvester Press, 1980), 141–55. Also available at http://www.princeton.edu/~hos/Mahoney/articles/beginnings/beginnings.html.
88. Joseph F. Grcar, "How Ordinary Elimination Became Gaussian Elimination," *Historia Mathematica* 38 (2012): 163–218; June Barrow-Green, "From Cascades to Calculus: Rolle's Theorem," in Eleanor Robson and Jacqueline A. Stedall eds., *The Oxford Handbook of the History of Mathematics* (Oxford: Oxford University Press, 2009), 737–54.
89. Mahoney, "The Beginnings of Algebraic Thought," 142–43.
90. This tendency is stressed by Jean Itard in his article "Michel Rolle" in the *DSB*.
91. "Avis aux *géomètres*," *J.S.* (1693): 336–38.
92. Bernoulli's "Solution . . . ," *J.S.* (1693): 405–8; Rolle's "Réponse à M. Bernoulli," *J.S.* (1693): 425–27; Bernoulli's "Réponse de M. Bernoulli," *J.S.* (1694): 32–34; Rolle's "Remarques sur la réponse . . . de M. Bernoulli," *J.S.* (1694): 77–80. For a summary of the dispute, see *BJB* 2:395.
93. The best accounts of Varignon's biography are Fontenelle's "Éloge de M. Varignon" from 1722, *O.F.* 7:19–32; Pierre Costabel, "Pierre Varignon," *DSB*; and J. J. O'Connor and E. F. Robertson, "Pierre Varignon," MacTutor History of Mathematics, http://www-history.mcs.st-andrews.ac.uk/Mathematicians/Varignon.html.
94. Sturdy, *Science and Social Status*, 218–20.
95. "Éloge de M. Varignon," *O.F.* 7:19–20.
96. One caveat to this claim is Varignon's "Démonstration de la possibilité de la présence réelle du corps de Jésus-Christ dans l'Eucharistie, conformément au sentiment des catholiques," a defense of the real presence of the body of Jesus Christ in the Eucharistic host, that was published posthumously in 1730 in a collection of short pieces on this topic, *Pièces fugitives sur l'eucharistie* (Geneva: Marc-Michel Bousquet, 1730), 8–23. It is not clear when Varignon wrote this essay, and he may

have written the text before 1687 while still a priest. But the fact that the argument was published alongside others by Malebranche, who Varignon only met after moving to Paris, and that the article presents its case in an axiomatic, geometrical style, suggests that this might be a later return of Varignon's earlier theological orientation.

97. "Éloge de M. Varignon," *O.F.* 6:20. On the abbé Saint-Pierre, see Joseph Drouet, *L'Abbé de Saint-Pierre: L'Homme et l'oeuvre* (Paris: Honoré Champion, 1912); Merle L. Perkins, *The Moral and Political Philosophy of the Abbe de Saint-Pierre* (Geneva: E. Droz, 1959); Nannerl O. Keohane, *Philosophy and the State in France: The Renaissance to the Enlightenment* (Princeton, NJ: Princeton University Press, 1980), 361–76; Thomas E. Kaiser, "The Abbé de Saint-Pierre, Public Opinion, and the Reconstitution of the French Monarchy," *Journal of Modern History* 55 (1983): 618–43; J. B. Shank, "The Abbé de Saint-Pierre and the Emergence of the 'Quantifying Spirit' in French Enlightenment Thought," in Mary Jane Parrine ed., *A Vast and Useful Art: The Gustave Gimon Collection on French Political Economy* (Stanford, CA: Stanford University Libraries, 2004), 29–47.
98. "Éloge de M. Varignon," *O.F.* 7:22.
99. "Démonstration générale de l'usage Poulies a Moussle par M. Varignon," *Nouvelles de la république des lettres* (1687): 487–98.
100. Pierre Varignon, *Projet d'une nouvelle méchanique, avec un examen de l'opinion de M. Borelli sur les propriétés des poids suspendus par des cordes* (Paris: Martin, Boudot, et Martin, 1687).
101. "Epitre à Messieurs de l'Académie Royale des Sciences," ibid.

INTRODUCTION TO PART TWO

1. Aiton, *The Vortex Theory of Planetary Motions*, 200.

CHAPTER FOUR

1. *J.S.* (August 1688): 153–54. On this review, see also Alan Gabbey, "Newton's 'Mathematical Principles of Natural Philosophy': A Treatise on Mechanics?," in P. M. Harman and Alan E. Shapiro eds., *The Investigation of Difficult Things: Essays on Newton and the History of the Exact Sciences in Honour of D. T. Whiteside* (Cambridge: Cambridge University Press, 1992).
2. *J.S.* (August 1688): 153.
3. Ibid., 154; emphasis in original.
4. Ibid.
5. See Gaukroger, *Emergence of a Scientific Culture*, ch. 11; Bos, *Redefining Geometri-*

cal Exactness; Domenico Bertoloni Meli, *Thinking with Objects: The Transformation of Mechanics in the Seventeenth Century* (Baltimore: Johns Hopkins University Press, 2006).

6. *Oeuvres de René Descartes*, Charles Adam and Paul Tannery eds., 12 vols. (Paris: Léopold Cerf, 1897–1913), 2:380. See also the preface to the French translation of the *Principles*, ibid., 9:5–11.
7. Christian Huygens, *Horologium oscillatorium sive de motu pendulorum.*
8. See especially I. Bernard Cohen, *The Newtonian Revolution, with Illustrations of the Transformation of Scientific Ideas* (Cambridge: Cambridge University Press, 1980); and George Smith, "Newton's *Philosophiae Naturalis Principia Mathematica*," *Stanford Encyclopedia of Philosophy* (Winter 2008), http://plato.stanford.edu/archives/win2008/entries/newton-principia/.
9. "*Philosophiae Naturalis Principia Mathematica*, Autore Is. Newton . . . ," *Philosophical Transactions of the Royal Society of London* 16, no. 186 (1686): 291–97.
10. Newton, *Principia*, 793.
11. Ibid.
12. Ibid.
13. "*Philosophiae Naturalis Principia Mathematica*, Autore Is. Newton . . . ," *Bibliothèque universelle et historique* 8 (1688): 363–75.
14. Ibid., 369. On Locke's authorship of this review, see James L. Axtell, "Locke's Review of the 'Principia,'" *Notes and Records of the Royal Society of London* 20, no. 2 (December 1965): 152–61.
15. Newton, *Principia*, 790.
16. On Cartesian vortical physics, see E. J. Aiton, *The Vortex Theory of Planetary Motions*; Pierre Brunet, *L'Introduction des théories de Newton en France au XVIIIe siècle* (Paris: Blanchard 1931); René Dugas, *A History of Mechanics*, J. R. Maddox trans. (New York: Dover, 1988); Daniel Garber, *Descartes' Metaphysical Physics* (Chicago: University of Chicago Press, 1992); Stephen Gaukroger, *The Emergence of a Scientific Culture: Science and the Shaping of Modernity 1210–1685* (Oxford: Oxford University Press, 2007), chs. 8, 9, 11; Paul Mouy, *Le Développement de la physique cartésienne, 1646–1712* (Paris: J. Vrin, 1934); André Robinet, *Malebranche, de l'Académie des sciences: L'Oeuvre scientifique, 1674–1715* (Paris: Vrin, 1970).
17. Brunet, *L'Introduction des théories de Newton*, v–vi.
18. See esp. Henry Guerlac, *Newton on the Continent* (Ithaca, NY: Cornell University Press, 1980), where Guerlac develops, among other arguments, the idea that Malebranchianism constitutes "the forgotten half-way house [in France] between Cartesianism and Newtonianism." See also Henry Guerlac, "Newton's Changing Reputation in the Eighteenth Century," in Raymond O. Rockwood ed., *Carl*

Becker's Heavenly City Revisited (Ithaca, NY: Cornell University Press, 1958); "Where the Statue Stood: Divergent Loyalties to Newton in the Eighteenth Century," in Earl R. Wasserman ed., *Aspects of the Eighteenth Century* (Baltimore: Johns Hopkins University Press, 1965).

19. Aiton, *The Vortex Theory of Planetary Motions*, 200.
20. *Principia*, 408.
21. William Whewell, *History of the Inductive Sciences from the Earliest to the Present Time* (London: John W. Parker and Son, 1837), 167; Guicciardini, *Newton on Mathematical Certainty and Method*, 235.
22. Huygens to Roemer, "Newtonii librum cui titulos Philosophiae Principia Mathematica non dubito quin videris, in quou obscuritas magna. Attamen multa acute inventa," *O.H.* 9:490. Guicciardini, *Reading the Principia*, 124.
23. *O.H.* 10:346.
24. Guicciardini, *Reading the Principia*, 180.
25. See as a classic example, Carl B. Boyer, *The History of the Calculus and Its Conceptual Development* (New York: Dover, 1959).
26. G. W. Leibniz, "Nova methodus pro maximis et minimis, itemque tangentibus, quae nec fractas nec irrationales quantitates moratur, et singulare pro illis calculi genus," *Acta Eruditorum* (October 1684): 476–83. Reprinted in *LMS* 5: 220–26.
27. J. F. Montucla, *Histoire des mathématiques, dans laquelle on rend compte de leurs progrès depuis leur origine jusqu'à nos jours*, 4 vols., 2nd ed. (Paris: Henri Agasse, 1799–1802), 3:6.
28. See esp. Guicciardini, *Newton on Mathematical Certainty and Method*, chs. 10.3, 11, 12.

CHAPTER FIVE

1. Blay, *La Naissance de la mécanique analytique*, 221.
2. Michel Blay, *Les Raisons de l'infini: Du monde clos à l'univers mathématique* (Paris: Gallimard, 1993), 173; emphasis in original. This work is translated as *Reasoning with the Infinite: From the Closed World to the Mathematical Universe*, M. B. DeBevoise trans. (Chicago: University of Chicago Press, 1998).
3. Jean-Sylvain Bailly, *Histoire de l'astronomie modern depuis la fondation de l'école d'Alexandrie, jusqu'à l'époque de M.D.CC.XXX* (Paris: Chez les Frères de Bure, 1779).
4. Ibid., 471–72.
5. Ibid., 472–73.
6. Joseph-Louis Lagrange, *Mécanique analytique* (Paris: La Veuve Desaint, 1788).

7. Ibid., 159.
8. M. Mathieu ed., *Histoire de l'astronomie au dix-huitième siècle par M. Delambre* (Paris: Bachelier, 1827), vi.
9. Isaac Newton, *Principes mathématiques de la philosophie naturelle*, translation of the Marquise du Châtelet reviewed by Alexis-Claude Clairaut, 2 vols. (Paris: Desaint, Saillant, et Lambert, 1759), 2:9.
10. Jean-Sylvain Bailly, *Eloge de Leibnitz qui a remportee le prix de l'Académie royale des sciences et des belles lettres de Prusse en 1768* (Paris: Debure, 1790).
11. Ibid., 201.
12. Ibid., 209–10.
13. Preface, *Analyse des infiniment petits, pour l'intelligence des lignes courbes* (Paris: Imprimerie Royale, 1696), xiii.
14. *O.M.* 19:577–78.
15. The fact that the Duc d'Orléans became Regent of France in 1715 when the legitimate Bourbon heir to the throne was still too young to rule indicates the supreme status of this house within the dynastic structures of Old Regime France.
16. "Éloge de M. le marquis de l'Hôpital," *O.F.* 6:97.
17. The earliest letter in the published correspondence of Malebranche is dated October 23, 1690, but it references an earlier letter that is lost. The tone of the letter indicates that the earlier letter might have initiated the correspondence. It also references letters to Huygens from the summer of 1690. *O.M.* 19:558. The first letter from l'Hôpital in the published Huygens correspondence is dated April 18, 1690. *O.H.* 9:401.
18. *O.M.* 19:577–78.
19. The early Bernoulli-Varignon correspondence is collected in *BJB*, vol. 1.
20. Ibid., 173. The appointment is noted in PVARS, June 17, 1693.
21. Sturdy, *Science and Social Status*, 249.
22. Leibniz's correspondence with Bernoulli, l'Hôpital, and Varignon is collected in *LMS*, vols. 3–4.
23. Huygens's correspondence from 1685 to 1695 is found in *O.H.*, vols. 9–10.
24. Newton's correspondence between 1688 and 1704, which contains few epistolary exchanges with these Continental mathematicians, is collected in H. W. Turnbull ed., *The Correspondence of Isaac Newton*, vol. 3, *1688–1694* (Cambridge: Cambridge University Press, 1961); and J. F. Scott ed., *The Correspondence of Isaac Newton*, vol. 4, *1694–1704* (Cambridge: Cambridge University Press, 1967).
25. The arrangement is noted in *BJB* 1:202.
26. G. W. Leibniz, "Nova methodus pro maximis et minimis . . . ," *Acta Eruditorum* (October 1684): 476–83.

27. Perhaps most interesting in this respect was the exchange of letters between Leibniz and his mathematical teacher and mentor Huygens, which began in the summer of 1690, and continued until Huygens's death in 1695. Huygens came to appreciate the value of the new calculus, but he never saw it as anything other than a convenient tool for problem solving, one that also raised troubling epistemological questions. For the relevant letters, see *O.H.* 9:450–52, 471–73, 496–69, 516–27, 532–40, 546–52, 555–58, 568–72, and 10:9–22, 49–52, 93–94, 109–12, 127–34, 139–42, 156–62, 182–91, 197–202, 221–30, 238–40, 260–63, 283–86, 296–304, 383–89, 425–31, 509–12, 538–43, 572–77, 609–15, 639–46, 649–51, 664–72, 675–84, 696–99, 714–18.
28. Kline, *Mathematical Thought*, 83–85, 112–15.
29. Ibid., 39–40.
30. See Douglas M. Jesseph, *Squaring the Circle: The War between Hobbes and Wallis* (Chicago: University of Chicago Press, 1999), 26–28.
31. Ibid., 25.
32. Bos, *Redefining Geometrical Exactness*, 237–53.
33. A stimulating popular account of the controversies provoked by infinitesimals is found in Amir Alexander, *Infinitesimal: How a Dangerous Mathematical Theory Shaped the Modern World* (New York: Scientific American/Farrar, Straus & Giroux, 2014).
34. My account is especially indebted to Kirsti Anderson, "Cavalieri's Method of Indivisibles," *Archive for History of Exact Sciences* 31(1985): 291–367; Blay, *Les Raisons de l'infini*; Antoni Malet, *From Indivisibles to Infinitesimals: Studies on Seventeenth-Century Mathematizations of Infinitely Small Quantities* (Bellaterra: Universitat Autonoma de Barcelona, 1996); Antoni Malet, "Barrow, Wallis, and the Remaking of Seventeenth Century Indivisibles," *Centaurus* 39 (1997): 67–92; François de Gandt, "Cavalieri's Indivisibles and Euclid's Canons," in Peter Barker and Roger Ariew eds., *Revolution and Continuity: Essays in the History and Philosophy of Early Modern Science* (Washington, DC: Catholic University of America Press, 1991), 157–82; and Paolo Mancosu, *Philosophy of Mathematics and Mathematical Practice* (Oxford: Oxford University Press, 1996), ch. 2. See also Enrico Giusti, *Bonaventura Cavalieri and the Theory of Indivisibles* (Bologna: Edizioni Cremonese, 1980).
35. Mancosu, 50–56. See also Jesseph, *Squaring the Circle*, 40–42.
36. Mancosu, ch. 2.
37. Anderson, *Cavalieri's Method of Indivisibles*, 355–58.
38. Ibid., 353–54.
39. Cited in Kline, *Mathematical Thought*, 339.

40. From the 1729 English translation by Andrew Motte, reprinted in Florian Cajori ed., *Sir Isaac Newton's Mathematical Principles of Natural Philosophy and His System of the World*, 2 vols. (Berkeley: University of California Press, 1962), 1:xvii.
41. See Bos, "Higher Order Differentials."
42. L'Hôpital, *Analyse des infiniment petits*, 2.
43. Voltaire, *Philosophical Letters*, Ernest Dilworth trans. (Indianapolis: Bobbs-Merrill, 1961), 79.
44. The epistemological dilemmas posed by the calculus are explored most fully in Kline, *Mathematics: The Loss of Certainty* (Oxford: Oxford University Press, 1980), 127–52.
45. The exchange of letters between Leibniz and Huygens in the 1690s (see note 27 above for these citations), and those between Huygens and l'Hôpital in the same years, where this discussion occurred, are found in *O.H.* 10:304–36, 312–13, 325–30, 342–46, 348–55, 437–44, 446–50, 451–53, 457–68, 481–85, 490–99, 544–46, 579–81, 621–26, 686–87, 711–13.
46. See Guicciardini, *Reading the Principia*; and *Newton on Mathematical Certainty and Method.*

CHAPTER SIX

1. My account of the Oratorian order and its schools in France is drawn primarily from Paul Lallemand, *Histoire de l'éducation dans l'Ancien Oratoire* (Paris: A. Bichat, 1888); and Pierre Costabel, "L'Oratoire de France et ses Collèges," in René Taton ed., *L'Enseignement et diffusion des sciences en France au XVIIIe siècle* (Paris: Hermann, 1986), part 3, 67–100. Also useful as a resource on Oratorian history before 1800 is Jean Félicissime Adry, *Bibliothèque des écrivains de l'oratoire, ou histoire littéraire de cette congrégation ou l'on trouve la vie & les ouvrages, tant imprimés que manuscrits, des auteurs qu'elle a produits depuis son origine en 1613, jusqu'à present* (Paris: [s.n.], 1790), the only known copy of which is found in B.N. Fonds Français 25681–86.
2. On Lamy, see François Girbal, *Bernard Lamy (1640–1715): Étude biographique et bibliographique* (Paris: Presses Universitaires de France, 1964).
3. A modern critical edition of this text is available as Bernard Lamy, *Entretiens sur les sciences*, François Girbal and Pierre Clair eds. (Paris: Presses Universitaires de France, 1966). Lamy is also the author of *Traité de mécanique* (Paris: A. Prelard, 1682); *Elémens des mathématiques, ou traité de la grandeur en general* (Paris: A. Prelard, 1689); *Les Elémens de géométrie, ou, De la mesure de corps* (Paris, 1685); and *Traité de perspective, où sont contenus les fondemens de la peinture* (Paris: Chez Anisson, 1701).

4. Aristide Marre has described the connections of two other Oratorians, Father Jean Prestet and Father Louis de Bizance (a converted Jew originally named Raphael Levi), to this mathematical network. See "Deux mathématiciens de l'Oratoire," in *Bullettino di bibliografia e di storia delle scienze matematiche e fisiche* (Bologna: Arnoldo Forni Editore, 1879), 12:21–24, 881–93.
5. The best modern critical edition of this work, edited by Geneviève Rodis-Lewis, is found in *O.M.*, vols. 1–3. My citations will be drawn from *The Search after Truth*, Thomas Lennon trans. and ed. (Cambridge: Cambridge University Press, 1997).
6. Malebranche, *De la recherche de la verité*, 1.
7. Ibid., 204.
8. Ibid.
9. Ibid., 209.
10. Ibid., 408.
11. Ibid., 409.
12. Ibid., 489.
13. Ibid., 484.
14. Ibid., 333.
15. Ibid., 421.
16. Ibid., 429.
17. Ibid., 435.
18. Ibid.
19. Ibid.
20. Ibid., 436.
21. Ibid.
22. *O.M.* 17-2:2–4, 16–17, 41, 162–63, 304–7.
23. Louis Carré, *Méthode pour la mesure des surfaces, la dimension des solides, leurs centres de pesanteur, de percussion et d'oscillation par l'application du calcul integral* (Paris: Jean Boudot, 1700).
24. The letter is found in "Sur huit letters inédites du P. Claude Jaquemet de l'Oratoire," in *Bullettino di bibliografia e di storia delle scienze matematiche e fisiche* (Bologna: Arnoldo Forni Editore, 1882), XV: 685–88.
25. Charles René Reyneau, *Analyse démontrée, ou la méthode de résoudre les problèmes des mathématiques . . . expliqué et démontrée* (Paris: J. Quillau, 1708); Charles René Reyneau, *La Science du calcul des grandeurs en général; ou Les Elémens des mathématiques* (Paris: J. Quillau, 1714).
26. Bernard Lamy, *La Rhétorique, ou l'art de parler* (Amsterdam: Paul Marret, 1699), 230–31.
27. Ibid., 231–32.

28. "Éloge de M. Renau," *O.F.* 6:488.
29. See Massimo Mazzotti, *The World of Maria Gaetana Agnesi, Mathematician of God* (Baltimore: Johns Hopkins University Press, 2012).
30. Jacques Ozanam, *Récréations mathématiques et physiques, qui contiennent plusieurs problèmes utiles [et] agreables, d'arithmétique, de géométrie, d'optique, de gnomonique, de cosmographie, de mécanique, de pyrotechnie, [et] de physique* (Paris: J. Jombert, 1698). Editions of this work were still being printed in 1770. On Ozanam and the influence of mathematical recreations more generally in this period, see Barbara Maria Stafford, *Artful Science: Enlightenment Entertainment and the Eclipse of Visual Education* (Cambridge, MA: MIT Press, 1994), esp. 45–47.
31. *O.F.* 1:515–20.
32. Leibniz, "Explication de l'Arithmétique binaire qui se sert des seuls caractères 0 et 1; avec des remarques sur son utilité, et sur ce qu'elle donne le sens des anciennes figures chinoises de Fo-Hi," *HARS* (1702), *Mem.*, 175–93; *Hist.*, 58–63.
33. "Ouvrage concernant l'Algebre," *Merc. gal.* (April 1697): 42–83; (May 1697): 164–212; (July 1697): 72–136; (October 1697): 77–82. The *Mercure* also published a "Discours sur l'Algebre" (February 1698): 67–122.
34. *Merc. gal.* (February 1698): 276–77.
35. A related culture was created in Britain around the periodical *The Ladies Diary*. On this topic see Shelley Costa, "The Ladies' Diary: Gender, Mathematics, and Civil Society in Early Eighteenth-Century England," *Osiris* 17 (2002): 49–73.
36. J. F. Montucla, *Histoire des mathématiques, dans laquelle on rend compte de leurs progrès depuis leur origine jusqu'à nos jours*, 4 vols., 2nd ed. (Paris: Henri Agasse, 1799–1802), 2:398.
37. See Fontenelle, "Éloge de M. Carré," in *O.F.* 6:249–56.
38. "Éloge de Reyneau," *O.F.* 7:137.
39. The literature on Descartes's philosophy is vast, but most helpful to my understanding has been François Azouvi, *Descartes et la France: Histoire d'une passion nationale* (Paris: Fayard, 2002); John Cottingham ed., *The Cambridge Companion to Descartes* (Cambridge: Cambridge University Press, 1992); Daniel Garber, *Descartes' Metaphysical Physics* (Chicago: University of Chicago Press, 1992); Stephen Gaukroger, *Descartes: An Intellectual Biography* (Oxford: Oxford University Press, 1995); Erica Harth, *Cartesian Women: Versions and Subversions of Rational Discourse in the Old Regime* (Ithaca, NY: Cornell University Press, 1992); Geneviève Rodis-Lewis, *Descartes: His Life and Thought*, Jame Marie Todd trans. (Ithaca, NY: Cornell University Press, 1998); Marleen Rozemond, *Descartes's Dualism* (Cambridge, MA: Harvard University Press, 1998).
40. Nicholas Jolley, "The Reception of Descartes' Philosophy," in Cottingham, *Cam-*

bridge Companion to Descartes, 393–423; Paul Mouy, *Le Développement de la physique cartésienne, 1646–1712* (Paris: J. Vrin, 1934); Francisque Bouillier, *Histoire de la philosophie cartésienne*, 2 vols. (Paris: Durand, 1868).

41. The strand of Cartesianism I have in mind is explored in Mihnea Dobre and Tammy Nyden eds., *Cartesian Empiricisms* (Dordrecht: Springer, 2013).
42. Jacques Rohault, *Traité de physique*, 2 vols. (Paris: Denys Thierry, 1671). On Rohault, see Mouy, esp. 108–38.
43. Jacques Rohault, *A System of Natural Philosophy: A Facsimile of the Edition and Translation [of Traité de physique] by John and Samuel Clarke published in 1723; With a New Introduction by L. L. Laudan* (New York: Johnson Reprint Company, 1969), author's preface (unpaginated).
44. Ibid.
45. Varignon, *Nouvelles conjectures sur la pesanteur* (Paris: J. Boudot, 1690). On this work see Koffi Maglo, "The Reception of Newton's Gravitational Theory by Huygens, Varignon, and Maupertuis: How Normal Science May Be Revolutionary," *Perspectives on Science* 11, no. 2 (2003): 135–69; and Gingras, 6–7.
46. Antoine Arnauld and Pierre Nicole, *Logic, or the Art of Thinking*, Jill Vance Buroker trans. and ed. (Cambridge: Cambridge University Press, 1996).
47. William Molyneux, *Dioptrica Nova: A Treatise on Opticks in Two Parts* (London: Benjamin Tooke 1692), iv. Cited in Charles J. McCracken, *Malebranche and British Philosophy* (Oxford: Oxford University Press, 1983), 9. As Jill Vance Buroker states in her introduction to the modern Cambridge translation: "The *Port-Royal Logic* was the most influential logic from Aristotle to the end of the nineteenth century. The 1981 critical edition lists 63 French editions and 10 English editions, one of which (1818) served as a text in courses at the Universities of Oxford and Cambridge." *Logic*, xxiii.
48. The reception of Malebranche's philosophy has not been as fully studied as that of Descartes's, but beyond McCracken see Stuart Brown, "The Critical Reception of Malebranche, from His Own Time to the End of the Eighteenth Century," in Steven Nadler ed., *The Cambridge Companion to Malebranche* (Cambridge: Cambridge University Press, 2000), 262–87.
49. *O.F.* 1:525–86.
50. Pierre Sylvain Régis, *Système de philosophie, contenant la logique, la métaphysique, le physique, et la morale* (Paris: Denys Thierry, 1690).
51. See André Robinet, *Malebranche et Leibniz: Relations personnelles présentées avec les textes complets des auteurs et de leurs correspondants* (Paris: J. Vrin, 1955).
52. "Éloge du père Malebranche," *O.F.* 6:337–60. It was one of the longest *éloges* that Fontenelle ever delivered.

53. Ibid., 354.
54. Ibid.
55. "Éloge de M. le marquis de l'Hôpital," *O.F.* 6:97.

INTRODUCTION TO PART THREE

1. "Règle générale pour toutes sortes de mouvements de vitesses quelconques variées à discretion," PVARS, July 5, 1698.
2. All these papers by Varignon, which were transcribed in the registers of the Academy but never published, are reproduced in Michel Blay, "Quatre mémoires inédites de Pierre Varignon consacrés à la science du mouvement," *Archives internationales d'histoire des sciences* (1989): 218–48.
3. See PVARS, February 28, 1699.
4. See for example George Berkeley, *The Analyst; or, A Discourse Addressed to an Infidel Mathematician. Wherein It Is Examined Whether the Object, Principles, and Inferences of the Modern Analysis Are More Distinctly Conceived, or More Evidently Deduced, Than Religious Mysteries and Points of Faith* (London: J. Tonson, 1734). On Berkeley, see Douglas Jesseph, *Berkeley's Philosophy of Mathematics* (Chicago: University of Chicago Press, 1993).

CHAPTER SEVEN

1. Guillaume-François Marquis de l'Hôpital, "Solution d'une problem posé dans le *Journal de Leipzig*," in *Mémoires de mathématique et de physique tirés des registres de l'Académie royale des sciences*, 2 vols. (Paris: Imprimeur Royal Jean Anisson, 1692–93), 1:97–101.
2. Abbé Catelan, *Principe de la science générale des lignes courbes, ou des principaux élémens de la géométrie universelle* (Paris: L. Roulland, 1691). [L'Hôpital], "Solution d'un problem que M. Beaune a proposé autrefois à M. Descartes, et que l'on trouve dans le 79 de ses lettres, tome 3. Par M. G***," *J.S.* (1692): 401–3.
3. [Catelan], "Réponse a quelques objections contre an ecrit intitulé *Principe de la science générale des lignes courbes*," *J.S.* (1692): 427–30; [l'Hôpital], "Nouvelles réflexions de M. G*** sur la réponse à quelques objections contre un ecrit intitulé *Principe de la science générale des lignes courbes*, inséré dans le 36 Journal de cette année," *J.S.* (1692): 486–90, 494–96.
4. *BJB* 2:24–25. On the Catelan episode more generally, see Costabel's discussion at *O.M.* 17:95–100.
5. *BJB* 2:36.

6. *Diverse ouvrages de mathématiques et physiques et de physiques: Par messieurs de l'Académie royale des sciences* (Paris: Imprimeur Royal Jean Anisson, 1693).
7. *BJB* 2:33.
8. Ibid., 10.
9. Ibid., 42.
10. Varignon, "L'Action de l'eau sur le fond d'un vaisseau," "Conjectures sur la durété," and "Dimension d'une espèce de coeur que forme d'une demi ellipse en tournat autour d'un de ses diametres obliques," in *Mémoires de mathématique et de physique*, 1:12–16, 62–64, 126–28.
11. Varignon, "Règles du mouvement en général," in *Mémoires de mathématique et de physique*, 1:190–95.
12. Varignon, "Règles du mouvement accéleré suivant toutes les proportions imaginables d'accélérations ordinnées," and "Application d'une règle générale du mouvement accéleré," in *Mémoires de mathématique et de physique*, 2:93–96, 107–12.
13. Varignon, "Règles du mouvement accéleré suivant toutes les proportions imaginables d'accélérations ordinnées," and "Application d'une règle générale du mouvement accéleré."
14. This episode is analyzed in detail in Clara Silvia Roero, "Leibniz and the Temple of Viviani: Leibniz's Prompt Reply to the Challenge and the Repercussions in the Field of Mathematics," *Annals of Science* 47 (1990): 423–43. On Viviani more generally, see Michael Segre, *In the Wake of Galileo* (New Brunswick, NJ: Rutgers University Press, 1991); Antonio Favaro, "Vincenzio Viviani," in Paolo Galuzzi ed., *Amici e corrispondenti di Galileo Galilei*, 3 vols. (Florence: Salimbeni, 1983), 2:1009–1163.
15. On the use of state diplomatic correspondence networks as a vehicle for sustaining seventeenth-century mathematical discussions, see Clara Silva Roero, "La matematica tra gli affari di Stato nel Granducato di Toscana alla fine del XVII secolo," *Bollettino di storia delle scienze matematiche* 11, no. 2 (1991): 85–142.
16. *Aenigma geometricum de miro opificio testudinis quadrabilis hemisphaericae*. Reproduced in Roero, "Leibniz and the Temple of Viviani," 424–25.
17. Cited in Roero, "Leibniz and the Temple of Viviani," 429.
18. On Viviani's mathematical philosophy, and its relationship with Leibniz's, see Silvia Mazzone and Clara Silvia Roero, *Jacob Hermann and the Diffusion of the Leibnizian Calculus in Italy* (Florence: Leo Olshiki, 1996); André Robinet, *G. W. Leibniz Iter Italicum (Mars 1689–Mars 1690)* (Florence: Leo Olshiki, 1988); André Robinet, *L'Empire leibnizien: La conquête de la chaire de mathématique de l'Université de Padoue* (Trieste: Lint, 1991); Clara Silvia Roero, "Viviani and Leib-

niz: Two Different Attitudes toward the Archimedean Tradition," *Studia Leibnitiana* (1990): 231–43.

19. Cited in Roero, "Leibniz and the temple of Viviani," 437.
20. See for example Torricelli's "De dimensione parabole," found in his *Opera geometrica* (Florence: Masse & de Landis, 1644), where he offers twenty-one different proofs that the area under a parabola is four-thirds the area of a triangle with the same base and height. On this work, see Amir Alexander, *Infinitesimal: How a Dangerous Mathematical Theory Shaped the Modern World* (New York: Scientific American/Farrar, Straus & Giroux, 2014), 104–11.
21. PVARS, April 9, 1692.
22. PVARS, March 13, 1694.
23. Bernard Nieuwentijt, *Considerationes circa analyseos ad quantitates infinitè parvas applicatæ principia, & calculi differentialis usum in resolvendis problematibus geometricis* (Amsterdam: Wolters, 1694).
24. PVARS, March 13, 1694.
25. PVARS, June 5, 1964.
26. Sturdy, 253.
27. Abbé Jean-Louis de Cordemoy, *Nouveau traité de toute l'architecture, utile aux entrepreneurs, aux ouvriers, et à ceux qui font bâtir* (Paris: Chez Jean-Baptiste Coignard, 1706). On Cordemoy and his work, see Robin Middleton, "The Abbé de Cordemoy and the Graeco-Gothic Ideal: A Prelude to Romantic Classicism," *Journal of the Warburg and Courtauld Institute* 25 (1962): 278–320, and 26 (1963): 90–123.
28. "Jean-Louis de Cordemoy," *Wikipedia*, https://en.wikipedia.org/w/index.php?title=Jean-Louis_de_Cordemoy&oldid=682450408.
29. Sauveur's overall career arc is surveyed in Sturdy, 249–54.
30. PVARS, May 2, 1693.
31. The paper is found at the end of Tome 14bis of PVARS, which contains the entries for the meetings between July 30, 1695, and March 20, 1696.
32. PVARS, February 22, 1696.
33. PVARS, May 12, 1696.
34. PVARS, February 4, March 10, April 7, and May 26, 1696.
35. Johann Caspar Eisenschmidt, *Diatribe de figura telluris, elliptico-sphaeroide* (Strasbourg: Johann Frederic Spoor, 1691).
36. On the eighteenth-century debate, see John L. Greenberg, *The Problem of the Earth's Shape from Newton to Clairaut: The Rise of Mathematical Science in Eighteenth-Century Paris and the Fall of "Normal" Science* (Cambridge: Cambridge University Press, 1995), and Mary Terrall, *The Man Who Flattened the*

Earth: Maupertuis and the Sciences in the Enlightenment (Chicago: University of Chicago Press, 2002), esp. chs. 3–5.

37. PVARS, August 18, 1696.
38. On the problem, see Kline, *Mathematical Thought*, 574–75; Jeanne Peiffer, "Le problème de la brachystochrone à travers les relations de Jean I Bernoulli avec≈L'Hôpital et Varignon," in Heinz-Jürgen Hess and Fritz Nagel eds., *Der Ausbau des Calculus durch Leibniz und die Brüder Bernoulli* (Stuttgart: Wiesbaden, 1989).
39. *Acta Eruditorum* (June 1696): 269.
40. *BJB* 1:333–34.
41. *BJB* 2:117–18.
42. Ibid.
43. *BJB* 1:347.
44. Nieuwentijt, *Considerationes circa analyseos ad quantitates infinitè.* On this work and the early debates it triggered, see Douglas M. Jesseph, "Leibniz on the Foundations of the Calculus: The Question of the Reality of Infinitesimal Magnitudes," *Perspectives on Science* 6, nos. 1, 2 (1998): 6–40.
45. "Remarque sur l'usage qu'on doit faire de quelques suppositions dans la méthode des infiniment petits," PVARS, February 23, 1697.
46. PVARS, January 26, 1697. Varignon's presentation was noted by du Hamel on January 19, 1697, but his paper was not transcribed. It was this exchange, combined with more analytical mathematical work presented to the Academy in early February by l'Hôpital, Varignon, and Lagny, that provoked de la Hire's "Remarks on the usage of the method of the infinitesimals" presented on February 23.
47. PVARS, July 20, 1697.
48. PVARS, August 3, 1697.
49. PVARS, January 9, 1697.
50. PVARS, March 13, 1697.
51. PVARS, March 16, 1697.
52. *BJB* 2:124.
53. One of the very few treatments of Rolle's broader mathematical thinking is found in J. F. Montucla, *Histoire des mathématiques, dans laquelle on rend compte de leurs progrès depuis leur origine jusqu'à nos jours*, 4 vols. (Paris: Charles Antoine Jombert, 1758), 2:361–67.
54. Michel Rolle, "Règles pour l'approximation des racines," in *Mémoires de mathématique et de physique*, 1:16.
55. Varignon, by contrast, contributed ten reports in all, and l'Hôpital three, although he entered the Academy in only 1693, when the monthly reports were reaching

their end, and contributed papers to three of the final five issues. The contributions of Rolle, Varignon, and l'Hôpital were dwarfed, however, by those of the royal astronomers, who contributed forty-six reports in all (de la Hire thirty and Cassini sixteen).

56. Bernoulli's "Solution . . . ," *J.S.* (1693): 405–8; Rolle's "Réponse à M. Bernoulli," *J.S.* (1693): 425–27; Bernoulli's "Réponse de M. Bernoulli," *J.S.* (1694): 32–34; Rolle's "Remarques sur la réponse . . . de M. Bernoulli," *J.S.* (1694): 77–80. For a summary of the dispute, see *BJB* 2:395.
57. Leibniz, "Responsio ad nonnullas difficultates a Dn. Bernardo Niewentiit circa methodum differentialem seu infinitesimalem motas," *Acta Eruditorum* (July 1695). Reprinted in *LMS* 5:321–26. On this reply, see Jesseph, 19–22.
58. "Considérations sur la difference qu'il y a entre l'Analyse ordinaire et le nouveau Calcul des Transcendantes," *J.S.* (1694). Reprinted in *LMS* 5:306–38.
59. *Merc. gal.* (April 1697): 47.
60. *Merc. gal.* (February 1698): 276–77.
61. PVARS, March 16, 1697.
62. "Éloge de M. l'abbé Gallois," *O.F.* 6:172.
63. Gallois's brief reports are found in *Mémoires de mathématique et de physique*, 1:113–20; 1:158–60; 2:49–64.
64. On this claim see *BJB* 3:11–15.
65. Guicciardini, *Reading the Principia*, 202.
66. L'Hôpital, *Analyse des infiniment petits*, i–ii.
67. Ibid., iii.
68. Ibid., iv.
69. Ibid., xiv.
70. *BJB* 2:96.
71. For l'Hôpital, see *O.H.* 10:346, and for Leibniz, *O.H.* 10:601.
72. *Acta Eruditorum* (June 1696): 269.
73. Cited in David Brewster, *Memoirs of the Life, Times, and Writings of Sir Isaac Newton*, 2 vols. (Edinburgh: Thomas Constable, 1855), 2:204.
74. "Epistola Missa ad Praenobilem Virum D.Carolum Mountague Armigerum, Scaccarii Regii apud Anglos Cancellarium, et Societatis Regiae Praesidem, in qua Solvuntur duo Problemata Mathematica a Johanne Barnoullo Mathematico Celeberrimo Proposita," *Philosophical Transactions of the Royal Society of London* 19 (1697): 384–89.
75. *BJB* 1:429–30.
76. See Goldgar, *Impolite Learning*.
77. Leibniz's letter is found in *O.H.* 10:719–22. On the relation of Huygens and

l'Hôpital to his journal, see *O.H.* 10:169–74, 216–18, 407–17, 437–41, 447–50, 694–95.

78. On Basnage de Beauval as journalist, see Jean Sgard, *Dictionnaire des journalistes (1600–1789)* (Oxford: Voltaire Foundation, 1999).
79. The correct answers, minus l'Hôpital's, were published in *Acta Eruditorum* (May 1697): 205–25.
80. *BJB* 2:189.

CHAPTER EIGHT

1. PVARS, February 4, 1699.
2. A useful chart summarizing these membership changes is found in Sturdy, 289–91.
3. PVARS, November–December 1699.
4. Structures and practices codified in 1699 include four formal ranks, those of *honoraire*, *pensionnaire*, *associé*, and *élève*; specifications regarding the number of members to be contained in each and their rights and responsibilities; a new corps of officers: a president and vice president appointed annually by the king to oversee the Academy, and a director and sous-director nominated by the Academy but approved by the king to preside over meetings; a perpetual treasurer to oversee finances, and a perpetual secretary to keep records of the Academy's business; a fixed meeting location (the Louvre); specific dates, times, and expected duration for the assemblies (Wednesdays and Saturdays from three to at least five o'clock except on holidays, when they will be held the next day); official annual vacation periods (September 8 through November 11, the fifteen days of Easter, the week of Pentecost, and the period from Christmas to Epiphany); a new requirement of Parisian residence for all academicians; and a new set of definitions, some rather strict, outlining the practices and duties expected of the institution and its members. The complete *règlements* are found in Léon Aucoc, *Lois, statuts, et règlements concernant les anciennes académies et l'Institut de 1635 à 1889* (Paris: L'Institut de France, 1889), LXXXIV–XCII.
5. *O.F.* 6:53. See also Fontenelle, *Histoire de l'Académie royale des sciences: Depuis son établissement en 1666 jusqu'à 1699*, 2 vols. (Paris, 1733), 2:81–83.
6. Joseph Bertrand, *L'Académie des sciences et les académiciens de 1666 à 1793* (Paris: Hertzl Relié, 1869); L. F. Alfred Maury, *Histoire de l'ancienne académie des sciences* (Paris, 1864).
7. Hahn, *Anatomy of a Scientific Institution.*
8. Ibid., 19–34.
9. Aucoc, *Lois*, LXXVII.

10. The subsequent history of academic publishing practices is studied in great detail in James E. McClellan, "Specialist Control: The Publications Committee of the Académie royale des sciences (Paris), 1700–1793," *Transactions of the American Philosophical Society* 93, no. 3 (December 2003).
11. Aucoc, *Lois*, LXXIV–XCII.
12. Michael Gordin, "The Importation of Being Earnest: The Early St. Petersburg Academy of Sciences," *Isis* 91, no. 1 (2000): 1–31.
13. Jürgen Habermas, *The Structural Transformation of the Bourgeois Public Sphere: An Inquiry into a Category of Bourgeois Society*, Thomas Burger and Frederick Lawrence trans. (Cambridge, MA: MIT Press, 1991), 19.
14. Ibid., 5–26.
15. Accounts of the Ancients-versus-Moderns struggle include Marc Fumaroli, "Les *abeilles* et les *araignées*," in A.-M. Lecoq ed., *La Querelle des anciens et des modernes* (Paris, 2001), 7–218; Hubert Gillot, *La Querelle des anciens et des modernes en France* (Paris: Honoré Champion, 1914); Noémi Hepp, *Homère en France au XVIIe siècle* (Paris: Klincksieck, 1968); Robert J. Nelson, "The Ancients and the Moderns," in Denis Hollier ed., *A New History of French Literature* (Cambridge, MA: Harvard University Press, 1989), 364–69; Stanley Rosen, *The Ancients and the Moderns: Rethinking Modernity* (New Haven, CT: Yale University Press, 1968); Levent Yilmaz, "La Querelle des anciens et modernes et sa posterité," *Intellectual News* (2002); Levent Yilmaz, *Le Temps moderne: Variations sur les anciens et les contemporains* (Paris: Gallimard, 2004); Larry F. Norman, *The Shock of the Ancient: Literature and History in Early Modern France* (Chicago: University of Chicago Press, 2011).
16. Joan DeJean, *Ancients against Moderns: Culture Wars and the Making of a Fin de Siècle* (Chicago: University of Chicago Press, 1997).
17. Charles Perrault, *Le Siècle de Louis XIV*, in *Mémoires, contes, et autres oeuvres de Charles Perrault*, Paul L. Jacob ed. (Paris: Gosselin, 1842), 320–36. The stanzas in question appear on 320–22.
18. DeJean, 44.
19. In England, a similar struggle occurred at roughly the same time and was dubbed the "Battle of the Books." On the English struggle, see Levine, *The Battle of the Books: History and Literature in the Augustan Age* (Ithaca, NY: Cornell University Press, 1981).
20. DeJean, 45–46.
21. One anonymous work from this period is titled "The Voice of the Public to M. de Scudéry concerning His Observations on *Le Cid*," DeJean, 35.
22. Ibid.

23. For a related argument about the origins of the public sphere in France, see Habermas, *The Structural Transformation of the Bourgeois Public Sphere*, 31–43.
24. DeJean, 57–66.
25. Cited in DeJean, 36.
26. Cf. Habermas, 31–43, 67–69.
27. James Pritchard, *In Search of Empire: The French in the Americas, 1670–1730* (Cambridge: Cambridge University Press, 2004), 238.
28. Cited in Bluche, 494. Saint-Simon, by contrast, thought Louis de Pontchartrain to be a model gentleman and statesman.
29. See Seymour L. Chapin, "Science in the Reign of Louis XIV," in Sonnino ed., *The Reign of Louis XIV*, 187; John C. Rule, "Jérôme Phélypeaux, Comte de Pontchartrain, and the Establishment of Louisiana, 1696–1715," in John F. McDermotte ed., *Frenchmen and French Ways in the Mississippi Valley* (Urbana: University of Illinois Press, 1969), 179–97.
30. Bluche, 226. He references the works by Frostin noted above in defending this claim. Even Geoffrey Symcox sees a continuity in the technical excellence of the French navy throughout the crisis in French naval policy that he documents. See Symcox, 39.
31. Kenneth J. Banks, *Chasing Empire across the Sea: Communications and the State in the French Atlantic, 1713–1763* (Montreal: McGill University Press, 2002).
32. Pontchartrain's neo-Colbertian initiatives in the domestic economy also shaped the eighteenth-century academy. See Schaeper, *The French Council of Commerce*, and David K. Smith, "'Au Bien du Commerce': Economic Discourse and Visions of Society in France" (PhD diss., University of Pennsylvania, 1995); and "Structuring Politics in Early Eighteenth-Century France: The Political Innovations of the French Council of Commerce," *Journal of Modern History* 74, no. 3 (September 2002): 490–537.
33. Niderst, *Fontenelle*, 179–81. Fontenelle and Jérôme de Pontchartrain began corresponding as early as 1691. Some of Pontchartrain's letters to Fontenelle are found in *O.F.* 6:11–23. Jérôme de Pontchartrain's other correspondence with men of letters is collected in G. Depping, "Lettres de Phélypeaux, comte de Pontchartrain, secrétaire d'état sous le règne de Louis XIV, à des littérateurs et à des amis de la littérature de son temps," Comité historique des arts et monuments, *Bulletin d'histoire, sciences, et lettres* 2 (1850): 52–64, 80–92.
34. For Pontchartrain's Ancient sympathies, see Depping, 59. For his defense of the Modern position, see Niderst, *Fontenelle*, 162.
35. Cited in Niderst, *Fontenelle*, 28.
36. The classic study of the culture of *préciosité* in seventeenth-century France is

Roger Lathuillère, *La Préciosité: Étude historique et linguistique* (Geneva: Droz, 1966). See also Harth, 79–90; Christoph Strosetzki, "Rhétorique de la conversation: Sa dimension littéraire et linguistique dans la société française du XVIIe siècle," Sabine Seubert trans., in *Biblio 17–20, Papers on French Seventeenth-Century Literature* (1984).

37. This history is drawn from the extensive survey of Fontenelle's early literary activities contained in Niderst, *Fontenelle*, esp. chs. 2–5.
38. Fontenelle's most famous and systematic pronouncement on the quarrel was published in 1688 under the title *Digression sur les anciens et modernes*. But other, smaller pieces published in the *Mercure galant*, such as his "Lettre sur *Eléonore d'Yvrée*," which appeared in September 1687, addressed this controversy as well. These and other texts from the period 1686–88 are collected in *O.F.*, vol. 2.
39. Niderst, *Fontenelle*, 36–37.
40. Jacques Rohault, *Traité de physique* (Paris, 1671).
41. See François Azouvi, *Descartes et la France: Histoire d'une passion nationale* (Paris: Fayard, 2002), chs. 2, 3; Stéphane Van Damme, *Descartes* (Paris: Presses de Sciences Po, 2002), ch. 1.
42. *Merc. gal.* (January 1686): 202–3.
43. *J.S.* (April 1686): 86.
44. For a fuller discussion of this point, see Shank, "Neither Natural Philosophy, nor Science, nor Literature."
45. Fontenelle had formed close ties with both Jérôme de Pontchartrain and Bignon from as early as 1691, and by 1693 both were actively involved in the affairs of the Academy of Sciences. Their extant correspondence reveals no explicit discussion of these questions, but it would not be surprising to learn that they in fact did exchange ideas concerning these matters before 1697. See Niderst, *Fontenelle*, 148; and Saunders, "Decline and Reform," 134–35.
46. *J.S.* (1692): 247–48.
47. Blay, *La Naissance de la mécanique analytique*, 221.
48. PVARS, January 19, 1692.
49. The passage is taken from René Dugas, *A History of Mechanics*, J. R. Maddox trans. (New York: Dover, 1988), 158.
50. Beyond Guicciardini, *Newton on Mathematical Certainty and Method*, which is the best available source on this topic, see Derek T. Whitesides, *The Mathematical Principles underlying Newton's Principia Mathematica* (Glasgow: University of Glasgow, 1970). A helpfully simplified and accessible explication of Newton's mathematical methods in the *Principia* is found in Dana Densmore, *Newton's* Principia: *The Central Argument* (Santa Fe, NM: Green Lion Press, 1996), xxxvii–xl, 17–91.

51. Bos, "Differentials," 4.
52. The two other general rules offered in this work captured the same general, algebraic relations for time and distance traveled.
53. PVARS, July 5, 1698.
54. Peter Galison, *Image and Logic: A Material Culture of Microphysics* (Chicago: University of Chicago Press, 1997), 19.
55. Lorraine Daston, *Classical Probability in the Enlightenment* (Princeton, NJ: Princeton University Press, 1988), 54.
56. Blay, *La Naissance de la mécanique analytique*, 161.
57. Joseph-Louis Lagrange, *Mécanique analytique* (Paris: La Veuve Desaint, 1788). Cited in ibid.
58. Blay, *Reasoning with the Infinite*, 129–30.
59. Ibid.
60. Daniel Garber is the best source on this topic, and Leibniz's extensive work on these questions is magisterially analyzed in *Leibniz: Body, Substance, Monad* (Oxford: Oxford University Press, 2011).
61. Gottfried Wilhelm Leibniz, "Brevis demon-stratio erroris memorabilis Cartesii et aliorum circa legem naturalem, secundum quam volunt a Deo eandem semper quantitatem motus con-servari; qua et in re mechanica abutuntur," *Acta Eruditorum* (1686): 161–63. A translation appears in *Gottfried Wilhelm Leibniz: Philosophical Papers and Letters*, Leroy E. Loemker trans., 2 vols. (Chicago: University of Chicago Press, 1956), 1:455–63. On the initial foundations of the *vis viva* controversy see Carolyn Iltis, "Leibniz and the Vis Viva Controversy," *Isis* 62, no. 1 (Spring 1971): 21–35.
62. On the eighteenth-century *vis viva* debate, see Thomas Hankins, "Eighteenth-Century Attempts to Resolve the *Vis Viva* Controversy," *Isis* 56 (1965): 281–97; Mary Terrall, "Vis viva Revisited," *History of Science* 42 (2004): 189–209; Pierre Costabel, "Le *De viribus viris* de R. Boscovich ou de la vertu des querelles de mot," *Archives internationales d'histoire des sciences* 14 (1961): 3–12; and L. L. Laudan, "The Vis Viva Controversy, a Post-Mortem," *Isis* 59, no. 2 (Summer 1968): 130–43.
63. Nicolas Malebranche, *Des loix de la communication des mouvements* (Paris: Pralard, 1692).
64. *BJB* 1:323–37.
65. *BJB* 2:168–72.
66. Ibid., 173.
67. Fontenelle, "Éloge de M. Varignon," *O.F.* 7:19–32.
68. For the evidence proving that Leibniz had read the *Principia* before writing this

piece, see Domenico Bertoloni Meli, *Equivalence and Priority: Newton versus Leibniz* (Oxford: Clarendon Press, 2002), ch. 5.

69. On this vortical mechanics, see Mouy, *Le Développement de la physique cartésienne*; Aiton, *The Vortex Theory of Planetary Motions.*
70. PVARS, July 5, 1698.
71. PVARS, August 9, 1698.
72. PVARS, September 6, 1698.
73. PVARS, June 8, 1697.
74. PVARS, September 6, 1698.
75. Varignon, "Méthode pour trouver un égalité divisée d'un angle rectiligne quelconque en autant parties égales qu'on voudra," PVARS, November 29, 1698; "Méthode pour prouver des courbes les longs desquelles un corps tombans les temps de chute en telle raison qu'on voudra . . . ," PVARS, January 24, 1699.
76. The new appointments and disciplinary classifications created in February 1699 are summarized in Sturdy, 289–92.
77. PVARS, February 28, 1699.
78. These are recorded in ibid.
79. *O.F.* 6:55.
80. PVARS, February 28, 1699.
81. Ibid.
82. Ibid.
83. Louis Carré, *Méthode pour la mesure des surfaces, la dimension des solides, leurs centres de pesanteur, de percussion et d'oscillation par l'application du calcul integral* (Paris: Jean Boudot, 1700). The academic approval of Carré's book by Varignon and Malebranche is recorded in PVARS, August 31, 1699.
84. PVARS, February 28, 1699.
85. Ibid.
86. PVARS, May 9, 1699.
87. PVARS, December 9, 1699.
88. PVARS, December 12, 1699.
89. PVARS, June 20, 1699.
90. I. Bernard Cohen, "Isaac Newton, Hans Sloane and the Académie royale des sciences," in *Mélanges Alexandre Koyré*, vol. 1, *L'Aventure de la sciences* (Paris: Hermann, 1964), 60–116.
91. PVARS, April 29, 1699.
92. Cited in Sturdy, 294.
93. For a sampling of Donneau de Visé's early coverage of the Academy of Sciences, see *Merc. gal.* (March 1699): 9–41; (May 1699): 1–15; (September 1699): 212–19; (November 1699): 179–84.

94. B.N. mss. Collection Clairambault.
95. *Merc. gal.* (May 1699): 9–11.
96. PVARS, May 9, 1699.
97. PVARS, end of Tome 18bis, June 5–December 22, 1699.
98. A picture with a textual description of the Academy's seating chart can be found in a set of manuscript annotations to the Academy's regulations kept by the astronomer Lalande. These notes document the internal institutional history of the Academy, and as such they are one of the few insider accounts of the actual internal practices of the institution. The notes themselves are in the Biblioteca Medicea Laurenziana in Florence, but a microfilm copy is available at Ac. Sci. Fontenelle also noted the seating arrangement in his public presentation of the 1699 regulations. See *O.F.* 6:67–68.
99. Aucoc, *Lois*, LXXXVII.
100. PVARS, August 7 and 11, 1699.

CHAPTER NINE

1. *Merc. gal.* (November 1700): 138–39.
2. PVARS, January 30, 1700. This paper was also published in *HARS* (1700), *Mem.*, 22–27.
3. Varignon, "Du mouvement en générale par toutes sortes de courbes; et des Forces centrales, tant centrifuges que centripètes, nécessaire aux corps qui les décrivent," in *HARS* (1700), *Mem.*, 83–101; and Varignon, "Des forces centrales, ou des pesanteurs nécessaire aux planètes pour faire décrire les orbes qu'on leur a supposé jusqu'ici," in *HARS* (1700), *Mem.*, 224–43.
4. *HARS* (1700), *Mem.*, 27.
5. Ibid., 84.
6. Later in the 1700s, Varignon applied his new mechanics to more complex problems of planetary motion, treating the orbits of Jupiter and Saturn in particular. Varignon also applied this same approach to the problem of *pesanteur*, claiming that the analytical expressions developed to account for movements according to "central forces" were applicable and made more general calculations regarding motions of this sort. Beginning in 1707 and continuing until 1711, Varignon likewise produced a series of papers focused on the motion of bodies in resisting media. This was the subject of Book II of Newton's *Principia*, and in what was by this date his paradigmatic practice Varignon applied the new tools of analysis to these propositions as well. In doing so, he cited Newton where appropriate but always claimed a more general approach. Varignon died in 1722 while engaged in a debate with an Italian rival who had challenged his work on central-force mechanics.

7. *HARS* (1700), *Hist.*, 99–100.
8. *HARS* (1700), *Mem.*, 22–27.
9. For a discussion of how the calculus is implicit in Newton's derivation, see Chandrasekhar, 161–64.
10. For Malebranche's fears about the impiety of more physicalist and naturalist conceptions of force, see *O.M.* 1:102.
11. "Demonstration de la possibilité de la Présence réelle du Corps de Jesus-Christ dans l'Eucharistie, conformément au sentiments des Catholiques," in *Pièces fugitives sur l'eucharistie* (Geneva: Marc-Michel Bousquet, 1730), 8–23.
12. On the philosophical problem of the material and spiritual transformations of the Eucharist in seventeenth-century France, see Tad Schmaltz, *Radical Cartesianism: The French Reception of Descartes* (Cambridge: Cambridge University Press, 2002).
13. Contestation over this issue in the Enlightenment would provide just such an occasion, and Voltaire's argument, built upon the statements of Malebranchian analysts such as Maupertuis and d'Alembert, that the metaphysics of Newtonian science rested precisely in its use of mathematics to eliminate metaphysics from science altogether, can be seen as a later articulation of ideas already present implicitly in the early decades of the eighteenth century. See Voltaire, *La Métaphysique de Newton* (1740).
14. A case for Newton's phenomenalism is made powerfully in three foundational works of the history of science: Léon Bloch, *La Philosophie de Newton* (Paris: F. Alcan, 1908); E. A. Burtt, *Metaphysical Foundations of Modern Physical Science* (London: Paul. Trench, Trubner, 1925); and Alexandre Koyré, *Newtonian Studies* (London: Chapman & Hall, 1965). See also Cohen, *The Newtonian Revolution*, 15–16.
15. PVARS, November 27, 1700.
16. PVARS, end of Tome 19, January 9–December 22, 1700.
17. Montucla offers an account of "Les Querelles suscitées au calcul différentiel" in the first edition of his *Histoire des mathématiques* (1758), 2:359–68. Recent work on the dispute about the validity of the calculus in France includes Blay, *La Naissance de la mécanique analytique*; and Mancosu, *The Philosophy of Mathematics*.
18. Ac. Sci., "Dossier Gouye."
19. A convenient table summarizing the new membership and rankings of the Royal Academy in 1700 is offered in Sturdy, 289–91.
20. Claude Bourdelin I (1620–99) was a chemist, trained as an apothecary, who joined the Academy as a part of the founding group in 1666. When he died in 1699, his seat passed to his son Claude II (1667–1711), who was then made an *associé anato-*

miste after the institutionalization of the reform. Claude II kept a personal diary of the academic meetings he attended from 1699 to 1709. The diary is found in B.N. mss. N.A. f. fr. 5148. The entries are dated, and my citations will be to "Bourdelin" noting the day, month, and year of the citation. On Claude Bourdelin II, see Sturdy, 292, 307–308.

21. Bourdelin, April 8, 1699.
22. Ibid., November 14, 1699.
23. Ibid., December 16, 1699.
24. Ibid., December 9, 1699.
25. Ibid., November 25, 1699.
26. PVARS, November 25, 1699.
27. Bourdelin, May 13, 1699.
28. Ibid., April 4, 1699.
29. Ibid., April 29 and May 9, 1699.
30. Ibid., July 11, 1699.
31. PVARS, July 18, 1699.
32. Bourdelin, July 18, 1699.
33. PVARS, August 5, 1699.
34. Bourdelin, August 5, 1699.
35. PVARS, December 9, 1699.
36. Ibid., December 12, 1699.
37. Bourdelin, December 12, 1699.
38. Ibid., February 27 and May 15, 1700.
39. Ibid., January 30, March 31, and May 29, 1700.
40. PVARS, January 20, 1700.
41. Ibid., January 23, 1700.
42. Bourdelin, January 23, 1700.
43. Ibid., June 20, 1699.
44. PVARS, July 17, 1700.
45. Bourdelin, July 17, 1700.
46. PVARS, July 21, 1700.
47. Bourdelin, July 21, 1700.
48. PVARS, July 21, 1700.
49. Bourdelin, August 7, 1700.
50. A summation of Rolle's views on infinitesimal analysis was eventually published as "Du nouveau système d'infini," *HARS* (1701), *Mem.*, 312–36.
51. PVARS, December 19, 1699.
52. Ibid., December 23, 1699.

53. Bourdelin, December 19, 1699.
54. Ibid., July 24, 1700.
55. According to the registers, the academic session for July 24, 1700 contained a letter read by Geoffroy and a paper by de la Hire, which was transcribed, on the mechanics of artillery bombs. Parent was noted as present in the attendance tally. PVARS, July 24, 1700. No paper such as this by Parent appears in the sessions earlier in the month.
56. *BJB* 2:352. The manuscript of Reyneau's record, "Extrait des réponses faites par M. Varignon en 1700–1701 aux objections que M. Rolle avait faites contre le calcul différentiel," is found at B.N. mss. f. fr. 25302, fol. 144–55. But the text has also been published in its entirety in *BJB* 2, Annexe IV, 351–76.
57. *HARS* (1701), *Mem.*, 312.
58. See l'Hôpital, *Analyse des infiniment petits*, 2.
59. *HARS* (1701), *Mem.*, 336.
60. Ibid.
61. *LMS* (letter to Varignon).
62. Bourdelin, November 27, 1700.
63. PVARS, August 11, 1700.
64. PVARS, November 27, 1700.
65. Varignon's paper, "Autre règles générale des forces central," is found at PVARS, January 29, 1701. His comment to Rolle is noted in Bourdelin, January 29, 1701. In the same entry, Bourdelin records that Varignon read a paper by l'Hôpital on the quadrature of the "Lunule d'Hippocrates de Chio," but the Academy registers say that l'Hôpital read the paper.
66. Bourdelin, February 12, 1701.
67. PVARS, February 19, 1701.
68. Rolle's interventions are noted in PVARS, November 27, 1700; December 1, 1700; March 12, 1701; and May 27, 1701. Varignon's interventions are noted in PVARS, July 21, 1700; August 7 and 11, 1700; January 29, 1701; and May 7, 1701. L'Hôpital's and abbé Gallois's interventions are noted respectively in PVARS, January 29 and February 19, 1701.
69. *BJB* 3:267, 317.
70. PVARS, February 23, 1701.
71. Bourdelin, February 23, 1701.
72. Bourdelin, May 7, 1701.
73. PVARS, May 23, 1701.
74. Fontenelle recorded Bignon's intervention in PVARS, May 27, 1701.
75. Louvois's report is found in B.N. mss. Collection Clairambault.

76. Ibid.
77. Ibid.
78. Bourdelin, June 4, 1701.
79. *BJB* 3:256–57.
80. PVARS, July 2, 1701.
81. Ibid., July 9, 1701.
82. Bourdelin, July 9, 1701. Varignon's intervention is noted without commentary in PVARS, July 9, 1701.
83. PVARS, August 1709.
84. Bignon's directive was recorded in PVARS, September 3, 1701.
85. *BJB* 2:352.
86. *J. T.* (January–February 1701): i–ii.
87. Ibid., iv–v.
88. Ibid., v–vi.
89. On the *Journal de Trévoux* in relation to the Republic of Letters, see Goldgar, *Impolite Learning*, 94, 99, 102–103.
90. "Examen du sentiment des Cartésiens sur la cause de la continuation du mouvement," *J. T.* (May–June 1701): 160–73; Cordemoy and Miron, "Réponse à l'examen du sentiment des Cartésiens sur la cause du mouvement," *J. T.* (May–June 1701): 321–29; "Eclaircissement de l'Auteur de la dissertation qui a pour titre *Examen du sentiment des Cartésiens etc.*," *J. T.* (May–June 1701): 329–34; "Dissertation sur la cause de la continuation du mouvement dans les corps qu'on appelle en Latin *projecta, missilia*, etc.," *J. T.* (May–June 1701): 334–45. Philippe de la Hire offered the Academy's judgment on this debate in a paper read before the Academy, PVARS, November 26, 1701. De la Hire's judgment was published in *J. T.* (February 1702) followed by a critical response, 46–56. This calmed the debate in the *Journal de Trévoux* for the time being, but the Academy offered one last pronouncement on the matter in 1703 publishing a synopsis of their official judgment and a description by Fontenelle of new work on this topic by Parent. See *HARS* (1701), *Hist.*, 14.
91. See for example *J. T.* (November 1707): 2027–39; (December 1707): 2170–74; (January 1708): 155–64; (June 1708): 1079–88; (July 1708): 1276–88; (December 1708): 2119–25. This sampling is typical of Laval's annual work throughout his lifetime.
92. I have not done a systematic study of these practices, but numerous indications that academicians considered the journal a viable alternative to academic publication are present from this period. See for example Louville's letters to Delisle from July 1718 (A.N., Archive de la Marine, 2JJ 60) in which he explores publication options at the journal. Similarly, Fontenelle noted Parent's use of the *Journal de Trévoux* as an organ for his ideas in his 1719 *éloge* of the academician. See *O.F.*

6:375. I suspect that a systematic study of academician publishing practices would reveal a host of other academicians such as Louville and Parent who used the *Journal de Trévoux* this way.

93. As Gaston Sortais writes: "One discovers among the Jesuits [in this period] two different attitudes. Some showed themselves more or less favorable to Cartesian or Malebranchist ideas; they constituted a tiny [*infime*] minority within the order. The others, comprising the great majority, opposed or rejected Cartesianism." *Les Cartésiennes chez les jésuites français au XVIIe et au XVIIIe siècle* (Paris: Beauchesne, 1929), 11.
94. George R. Healy, "Mechanistic Science and the French Jesuits: A Study of the Responses of the Journal de Trévoux (1701–1762) to Descartes and Newton" (PhD diss., University of Minnesota, 1956), 26–30. See also Jolley, "The Reception of Descartes' Philosophy."
95. Mouy, *Le Développement de la physique cartésienne*, 170. Cf. Bouillier, *Histoire de la philosophie cartésienne*, 1:571–91.
96. I take the term "Baroque science" from Gunnar Eriksson's study of Olaus Rudbeck's *Atlantica*, titled *The Atlantic Vision: Olaus Rudbeck and Baroque Science* (Canton, MA: Science History Publications, 1994), 149–66. I also use the term here in a way different from its luminous and incisive use in Ofer Gal and Raz-Chen Morris, *Baroque Science* (Chicago: University of Chicago Press, 2012).
97. Cited in Healy, 56.
98. The pre-seventeenth-century roots of this wondrous conception of natural philosophy are explored in Lorraine Daston and Katherine Park, *Wonders and the Order of Nature, 1150–1750* (New York, 1998).
99. Other sources on French Jesuit philosophy and science in this period include, Jeffrey Burson, "Abdication of Legitimate Heirs: The Use and Abuse of Locke in the Jesuit *Journal de Trévoux* and the Origins of Counter-Enlightenment, 1737–1767," *Studies on Voltaire and the Eighteenth Century* 2005:297–327; Catherine Northeast, *The Parisian Jesuits and the Enlightenment 1700–1762* (Oxford: Voltaire Foundation, 1991); Donald Schier, *Louis Bertrand Castel, Anti-Newtonian Scientist* (Cedar Rapids, IA: Totch Press, 1941); Stéphane Van Damme, *Le Temple de la sagesse: Savoirs, écriture, et sociabilité urbaine (Lyon XVIIe–XVIIIe siècle)* (Paris: Éditions de l'École des Hautes Études en Sciences Sociales, 2005).
100. See Healy, 30–66; Jolley; Mouy, 168–72; Bouillier, 1:571–91.
101. Cited in Healy, 51.
102. For an example of these views, see the review of Malebranche's *De la recherche de la verité*, *J.T.* (July–August 1701): 3–14. See also Rodolphe du Tertre, S.J., *Réfutation d'un nouveau système de métaphysique proposé par le Père Malebranche* (Paris, 1715).

103. *J. T.* (May–June 1701): 160–73.
104. A good window into these wider Jesuit preoccupations is afforded by Father Tournemine's anti-Malebranchist writings on the union of the body and the soul and the reactions they engendered. See Tournemine, "Conjectures sur l'union de l'ame et du corps," *J. T.* (May 1702): 864–75; Languet de Montigny, "Lettre à de Tournemine," *J. T.* (September 1703): 1840–57; Tournemine, "Réponse aux objections proposé contre son système de l'ame et du corps," *J. T.* (September 1703), 1857–70.
105. See for example "Extrait d'un lettre de Leibnitz touchant la génération de la glace, et touchant la démonstration cartésienne de l'existence de Dieu par Lamy," *J. T.* (September–October 1701): 200–207.
106. This conceptual uncertainty allowed even Jesuits to attempt a divorce of Descartes's physics and metaphysics. The Jesuit Father Rapin, for example, declared that "[Descartes's] physics is one of the most subtle and the most accomplished of the modern physics, and, when one views it rightly, one sees a more ordered [*réglé*] body of doctrines than either Galileo's or the English." But he also quickly distanced himself from any similar praise for Descartes's metaphysics. More secular thinkers attempted similar divorces for different reasons in this period. Rapin's statements are cited in Bouillier, 1:585.
107. *J. T.* (January–February 1701): 147–61.
108. Ibid., 148.
109. "Nouvelle méthode pour determiner aisément les rayons de la developpé dans toute sorte de courbe algebraique," *Acta Eruditorum* (November 1700).
110. *J. T.* (May–June 1701): 223; emphasis in original.
111. Ibid., 224.
112. "Il ne suffit pas en Géométrie de conclure vray, il faut voir évidemment qu'on le conclut bien." Ibid., 233–34; emphasis in original.
113. On Guldin's opposition to Cavalieri's geometry of indivisibles, see Paolo Mancosu and Enzio Vailati, "Torricelli's Infinitely Long Solid and Its Philosophical Reception in the Seventeenth Century," *Isis* 82 (1991): 50–70. Gregory de Saint-Vincent, who actually deployed Cavalieri's methods in some of his work, was less a critic of analytical mathematics than an advocate for the power of traditional geometry to supersede the new methods. He became, therefore, one of the seventeenth century's most successful "circle squarers," convincing himself that he had developed a synthetic proof of a square that exactly equaled a circle in magnitude. The eighteenth-century French historian of mathematics Montucla said of him, "no one ever squared the circle with so much ability or (except for his principal object) with so much success." On Gregory de Saint-Vincent, see H. van Looy, "Chronology and Analysis of the Mathematical Manuscripts of Gregory of Saint Vincent (1584–1667)," *Historia Mathematica* 11 (1984): 57–75. On Jesuit math-

ematical work more generally see Mordechai Feingold ed., *The New Science and Jesuit Science: Seventeenth-Century Perspectives* (New York: Springer 2003); Antonella Romano, *La Contre-réforme mathématique: Constitution et diffusion d'un culture mathématique jésuite à la Renaissance (1540–1640)* (Rome: Bibliothèque des Écoles françaises d'Athènes et Rome, 1999).

114. Amir Alexander, *Infinitesimal: How a Dangerous Mathematical Theory Shaped the Modern World* (New York: Scientific American/Farrar, Straus, and Giroux, 2014).
115. On Gregory de Saint-Vincent's infinitesimalist mathematics, see Margaret E. Baron, *The Origins of the Infinitesimal Calculus* (Oxford: Pergamon Press, 1969), 135–47. Louis Bertrand Castel, a later extension of this Jesuit opposition to modern analysis and editor at the *Journal de Trévoux* from 1720 to 1742, was a particular champion of Gregory de Saint-Vincent.
116. Cited in Healy, 72–73.
117. Ibid., 70.
118. In 1706, Varignon exchanged barbs with a critic of his new analytical approach to hydrodynamics in the pages of the *Journal de Trévoux*, and other such debates occurred throughout the eighteenth century as well. See *J. T.* (January 1706): 167; and Varignon, "Des projections faites dans un milieu sans resistance . . . ," *HARS* (1706), *Mem.*, 69–85.
119. *J. T.* (September–October 1701): 219–20.
120. Ibid.
121. "Mémoire de Mr. Leibnitz touchant son sentiment sur le calcul différentiel," *J. T.* (November–December 1701): 270–72.
122. See note 45 above.
123. *J. T.* (November–December 1701): 270.
124. Ibid.
125. Ibid., 271.
126. Ibid., 271–72.
127. *BJB* 2:308–11.
128. *LMS* 4:89–90.
129. "Extrait d'une letter de M. Leibniz à M. Varignon contenant l'explication de ce qu'on a rapporté de lui dans les Mémoires de Trévoux de novembre et décembre derniers," *J.S.* (March 1702): 183–86.
130. *BJB* 2:312–13.
131. *J. T.* (Supplement, 1701): 1–15.
132. *J. T.* (Supplement, 1702): xxviii–lx.
133. *J. T.* (February 1702): 46–56.
134. Ibid., i–vi.

135. *J.T.* (May 1702): i–xii.
136. "Règle et remarques pour la problème général des tangents, par M. Rolle de l'Académie royale des sciences," *J.S.* (April 1702): 239–54.
137. *J.S.* (January 1702): i–iii. Bignon's reforms at the *Journal des savants* are discussed most fully in Birn, "Le Journal des Savants sous l'Ancien Régime," and Bléchet, "Un précurseur de l'Encyclopédie au service de l'État: l'abbé Bignon."
138. *BJB* 2:314–15.
139. Ibid., 318–19.
140. Ibid., 322–25.
141. Ibid.
142. On Saurin's life, see Fontenelle, "Éloge de M. Saurin," *O.F.* 7:271–85; Bernard de Casaban, "Joseph Saurin, membre de l'Académie royale des sciences de Paris (1655–1737)," *Mémoires de l'Académie de Vaucluse*, 6e série, 1 (1968): 187–310.
143. The fullest account of this history is in Casaban, 208–25.
144. *O.F.* 7:276.
145. Ibid., 279.
146. See ibid., 279–80; Casaban, 225–55.
147. *O.F.* 7:280–81.
148. Ibid., 283.
149. On de la Motte, see Roger Marchal, *Madame de Lambert et son milieu* (Oxford: Voltaire Foundation, 1991), esp. 215–21, 245–48, 262–73, 276–82, and passim.
150. This relationship worked both ways. PVARS from December 22, 1706, records, for example, that "M. de la Motte visited and read an ode to the Academy of Sciences dedicated to M. Bignon that was received with great approval by the company."
151. "Réponse à l'écrit de M. Rolle de l'ac. R. des Sc. inséré dans le Journal de 13 avril 1702 . . . Par M. Saurin," *J.S.* (August 1702): 519–34.
152. *BJB* 2:323–34.
153. *BJB* 3:107.
154. *J.S.* (July 1703): 479–80.
155. *BJB* 3:150–51.
156. The precise citations to all these articles are found in ibid., notes 11–13.
157. PVARS, August 8, 1705.
158. *LMS* 4:131.
159. PVARS, January 6, 1706.
160. *Nouvelles de la république des lettres* (January 1706): 120.
161. *BJB* 3:200–201.
162. PVARS, February 15, 1708.
163. Bourdelin, August 17, 1709.
164. PVARS, August 23, 1706.

CHAPTER TEN

1. See for example Voltaire's "Lettre XXIV—Sur les Académies," with its favorable comparison of the Académie Royale des Sciences with respect to other royal scientific institutions such as the Royal Society of London. *Lettres philosophiques* (Amsterdam: E. Lucas, 1734), 275–88.
2. In 1700, for example, the editor of *Les Connaissances des temps*, the annual almanac of astronomical and meteorological data published by the Royal Observatory, used the preface of this work to issue some rather pointed statements about a scientific rival. The Academy's minutes of December 7, 1700 report this infraction of regulation XXVI and similarly record the punishment levied by the Academy (censure before the Academy and public retraction). Other controversies were handled in the same way, and without exception, they generated the same neutral reporting in the Academy's minutes. An account of the controversy with *Les Connaissances des temps* is found in the Lalande diary, B.N. mss. f. fr. 12274. Cf. *HARS* (1700), *Hist.*, 113–14.
3. Alain Viala, *La Naissance de l'écrivain: Sociologie de la littérature à l'âge classique* (Paris: Editions de Minuit, 1985).
4. James E. McClellan, "Specialist Control: The Publications Committee of the Académie royale des sciences (Paris), 1700–1793," *Transactions of the American Philosophical Society* 93, no. 3 (December 2003).
5. Le Chevalier Louville to Joseph-Nicolas Delisle, July 4, 1718, Delisle correspondence, A.N., Archive de la Marine, 2JJ 60 (hereafter cited Delisle, A.N.). A useful two-volume manuscript catalogue of this correspondence prepared by Guillaume Bigourdan is also found at Obs., Mss. 1029a. The issues and contests attendant to academic publication have recently been treated in wonderful detail by James McClellan. See "Specialist Control."
6. Delisle to Louville, July 8, 1718, "Correspondence de J. N. Delisle," Archive de la Marine, A.N. 3 38 96.
7. Ibid.
8. Louville to Delisle, July 13, 1718, "Correspondence de J. N. Delisle."
9. Delisle to Louville, July 18, 1718, "Correspondence de J. N. Delisle."
10. *O.F.* 2:10.
11. A good discussion of Fontenelle's *éloges* is found in Charles B. Paul, *Science and Immortality: The Éloges of the Paris Academy of Sciences (1699–1791)* (Berkeley: University of California Press, 1980).
12. Bourdelin, April 21, 1700.
13. PVARS, April 6, 1701.

14. Keith Michael Baker, *Condorcet: From Natural Philosophy to Social Mathematics* (Chicago: University of Chicago Press, 1974), 1–3.
15. Marchal, *Madame de Lambert et son milieu*, 211.
16. PVARS, April 6, 1701.
17. Ibid.
18. *Histoire et mémoires de l'Académie royale des sciences depuis 1666 jusqu'en 1699*, 11 vols. in 13 (Paris: La Compagnie des Libraires, 1729–34).
19. *O.F.* 6:53–54.
20. *Merc. gal.* (April 1701): 107.
21. Bourdelin, April 6, 1701.
22. *Merc. gal.* (April 1701): 109.
23. *J.T.* (April 1702): 138–44.
24. The "Preface" was first published in *HARS* (1699), *Hist.*, i-xix. My citations will be to the version of the text titled *Préface sur l'utilité des mathématiques et de la physique et sur les travaux de l'Académie des sciences*, found in *O.F.* 6:37–53. This version is identical to the original version published in the Academy's *histoire* save for an introductory paragraph included in the original that explained the nature and purpose of the academic publication itself.
25. *Merc. gal.* (May 1699): 5–6.
26. Cited in Margaret C. Jacob, *Strangers Nowhere in the World: The Rise of Cosmopolitanism in Early Modern Europe* (Philadelphia: University of Pennsylvania Press, 2006), 58.
27. *O.F.* 6:37.
28. Ibid., 39–40.
29. *Merc. gal.* (August 1701): 283–84.
30. Ibid., 285.
31. *Merc. gal.* (July 1702): 125.
32. *O.F.* 6:39.
33. Ibid., 49–50.
34. Ibid.
35. *HARS* (1699), *Hist.*, 15.
36. Ibid., 117.
37. Ibid.
38. *O.F.* 6:42–43.
39. Ibid., 44.
40. Ibid., 40.
41. Ibid., 41.
42. Ibid., 43–44.

43. Ibid., 44.
44. Ibid.
45. Indeed, Fontenelle was continuing the tradition illuminated by Matthew L. Jones in *The Good Life in the Scientific Revolution: Descartes, Pascal, Leibniz, and the Cultivation of Virtue* (Chicago: University of Chicago Press, 2006).
46. In his study of the Academy publications from the first decades of the eighteenth century, Christian Licoppe sees a bias toward what calls a "discourse of clarity" in the Academy. This discourse is exactly the Fontenellian strand of Cartesianism described here, and its prevalence, I would argue, is due at least in part to the new public orientation of the Academy and Fontenelle's role in it after 1699. Christian Licoppe, *La Formation de la pratique scientifique: Le discours de l'experience en France et en Angleterre, 1630–1820* (Paris: La Découverte, 1996).
47. On Parent, see Fontenelle, "Éloge de M. Parent," *O.F.* 6:371–77; Sturdy, 291, 297, 299–300; John L. Greenberg, *The Problem of the Earth's Shape from Newton to Clairaut: The Rise of Mathematical Science in Eighteenth-Century Paris and the Fall of "Normal" Science* (Cambridge: Cambridge University Press, 1995), 117, 245; and J. Morton Briggs, "Antoine Parent," *DSB*.
48. In 1701, Leibniz approached Fontenelle seeking an assistant to help him with his projects in binary arithmetic. When the Academy secretary sought l'Hôpital's counsel, the marquis recommended Parent. Leibniz, however, was unable to arrange a salaried position for him at the Berlin Academy, and this arrangement then dissolved. Greenberg, 117.
49. Louvois, B.N. mss. Collection Clairambault 566, f. 238.
50. Bourdelin, May 13, 1699; January 9 and 16, 1700.
51. *BJB* 2:317. Varignon offers an even scathing attack on Parent in *BJB* 3:265–66, and overall his correspondence with Bernoulli is spiced with derogatory comments about Parent.
52. Sturdy, 300.
53. *O.F.* 6:373.
54. This lone copy of the March 1703 edition is found at B.N. V 18884. On the history of this journal, see Anne-Marie Chouillet, "Recherches de mathématique et de physique (1703, 1705)," in Jean Sgard ed., *Dictionnaire des journaux 1600–1789*, 2 vols. (Oxford: Voltaire Foundation, 1991), 2:1047 and see Anne-Marie Chouillet, "Parent, Antoine," in Sgard, *Dictionnaire des journalistes (1600–1789)*, 770–71. The extant copies of the 1705 edition are found at B.N. R 14115–16.
55. Chouillet, who has studied not only the extant copies but the references to the other issues found in the periodical literature and elsewhere, lists the following names: Newton, Huygens, du Hamel, Leibniz, Sauveur, l'Hôpital, Tournemine, Saurin, Varignon, Lamy, and de la Hire. *Dictionnaire des journaux*, 1047.

56. *J.S.* (April 1703): 245–51.
57. Ibid., 246.
58. Ibid., 251.
59. Bourdelin, April 21, 1703.
60. Bourdelin and PVARS, June 23, 1705.
61. Ibid., June 27 and July 1, 1705.
62. Ibid., July 18, 1705.
63. *HARS* (1699), *Hist.*, 66–67.
64. Ibid., 70.
65. Ibid., 71.
66. *HARS* (1700), *Hist.*, 78–101.
67. Ibid., 78–80.
68. Ibid., 80.
69. Ibid., 81.
70. Ibid., 89.
71. Ibid., 93–94.
72. *Éléments de la géométrie de l'infini* (Paris: Imprimerie Royale, 1727). A facsimile reproduction edited with a critical introduction by Michel Blay and Alain Niderst is available as Fontenelle, *Éléments de la géométrie de l'infini* (Paris: Klincksieck, 1995). The work is also found in *O.F.*, vol. 8. On the context of this work and its publication, see also Shank, *The Newton Wars and the Beginning of the French Enlightenment*, ch. 3.
73. *HARS* (1704), *Hist.*, 104.
74. Descartes to Mersenne, November 15, 1638. Published in John Cottingham et al. eds. and trans., *The Philosophical Writings of Descartes*, 3 vols. (Cambridge: Cambridge University Press, 1991), 3:128–29; Antoine Arnauld and Pierre Nicole, *La Logique, ou l'art de penser* (Paris: C. Savreux, 1662), esp. part 4, "On Method."
75. *HARS* (1704), *Hist.*, 105.
76. Ibid., 115.
77. *HARS* (1706), *Hist.*, 59–62.
78. *HARS* (1704), *Hist.*, 47.
79. Ibid.
80. Ibid., 51.
81. Ibid., 53. The *Dictionnaire de l'Académie française* (Paris, 1694) defines *grues* as "large machines used to raise heavy stones for use in building."
82. *HARS* (1705), *Hist.*, 74.
83. Ibid.
84. Ibid., *Avertissement.*
85. Bourdelin, December 10 and 17, 1701.

86. "Essai d'une methode pour trouver les rayons des developpees, les tangentes, les quadratures et les rectications de plusieurs courbes sans y supposer aucune grandeur infiniment petite. Par Mr de Tschirnhausen." *HARS* (1701), *Mem.*, 394–410.
87. *HARS* (1702), *Hist.*, 53.
88. Ibid.
89. For a discussion of this practice, see Blay, *La Naissance de la mécanique analytique*, 49–60.
90. *HARS* (1702), *Hist.*, 61–64.
91. *HARS* (1701), *Hist.*, 87–89.
92. Varignon notes the appearance of the volume in a letter to Bernoulli of February 16, 1704. *BJB* 3:113.
93. *HARS* (1701), *Hist.*, 87.
94. Ibid.
95. Ibid.
96. Ibid., 88.
97. Ibid.
98. Ibid.
99. Ibid.
100. See *LMS* 4:94.
101. *O.F.* 6:479–86.
102. *HARS* (1701), *Hist.*, 88–89.
103. *O.F.* 6:374.
104. *HARS* (1701), *Hist.*, 89.
105. *O.F.* 6:95–108.
106. Ibid., 109–22, 167–75, 249–56.
107. The first of Blondel's "letters to a friend informing him of what happened at the public assembly of the Academy of Sciences" appeared in *J.T.* (January 1702): 143–62. He reported on each of the public assemblies after this date until his self-imposed retirement in 1710. Blondel also published accounts of the public assemblies of the Academy of Inscriptions and Belles-Letters in the same journal.
108. In its May 1702 review of the *histoire* of 1699 that included Fontenelle's report on the calculus debate in the Academy, the journal passed over the entire subject without comment. Further reviews of the Academy's publications, at least in the period before 1720, were similarly nonpolemical. See *J.T.* (May 1702): 133–34.
109. "Solution de la principale difficulté proposée par M. Huygens contre le système de M. Descartes sur la cause de la pesanteur," *J.S.* (January 1703): 38–47. On the significance of this paper in relation to the vortical theory of celestial mechanics, see Brunet, *L'Introduction des théories de Newton*, 20–21. Parent published a

critique of this paper in his *Recherches de mathématiques et de physique* of the same year, and another was published by La Montre in *J. T.* (February 1703): 498–502.

110. See, for example, *J. T.* (1701), appendix, 1–15; and *J. T.* (1702), supplement, xviii–lx.
111. PVARS, August 21, 1706.
112. PVARS, January 29, 1707.
113. Bourdelin, January 29, 1707.
114. PVARS, February 19, 1707.
115. Bourdelin, February 19, 1707.
116. PVARS, May 7 and 18, 1707.
117. The academy only had three *pensionnaire* geometers. Before May 1707, they had been Rolle, Varignon, and Gallois. After that, they were Varignon, Saurin, and Rolle.
118. Very little is known about this person, and what is known is found in Ac. Sci., "Dossier Bomie."
119. PVARS, August 6, 1707.
120. Ibid.
121. I have not been able to determine who precisely invited Bomie to become his student in the Academy. I will discuss the precise relationship between the astronomical community and analytical mechanics in the next chapter.
122. *HARS* (1706), *Hist.*, 56–68.
123. *HARS* (1707), *Hist.*, 97–103.
124. Bomie, "Des forces centripètes et centrifuges, considerée en générale dans toutes sortes des courbes et en particulier dans le cercle," *HARS* (1707), *Mem.*, 477–87. In this paper, Bomie makes explicit reference to Newton's *Principia*, particularly proposition I.4. Bomie's appearance at the spring public assembly is noted in PVARS, April 18, 1708. His presentation was reviewed by Blondel in *J. T.* (July 1708): 1424–28. But perhaps not surprisingly given its content, the *Mercure galant* ignored Bomie's paper altogether in its review of this public assembly, focusing instead on Fontenelle's *éloge* of the popular scientific lecturer Régis and on a paper offered by the anatomist Littré. *Merc. gal.* (May 1708): 259–62.
125. On Nicole, see Greenberg, 232–42, 308–11; and Ac. Sci., "Dossier Nicole."
126. Reyneau, *Analyse démontrée, ou la méthode de résoudre les problèmes des mathématiques* (Paris, 1708).
127. *Encyclopédie, ou Dictionnaire raisonné des sciences, des arts, et des métiers, par une société de gens de lettres*, 17 vols. (Paris: Briasson, David, Le Breton, Durand, 1751–65), art. "Analyse."
128. *BJB* 3:256–57.
129. PVARS, September 7, 1709; and *HARS* (1709), *Mem.*, 257–58.

130. PVARS, August 17, 1709.
131. PVARS, June 9, 1708.

CHAPTER ELEVEN

1. On Clairaut's work, see Curtis Wilson, "Clairaut's Calculation of the Eighteenth-Century Return of Halley's Comet," *Journal for the History of Astronomy* 24 (1993): 1–15.
2. The historian of mathematics Charles Bossut thought that this celebrity in fact sent Clairaut to the grave early in 1765 at only fifty-two years of age. As he wrote, "[Clairaut's success] gave him an existence in society [*le grand monde*] and a consideration that talent alone does not provide. Unfortunately for the sciences, he gave himself over to the general impression. . . . Wanting to join pleasures to his work, he attended dinners and evenings driven by a lively taste for women. He lost rest, health, and finally his life at the age of fifty-two." Charles Bossut, *Histoire générale des mathématiques depuis leur origine jusqu'à l'année 1808*, 2 vols. (Paris: F. Louis, 1810), 2:428–29.
3. On this development, see Niccolò Guicciardini, *The Development of Newtonian Calculus in Britain, 1700–1800* (Cambridge: Cambridge University Press, 1989).
4. On the relationship of the Royal Society to mathematics between its founding and 1750, see Barbara J. Shapiro, *The Culture of Fact: England 1550–1720* (Ithaca, NY: Cornell University Press, 2003), esp. 155–58.
5. See J. M. Dubbey, "The Introduction of Differential Notation into Great Britain," *Annals of Science* 19 (1963): 35–48, and more comprehensively Jeremy Gray and Karen Hunger Parshall eds., *Episodes in the History of Modern Algebra (1800–1950)* (Providence, RI: American Mathematical Society, 2007), esp. chs. 2, 3.
6. George Berkeley, *The Analyst; or, A Discourse Addressed to an Infidel Mathematician. Wherein It Is Examined Whether the Object, Principles, and Inferences of the Modern Analysis Are More Distinctly Conceived, or More Evidently Deduced, Than Religious Mysteries and Points of Faith* (London: J. Tonson, 1734).
7. "M. Varignon spoke about the force that directs [*fait tendre*] all the planets toward the sun, or what is called their *pesanteur par rapport au soleil.* He demonstrated geometrically that there must be such a force in order to describe [*décrire*] the oval orbits that they trace." *Merc. gal.* (November 1700): 138.
8. Bourdelin, March 31, 1700.
9. The most thorough treatments of the vortical program are found in Aiton, *The Vortex Theory of Planetary Motions*; and Brunet, *L'Introduction des théories de Newton.*

10. Villemot, *Nouveau système ou nouvelle explication du mouvement des planètes* (Lyon: Louis Declaustre, 1707). For a discussion of Villemot's work, see Brunet, *L'Introduction des théories de Newton*, 23–25, 33–39, 40–42; and Aiton, *The Vortex Theory of Planetary Motions*, 152–72, 177–82, and 188–93.
11. This letter and a number of other papers associated with this work are found in the papers of Father Sebastien Truchet, A.N., M 849.
12. Ibid.
13. *J.S.* (June 1707): 406–18.
14. Truchet, A.N.
15. Ibid.
16. Ibid.
17. Ibid.
18. Ibid.
19. *HARS* (1707), *Hist.*, 97–103.
20. Truchet, A.N.
21. "Solution de la principale difficulté proposée par M. Huygens contre le système de M. Descartes sur la cause de la pesanteur," *J.S.* (January 1703): 24–36.
22. See Aiton, *The Vortex Theory of Planetary Motions*, 171–76.
23. PVARS, April 10, 1709.
24. *Merc. gal.* (April 1709): 176–84. This excerpt captures the flavor the review throughout: "M. Fontenelle, having ceased to speak, or in other words having ceased to earn himself admiration, gave the podium to M. Saurin whose extensive genius is widely known . . . and who never speaks publicly or in the Academy without receiving great applause." See also Blondel's more objective and extensive account in *J.T.* (September 1709): 1553–66.
25. Saurin, "Examen d'une difficulté considérable proposée par M. Huygens contre le système cartésien sur la cause de la pesanteur," *HARS* (1709), *Mem.*, 131–39.
26. See Aiton, 170–77; and Brunet, 25–29.
27. *O.M.* 2:321–45. On this work, see Robinet, *Malebranche, de l'Académie des sciences*, 206–10, 215–29; and Aiton, 177–80.
28. E. J. Aiton gives the following illustration of how this mechanism can be conceived: "It is analogous to the explanation that we would give of the rise to the surface of a sponge placed in water; the water displaces the sponge upwards because, although it passes easily through the sponge, the water has more force (weight) then the sponge has to descend," 176. Further reflections on another eighteenth-century ether-gravity theory are found in Matthew R. Edwards ed., *Pushing Gravity: New Perspectives on Le Sage's Theory of Gravitation* (Montreal: Apeiron, 2002).
29. Bourdelin, April 4, 1699.
30. *HARS* (1700), *Hist.*, 98.

31. Ibid., 99–100.
32. *HARS* (1700), *Mem.*, 316.
33. As a young man, Varignon had developed his own quasi-Cartesian vortical account of *pesanteur* and presented it to the Academy. But his work was widely criticized, and Varignon abandoned the project, never to return to vortical mechanics.
34. *HARS* (1700), *Hist.*, 97.
35. *HARS* (1703), *Hist.*, 73.
36. *HARS* (1705), *Hist.*, 94–95.
37. *HARS* (1706), *Hist.*, 56–57.
38. *HARS* (1709), *Mem.*
39. I offer a detailed account of the rise of the mid-eighteenth-century "Newton Wars" in *The Newton Wars and the Beginning of the French Enlightenment.* See esp. chs. 2 and 3.
40. On the common understanding of the *Principia* that linked European and English savants before 1705, see Guicciardini, *Reading the Principia*, ch. 7.
41. See Margaret Jacob, *The Newtonians and the English Revolution 1689–1720* (Ithaca, NY: Cornell University Press, 1976); Margaret Jacob, *The Radical Enlightenment: Pantheists, Freemasons, and Republicans* (London: Allen & Unwin, 1981); and Jeffrey Wigelsworth, *Deism in Enlightenment England: Theology, Politics, and Newtonian Public Science* (Manchester, UK: Manchester University Press, 2009).
42. *The Newton Wars*, ch. 3.
43. See Mary Terrall, *The Man Who Flattened the Earth: Maupertuis and the Sciences in the Enlightenment* (Chicago: University of Chicago Press, 2002).
44. See Shank, *The Newton Wars*, 282–83.
45. I draw the details of this story from Thomas L. Hankins, *Jean d'Alembert: Science and the Enlightenment* (Oxford: Oxford University Press, 1970), ch. 3.
46. PVARS, January 20, 1748.
47. Hankins, *Jean d'Alembert*, 32–42.

Index